Partial Differential Equations and Geometry

PURE AND APPLIED MATHEMATICS

A Program of Monographs, Textbooks, and Lecture Notes

Contributions to *Lecture Notes in Pure and Applied Mathematics* are reproduced by direct photography of the author's typewritten manuscript. Potential authors are advised to submit preliminary manuscripts for review purposes. After acceptance, the author is responsible for preparing the final manuscript in camera-ready form, suitable for direct reproduction. Marcel Dekker, Inc. will furnish instructions to authors and special typing paper. Sample pages are reviewed and returned with our suggestions to assure quality control and the most attractive rendering of your manuscript. The publisher will also be happy to supervise and assist in all stages of the preparation of your camera-ready manuscript.

LECTURE NOTES
IN PURE AND APPLIED MATHEMATICS

1. *N. Jacobson,* Exceptional Lie Algebras
2. *L.-Å. Lindahl and F. Poulsen,* Thin Sets in Harmonic Analysis
3. *I. Satake,* Classification Theory of Semi-Simple Algebraic Groups
4. *F. Hirzebruch, W. D. Newmann, and S. S. Koh,* Differentiable Manifolds and Quadratic Forms
5. *I. Chavel,* Riemannian Symmetric Spaces of Rank One
6. *R. B. Burckel,* Characterization of C(X) Among Its Subalgebras
7. *B. R. McDonald, A. R. Magid, and K. C. Smith,* Ring Theory: Proceedings of the Oklahoma Conference
8. *Y.-T. Siu,* Techniques of Extension of Analytic Objects
9. *S. R. Caradus, W. E. Pfaffenberger, and B. Yood,* Calkin Algebras and Algebras of Operators on Banach Spaces
10. *E. O. Roxin, P.-T. Liu, and R. L. Sternberg,* Differential Games and Control Theory
11. *M. Orzech and C. Small,* The Brauer Group of Commutative Rings
12. *S. Thomeier,* Topology and Its Applications
13. *J. M. López and K. A. Ross,* Sidon Sets
14. *W. W. Comfort and S. Negrepontis,* Continuous Pseudometrics
15. *K. McKennon and J. M. Robertson,* Locally Convex Spaces
16. *M. Carmeli and S. Malin,* Representations of the Rotation and Lorentz Groups: An Introduction
17. *G. B. Seligman,* Rational Methods in Lie Algebras
18. *D. G. de Figueiredo,* Functional Analysis: Proceedings of the Brazilian Mathematical Society Symposium
19. *L. Cesari, R. Kannan, and J. D. Schuur,* Nonlinear Functional Analysis and Differential Equations: Proceedings of the Michigan State University Conference
20. *J. J. Schäffer,* Geometry of Spheres in Normed Spaces
21. *K. Yano and M. Kon,* Anti-Invariant Submanifolds
22. *W. V. Vasconcelos,* The Rings of Dimension Two
23. *R. E. Chandler,* Hausdorff Compactifications
24. *S. P. Franklin and B. V. S. Thomas,* Topology: Proceedings of the Memphis State University Conference
25. *S. K. Jain,* Ring Theory: Proceedings of the Ohio University Conference
26. *B. R. McDonald and R. A. Morris,* Ring Theory II: Proceedings of the Second Oklahoma Conference
27. *R. B. Mura and A. Rhemtulla,* Orderable Groups
28. *J. R. Graef,* Stability of Dynamical Systems: Theory and Applications
29. *H.-C. Wang,* Homogeneous Banach Algebras
30. *E. O. Roxin, P.-T. Liu, and R. L. Sternberg,* Differential Games and Control Theory II
31. *R. D. Porter,* Introduction to Fibre Bundles
32. *M. Altman,* Contractors and Contractor Directions Theory and Applications
33. *J. S. Golan,* Decomposition and Dimension in Module Categories
34. *G. Fairweather,* Finite Element Galerkin Methods for Differential Equations
35. *J. D. Sally,* Numbers of Generators of Ideals in Local Rings
36. *S. S. Miller,* Complex Analysis: Proceedings of the S.U.N.Y. Brockport Conference
37. *R. Gordon,* Representation Theory of Algebras: Proceedings of the Philadelphia Conference
38. *M. Goto and F. D. Grosshans,* Semisimple Lie Algebras
39. *A. I. Arruda, N. C. A. da Costa, and R. Chuaqui,* Mathematical Logic: Proceedings of the First Brazilian Conference
40. *F. Van Oystaeyen,* Ring Theory: Proceedings of the 1977 Antwerp Conference

41. *F. Van Oystaeyen and A. Verschoren,* Reflectors and Localization: Application to Sheaf Theory
42. *M. Satyanarayana,* Positively Ordered Semigroups
43. *D. L. Russell,* Mathematics of Finite-Dimensional Control Systems
44. *P.-T. Liu and E. Roxin,* Differential Games and Control Theory III: Proceedings of the Third Kingston Conference, Part A
45. *A. Geramita and J. Seberry,* Orthogonal Design: Quadratic Forms and Hadamard Matrices
46. *J. Cigler, V. Losert, and P. Michor,* Banach Modules and Functors on Categories of Banach Spaces
47. *P.-T. Liu and J. G. Sutinen,* Control Theory in Mathematical Economics: Proceedings of the Third Kingston Conference, Part B
48. *C. I. Byrnes,* Partial Differential Equations and Geometry: Proceedings of Park City Conference

Other Volumes in Preparation

Partial Differential Equations and Geometry

PROCEEDINGS OF THE PARK CITY CONFERENCE

Edited by
CHRISTOPHER I. BYRNES
Harvard University
Cambridge, Massachusetts

MARCEL DEKKER, INC. New York and Basel

Library of Congress Cataloging in Publication Data
Main entry under title:

Partial differential equations and geometry.

(Lecture notes in pure and applied mathematics ; 48)
Papers presented at a converence held from
Feb. 19-23, 1977 at Park City, Utah.
Included bibliographies and index.
1. Differential equations, Partial--Addresses,
essays, lectures. 2. Geometry, Differential--Addresses,
essays, lectures. I. Byrnes, Christopher I.,
[Date]
QA374.P35 515'.353 79-12725
ISBN 0-8247-6775-6

MARCEL DEKKER, INC.

270 Madison Avenue, New York, New York 10016

Current printing (last digit)
10 9 8 7 6 5 4 3 2 1

PRINTED IN THE UNITED STATES OF AMERICA

PREFACE

The conference on Partial Differential Equations and Geometry, of which these proceedings are a record, was held from February 19-23, 1977 at Park City, Utah. It was organized by Morris Kalka, Hugo Rossi, Joseph Taylor, and the editor and was partially supported by the National Science Foundation and partially supported by the University of Utah, to which formal thanks are due.

The purpose of the conference was to review some recent, exciting developments in the terrain shared by geometry and differential equations; such as the study of analytic and topological invariants of manifolds, the interactions of differential analysis and geometry with the study of stochastic processes on riemannian manifolds, the study of partial differential equations arising in complex analysis and geometry, and the applications of geometry, both algebraic and differential, to the study of nonlinear equations. Three one-hour addresses were presented each day by the invited lecturers. In addition to those indicated in the table of contents, it was also a pleasure to have Professors Cheeger, Goldschmidt, Helgason, Kazhdan, and Weinstein report on survey work which is recorded elsewhere. The abstracts, as well as some of the longer papers, provide a partial record of the afternoon three hour sessions for contributed papers.

In closing, I take this opportunity to thank the organizing committee and, especially, the (roughly) 100 participants for providing the atmosphere and the mathematics which made the conference interesting. It is also a pleasure to thank Ann Reed, the conference secretary, and Ricki Rossi, the proceedings secretary, for their invaluable assistance.

Christopher I. Byrnes

CONTRIBUTORS

KINETSU ABE, University of Connecticut, Storrs, Connecticut

F. J. ALMGREN, JR., Princeton University, Princeton, New Jersey

ERIC BEDFORD, Princeton University, Princeton, New Jersey

JOHN K. BEEM, University of Missouri, Columbia, Missouri

ROBERT P. BUEMI, Massachusetts Institute of Technology, Cambridge, Massachusetts

DANIEL M. BURNS, JR., Princeton University, Princeton, New Jersey

T. E. DUNCAN, University of Kansas, Lawrence, Kansas

PAUL E. EHRLICH, University of Missouri, Columbia, Missouri

ROBERT B. GARDNER, University of North Carolina, Chapel Hill, North Carolina

B. GAVEAU, Université de Lille, Lille, France

PETER GILKEY, Princeton University, Princeton, New Jersey

V. GUILLEMIN, Massachusetts Institute of Technology, Cambridge, Massachusetts

GARY A. HARRIS, University of Kentucky, Lexington, Kentucky

REESE HARVEY, Rice University, Houston, Texas

ROBERT HERMANN, Rutgers University, New Brunswick, New Jersey

HANS-CHRISTOPH IMHOF, University of California, Berkeley, California

HOWARD JACOBOWITZ, Rice University, Houston, Texas

H. BLAINE LAWSON, JR., University of California, Berkeley, California

PAUL MALLIAVIN, Institut Mittag-Leffler, Djursholm, Sweden

H. P. McKEAN, Courant Institute of Mathematical Sciences, New York University, New York, New York

LINDA PREISS ROTHSCHILD, University of Wisconsin, Madison, Wisconsin

JEDRZEJ SNIATYCKI, University of Calgary, Calgary, Alberta, Canada

STANLY STEINBERG, University of New Mexico, Albuquerque, New Mexico

S. STERNBERG, Harvard University, Cambridge, Massachusetts

R. C. SWANSON, University of Missouri, Columbia, Missouri

DAVID S. TARTAKOFF, Institute for Advanced Study, Princeton, New Jersey

B. A. TAYLOR, University of Michigan, Ann Arbor, Michigan

JEAN E. TAYLOR, Princeton University, Princeton, New Jersey

MICHAEL E. TAYLOR, Rice University, Houston, Texas

JOSEPH A. WOLF, University of California, Berkeley, California

MARCUS WRIGHT, University of Kentucky, Lexington, Kentucky

SHING-TUNG YAU, University of California, Los Angeles, California

STEVEN ZUCKER, Rutgers University, New Brunswick, New Jersey

CONTENTS

*Invited Lecturer

*Invited lecturer

---

*Invited lecturer

Partial Differential Equations and Geometry

ABSTRACTS

SOME DIFFERENTIAL GEOMETRIC STRUCTURES ON THE GENERALIZED BRIESKORN MANIFOLDS AND THEIR PRODUCTS*

Kinetsu Abe

Introduction

The purpose of this article is to announce that the generalized Brieskorn manifolds (defined below) admit some differential geometric structures. These structures are quite natural in the sense that they possess characteristics of the Hopf fibration except possibly for regularity. The proofs of these results and further generalization will apear elsewhere in Abe (a) (to appear) and Abe (b) (to appear).

Preliminaries

Let V be an analytic subvariety of $\mathbb{C}^{n+1}$. Assume that there exists a $\mathbb{C}$-action of V induced from the $\mathbb{C}$-action of $\mathbb{C}^{n+1}$ given by

$$t(Z_0,\ldots,Z_n) = \left(e^{2\pi q_0 t} Z_0,\ldots,e^{2\pi q_n t} Z_n\right),$$

where $t \in C$ and $q_0,\ldots,q_n$ are positive real numbers.

If we assume that the origin of $\mathbb{C}^{n+1}$ is either an isolated singular point of V or a regular point of V, we have that the intersection $\Sigma = V \cap S$ of V and the unit sphere S of $\mathbb{C}^{n+1}$ is a closed and smooth manifold. Σ is in general highly connected except when the dimension of Σ is 3. If $q_0,\ldots,q_n$ are rational numbers, Σ has a S^1-action induced from the $\mathbb{C}$-action on V. This Σ will be referred to as a generalized Brieskorn manifold for the obvious reason. The Brieskorn manifolds and those which are obtained from the weighted homogeneous polynomials are typical examples; see Brieskorn (1966) and Milnor (1968) for example.

* Research partially supported by NSF Grant MCS 76-07152.

We shall talk about contact structures on Σ and complex structures on products of Σ's. As for the definition and the related matters on contact structures, we refer to Sasaki (1968). Let (ϕ,ξ,η) be an almost contact structure on a manifold M. Then (ϕ,ξ,η) is called regular (non-regular) if ξ gives rise to a regular (non-regular) foliation on M. (ϕ,ξ,η) is called normal if the torsion tensor $T\phi(X,Y) = [X,Y] + \phi[\phi X,Y] + \phi[X,\phi Y] - [\phi X,\phi Y] - (X\eta(Y) - Y\eta(X))\xi$ vanishes everywhere. Next, let ω be a contact form on M. Then, ω is called regular (non-regular) or normal if the associated almost contact structure under an appropriate Riemannian metric on M is regular (non-regular) or normal, respectively.

Statement of Results

THEOREM 1. Let Σ be a generalized Brieskorn manifold. Then

(a) There exists one-parameter families of almost contact structures on Σ. Most of them are non-regular, and have closed curves as the leaves of associated foliations. Furthermore, one of them is normal.

(b) There exists a one-parameter family of contact structures on Σ. This one-parameter family connects the contact structure constructed in Abe (1975) and a normal contact structure with closed curves as its leaves of the associated foliation.

For instance, a result of Brieskorn (1966) combined with Theorem 1 gives us that every Brieskorn sphere (exotic or standard) admits infinitely many seemingly different contact structures. Indeed, we can construct infinitely many mutually distinct normal contact structures on a class of exotic spheres and on all the odd dimensional standard spheres. Here we say that two contact structures ω_1 and ω_2 are distinct if there is no diffeomorphism f of M such that $f^*\omega_1 = \omega_2$. (Usually, ω_1 and ω_2 are said to be the same if there is a positive real-valued function g on M such that $f^*\omega_1 = g\omega_2$.) Some results concerning classification of these structures have been obtained. The following is one of them.

THEOREM 2. Let $P_1(Z) = Z_0 + Z_1^{a_1} + Z_2^{a_2}$ and $P_2(Z) = Z_0 + Z_1^{b_1} + Z_2^{b_2}$ be two polynomials with $a_2 \geqslant a_1 > 0$ and $b_2 \geqslant b_1 > 0$. Let ω_1 and ω_2 be the normal contact forms on S^3 corresponding to P_1 and P_2, respectively. Then $\omega_1 \nsim \omega_2$ if either $a_1/(a_1,a_2) \nsim b_1/(b_1,b_2)$ or $a_2/(a_1,a_2) \nsim b_2/(b_1,b_2)$, where (i,j) = g.c.d. of i and j.

THEOREM 3. Let B_p be the pth Betti number of Σ. Then

$$B_p \equiv 0 (\text{mod } 2) \quad \text{if} \quad p \equiv 1 (\text{mod } 2) \quad \text{and} \quad 1 \leqslant p \leqslant \left[\frac{\dim \Sigma}{2}\right],$$

and

$$B_p \equiv 0 \quad \text{if} \quad p \equiv 0 (\text{mod } 2) \quad \text{and} \quad \left[\frac{\dim \Sigma}{2}\right] < p \leqslant 2\left[\frac{\dim \Sigma}{2}\right].$$

Using Theorem 1 and the theorem of Newlander-Nirenberg ((1957), we have

THEOREM 4. $\Sigma_1 \times \Sigma_2$ admits a complex structure, where Σ_1 and Σ_2 are two generalized Brieskorn manifolds. In fact, the product of any even number of generalized Brieskorn manifolds admits a complex structure.

THEOREM 5. $\Sigma_1 \times \Sigma_2$ admits a foliation whose leaves are complex curves. Furthermore, all the analytic subvarieties of $\Sigma_1 \times \Sigma_2$ are also foliated by leaves of the foliation.

As a special case, we have

THEOREM 6. If the normal contact structures on Σ_1 and Σ_2 have closed curves as the leaves of their associated foliations, then $\Sigma_1 \times \Sigma_2$ is a total space of an analytic fibration over an analytic space. The fibers are elliptic curves. Furthermore, all the analytic subvarieties of $\Sigma_1 \times \Sigma_2$ admit a fibration induced from that of $\Sigma_1 \times \Sigma_2$.

Note here that the condition imposed on the normal contact structures on Σ_1 and Σ_2 above is not so artificial. For example, all the original Brieskorn manifolds and those which obtained from weighted homogeneous polynomials satisfy the condition.

The complex structures given in Theorem 4 are closely related to those of Hopf (1948), Calabi-Eckmann (1953) and Brieskorn-Van de Ven (1968).

As in the Calabi-Eckmann (1953), most of these complex structures do not admit a Kählerian structure; therefore, they are not algebraic. Finally, we can show that many of these products admit infinitely many seemingly different complex structures. The following are some of the more refined results.

THEOREM 7. $\Sigma_1 \times \Sigma_2$ admit infinitely many distinct complex structures if one of Σ_1 or Σ_2 is either a certain exotic sphere or any odd dimensional standard sphere or a generalized lens space. Furthermore, $S^1 \times S^2 \times \cdots \times S^{4k-1}$ $(k = 1,2,\cdots)$ admits infinitely many distinct complex structures.

THEOREM 8. Let $P_1(Z)$ and $P_2(Z)$ be given as in Theorem 2. Then the associated complex structures on $S^1 \times S^3$ with P_1 and P_2 are distinct if either $a_1/(a_1,a_2) \nmid b_1/(b_1,b_2)$ or $a_2/(a_1,a_2) \nmid b_2/(b_1,b_2)$ holds.

REFERENCES

Abe, K., and Erbacher, J. (1975). Non-regular contact structures on Brieskorn manifolds. Bull. Amer. Math. Soc. 81.

Abe, K. (to appear). On a generalization of Hopf fibrations I and II.

Abe, K. (to appear). On a class of Hermitian manifolds.

Brieskorn, E. (1966). Beispiele zur Differential Topologie von Singularitaten. Invention Math. 2: 1-44.

Brieskorn, E., and Von de Ven, A. (1968). Some complex structures on products of homotopy spheres. Topology 7: 389-393.

Calabi, E., and Echmann, B. (1953). A class of compact complex manifolds which are not algebraic. Ann. of Math. 58: 494-500.

Hopf, H. (1948). Zur Topologie der komplexen Mannigfaltigkeiten. In Studies and Essays presented to R. Courant, New York.

Milnor, J. (1968). Singular points of complex hypersurfaces. Princeton University Press.

Newlander, A., and Nirenberg, L. (1957). Complex analytic coordinates in almost complex manifolds. Ann. of Math. 65: 391-404.

Sasaki, S. (1968). Almost contact manifolds. Lecture Notes, I, II and III. Tohoku University.

University of Connecticut
Storrs, Connecticut

MULTIPLE VALUED SOLUTIONS TO VARIATIONAL PROBLEMS AND THE REGULARITY OF MASS MINIMIZING INTEGRAL CURRENTS

F. J. Almgren, Jr.

The generalized branching structure of some of the singularities in mass minimizing integral currents in general dimensions and codimensions can be modelled by multiple valued functions minimizing a generalized Dirichlet integral. Following several examples derived from complex algebraic varieties, the structure of such functions is discussed and is applied to the almost everywhere regularity of such currents.

Princeton University
Princeton, New Jersey

CONFORMAL DEFORMATIONS OF RICCI CURVATURE ON GLOBALLY HYPERBOLIC LORENTZ MANIFOLDS*

John K. Beem and Paul E. Ehrlich

Let (M,g) be a time oriented Lorentz manifold. This means that M is a connected n-dimensional smooth manifold with a Lorentz metric of signature $(-,+,\cdots,+)$ and a time orientation. A tangent vector v is

*This research was funded by a grant from the Research Council of the Graduate School, University of Missouri-Columbia.

nonspacelike (resp. timelike) if $v \neq 0$ and $g(v,v) \leq 0$ (resp. $g(v,v) < 0$). A future directed nonspacelike (resp. timelike) curve is a smooth curve whose tangent vector is always nonspacelike (resp. timelike). A point q is said to be in the causal future $J^+(p)$ of p if there is a future directed nonspacelike curve from p to q. The causal past $J^-(p)$ of p is defined using past directed nonspacelike curves. It is known that if g' is another Lorentz metric on M, then the causal future and causal past sets of M with respect to g' coincide with those of g iff the two metrics are conformal. Thus we will restrict our attention to the set $C(M,g)$ of Lorentz metrics on M conformal to g.

When $n = 4$, the Ricci tensor is related to the energy momentum tensor T by the Einstein equations. Thus the curvature condition which we study has interesting physical interpretations. Assuming the cosmological constant is zero, it can be shown that the curvature condition $Ric(g)(v,v) \geq 0$ for all nonspacelike v is equivalent to the energy condition $T(v,v) - g(v,v)trace(T)/2 \geq 0$ for all nonspacelike v. The energy condition has been used by physicists in studying cosmology while the corresponding curvature condition has been used in proving important singularity theorems, cf. Hawking and Ellis (1973). Furthermore, Lee (1975) has investigated compact pre-space-times and has shown that the Lorentz metric on such a manifold can always be chosen such that the condition $Ric(g)(v,v) > 0$ for all nonspacelike v cannot hold.

Given a riemannian manifold, the compactness and geodesic convexity of any sufficiently small closed metric ball can be used to study the local geometry. Analytically, this local convexity is equivalent to that of the function $(d(p,.))^2$ where $d(p,.)$ denotes the metric distance from p. The second author found these properties to be crucial in studying local metric deformations of the Ricci and sectional curvature of a given riemannian metric (Ehrlich (1975) and (1976), cf. Aubin (1970)). In Ehrlich (1976) in particular, given any riemannian metric g and any g-convex metric ball centered at p, a standard conformal deformation $g(t)$ of g was constructed using the g-distance function $(d(p,.))^2$.

The most obvious difficulties in trying to deform the Ricci tensor of a given Lorentzian metric by the same construction arise from the basic differences between local riemannian and Lorentzian geometry. The Lorentzian distance function $d(p,.)$ from an arbitrary point $p \in (M,g)$ is defined to be zero in $M - J^+(p)$ and is calculated for $q \varepsilon J^+(p)$ by taking the l.u.b. of future directed piecewise nonspacelike curves from p to q. The sets $\{q \varepsilon M: d(p,q) = \varepsilon > 0\}$ which are thus contained in $J^+(p)$ are neither topological spheres nor even compact in general. Thus, the Lorentzian distance function itself is not suitable for constructing the standard deformation of the Ricci tensor. In place of the Lorentzian distance, we utilize distance functions defined using fixed normal coordinate charts.

From now on, we consider globally hyperbolic Lorentz manifolds (M,g) of dimension $\geqslant 3$ satisfying the curvature condition $Ric(g)(v,v) \geqslant 0$ for all nonspacelike $v \varepsilon TM$. A globally hyperbolic Lorentz manifold is a Lorentz manifold satisfying strong causality and the condition $J^+(p) \cap J^-(q)$ is compact for all $p,q \varepsilon M$. (Strong causality means that every point has arbitrarily small neighborhoods which each nonspacelike curve intersects at most once.) If (M,g) is globally hyperbolic, then (M,g') is globally hyperbolic for all $g' \varepsilon C(M,g)$. Geroch (1970) has shown that all globally hyperbolic Lorentz manifolds (M,g) are metric products $M = R \times S$ where S is an $(n-1)$-dimensional spacelike submanifold of M such that $\{t\} \times S$ is a Cauchy surface of M for all $t \varepsilon R$. Further, there is a smooth function $h: M \to R$ whose gradient is nonvanishing and timelike at all points of M and moreover, $h^{-1}(t) = \{t\} \times S$ for all $t \varepsilon R$. Such a function h is called a globally hyperbolic time function.

Definition: The Lorentz manifold (M,g) satisfies the strict curvature condition on a subset N of M iff $Ric(g)(v,v) > 0$ for all nonspacelike tangent vectors v with $\pi(v) \in N$.

We prove

THEOREM A: Let (M,g) be a globally hyperbolic Lorentz manifold of dimension $\geqslant 3$ satisfying the curvature condition $Ric(g)(v,v) \geqslant 0$ for all nonspacelike $v \in TM$. Suppose (M,g) admits a globally hyperbolic time function h such that the strict curvature conditions holds for all nonspacelike tangent vectors v attached to some Cauchy hypersurface $h^{-1}(t_0)$. Then, there is a metric g' for M conformal to g such that the strict curvature condition $Ric(g')(v,v) > 0$ holds for all nonspacelike $v \in TM$.

The proof of Theorem A involves: (I) constructing a standard local deformation with support in a ball-like set using normal coordinates adapted to the causality structure, and (II) using global hyperbolicity to obtain natural compact subsets related to h having regions satisfying the strict curvature condition. We first discuss (II).

Fix $t_0 \in R$ and let p be any point in M with $h(p) > t_0$. Set $J(p,t_0) = J^+(p) \cap h^{-1}[t_0,\infty)$. Then $J(p,t_0)$ is compact and $N(p,t_0) = \mathrm{Int}(J(p,t_0))$ is globally hyperbolic. Thus, we may fix a globally hyperbolic time function $h_1 : N(p,t_0) \to R$ inducing the same time orientation on $N(p,t_0)$ as h. Typical of what is proved in (II) is:

Proposition B: For any $t \in R$ and $\varepsilon > 0$, the <u>edge</u>

$$E(t,\varepsilon,h_1) = \mathrm{cl}(h_1^{-1}[t - \varepsilon, t + \varepsilon]) - h_1^{-1}[t - \varepsilon, t + \varepsilon]$$

is contained in the Cauchy surface $h_1^{-1}(t_0)$. (Here the closure is taken in M.)

Proposition C: Suppose (M,g) satisfies the strict curvature condition on $h^{-1}(t_0)$. Then, given any $\alpha \in R$, there is a metric g' conformal to g such that (M,g') satisfies the strict curvature condition on $h^{-1}(t_0) \cup h_1^{-1}(-\infty,\alpha]$.

We now sketch the construction (I) of the standard local deformation used to enlarge the region of positive Ricci curvature for a Lorentz manifold (M,g) satisfying the curvature condition. Fix $p \in M$ and let h be any smooth real valued function defined near p such that ∇h is timelike in a neighborhood of p. Construct normal coordinates $x = (x^1,\ldots,x^n)$ on an open neighborhood U of p from an orthogonal basis for T_pM whose first basis vector has direction $-\nabla h(p)$. Given a fixed $q \in U$ and any real constant $L > 0$, define a smooth function $\rho = \rho(L,q) : U \to R$ by

$$\rho(m) = L^2 - \textstyle\sum_i (x^i(m) - x^i(q))^2, \quad m \in U$$

For L sufficiently small, the ball-like set

$$B(L,q) = \{m \in M; \rho(L,q)(m) > 0\}$$

is contained in U. Deform the metric g on B(L,q) by

$$q(t) = \begin{cases} e^{-2t\rho^5} g & \text{on } B(L,q) \\ g & \text{on } M - B(L,q) \end{cases}$$

For suitably chosen L and q, the new metrics g(t) with $0 < t \leqslant 1$ will all satisfy the strict curvature condition on $B(L,q) \cap h^{-1}[h(p),\infty)$. Intuitively, points near the north and south polar caps of b(L,q) satisfy the strict curvature condition for the deformed metrics.

REFERENCES

Aubin, T. "Métriques Riemannienes et Courbure." J. Diff. Geometry, 4 (1970), pp. 383-424.

Ehrlich, P. "Local convex deformations and sectional curvature." Archiv der Mathematik, 26 (1975), pp. 432-435.

Ehrlich, P. "Metric deformations of curvature I: local convex deformations." Geometriae Dedicata, 5 (1976), pp. 1-23.

Geroch, R. "The domain of dependence." J. Math. Physics, 11 (1970), pp. 437-439.

Hawking, S. and Ellis, G. The large scale structure of space-time. Cambridge monographs on mathematical physics. Cambridge U. Press (1973).

Lee, K. "Another possible abnormality of compact space-time." Canad. Math. Bull., 18 (1975), pp. 695-697.

The article on which this abstract was based is:

Beem, John K., and Ehrlich, Paul E. "Conformal deformations, Ricci curvature and energy conditions on globally hyperbolic space-times." To appear in Proc. Cambridge Phil. Soc.

University of Missouri
Columbia, Missouri

AN OBSTRUCTION TO CERTAIN NON-INTEGRABLE 2-PLANE FIELDS*

Robert P. Buemi

Motivated by Bott's vanishing Theorem on the vanishing of certain characteristic classes of integrable Pfaffian systems, we have studied, in this paper, a generalization of completely integrable Pfaffian systems, namely, the constant rank derived flag systems and have asked if there is any obstruction to the existence of such systems. In this paper, we have shown that a necessary condition for a manifold to admit a system of type (0,1,...,1.2) (for manifolds of dimension at least 4) or of type (1,...,1,2) (for manifolds of dimension at least 5) is the vanishing of the entire Pontryagin ring of the manifold.

Massachusetts Institute of Technology
Cambridge, Massachusetts

SOLUTIONS OF MONGE AMPERE EQUATIONS BY OPTIMAL CONTROL METHODS AND APPLICATIONS.

B. Gaveau

§1. In one complex variable, a function is the real part of a holomorphic function, if and only if it is harmonic (at least locally); this is why the theory of analytic functions of one complex variable often reduces to

*Published in Topology, 16, pp. 173-176.

the study of harmonic functions (for example, distributions of zeroes, growth questions, divisors on Riemann surfaces and abelian integrals via Dirichlet principle...). In several complex variables, the real part of a holomorphic function is harmonic for usual potential theory in $\mathbb{C}^n$, but the converse is false; this remark was first used by Poincaré to construct holomorphic functions; it is also used to solve $\bar{\partial}$ equations. More generally, P. Malliavin has noticed that a real part of a holomorphic function is harmonic for every Kähler metric, and he has applied this remark to study boundary behavior and distribution of zeroes using Kähler metrics adapted to the problem considered. (See P. Malliavin, 1969, 1974 and 1975.) As an example of the interactions of several potential theories, let us prove the following easy result:

Let B be the unit ball in $\mathbb{C}^n$, f holomorphic in B, $V = f^{-1}(0)$ its zeroes; call θ the angle between the complex tangent to V and the radial vector. Suppose that V is such that

$$\int_V d(z,\partial B)d\sigma(z) < +\infty \tag{1}$$

where $d\sigma$ is the euclidean area of V and $d(z,\partial B)$ is the distance from z to the boundary ∂B. Then

$$\int_V \cos^2 \theta \, d\sigma(z) < +\infty \tag{2}$$

The proof goes as follows: for any Kähler metric ds^2 in the ball one has

$$\int_{\partial B_\varepsilon} \log|f(\zeta)|d\mu(\zeta) = \log|f(0)| + \int_{V\cap B_\varepsilon} g_\varepsilon(0,\zeta)d\sigma_{ds^2}(\zeta) \tag{3}$$

where B_ε is the ball of radius $1 - \varepsilon$, g_ε is the Green's function of the metric for B_ε, $d\mu$ is the harmonic measure of 0 in ∂B, $d\sigma_{ds^2}$ the area element of V for the ds^2. Apply this for ds^2 the euclidean metric; the hypothesis (1) and the Riesz formula (3) imply

$$\sup \int_{\partial B_\varepsilon} \log|f(\zeta)|d\sigma(\zeta) < +\infty$$

because of the fact that the euclidean Green's function for the ball is like $d(z,\partial B)$ near ∂B (but f is not necessarily Nevanlinna). Now apply (3) with ds^2 the Bergmann metric, then

$$\int_V g(0,\zeta)d\sigma_{Berg}(\zeta) < +\infty$$

and by Malliavin's (1974) estimates this implies (2).

§2. Now, it is very elementary to prove that a function is plurisubharmonic (resp. pluriharmonic) if and only if it is subharmonic (resp. harmonic) for every elliptic operator of the form

$$\Delta_a = \sum a_{ij}(z) \frac{\partial^2}{\partial z_i \partial \bar{z}_j} \tag{4}$$

where a_{ij} is any hermitian matrix which is $\geqslant 0$. It seems that in some questions of function algebras and of envelopes of holomorphy, one is lead to use, at one time, all Kähler potential theories. In fact, this is a rather general phenomenon; more or less, each time there is a first order elliptic partial differential system, one can think to simultaneously consider all potential theories. For example, in topology, the best result should be obtained by considering all metrics or all connections at one time. (See B. Gaveau, 1977.)

§3. Consider now the complex Monge Ampère equations

$$\begin{cases} (i\partial\,\bar{\partial}u)^{\Lambda^n} = \left(\frac{2}{n}\, f\right)^n & \text{in } D \quad f^n \geqslant 0 \\ u = \phi & \text{in } \partial D \\ u & \text{plurisubharmonic} \end{cases} \tag{5}$$

You can view the $(i\partial\bar{\partial}u)^{\Lambda^n}$ as determinant of the complex hessian of u. But one has the following elementary equality

$$\frac{n}{2}(\det H)^{1/n} = \frac{1}{2}\,\mathrm{Inf}(\sum a_{ij}h_{ij}) \tag{6}$$

where $H = (h_{ij})$ is a positive hermitian matrix $n\times n$, and the infimum is taken over the class of all positive hermitian matrices $n\times n$ with $\det a \geqslant 1$. Using (6), the complex Monge Ampère problem (5) can be rewritten as

$$\begin{cases} \mathrm{Inf}\Delta_a u = f & \text{in } D \\ u = \phi & \text{in } \partial D \\ u & \text{plurisubharmonic} \end{cases} \tag{7}$$

where Δ_a is defined by (4) and the infimum is taken over the class of positive hermitian n×n matrices a(z) with det a ⩾ 1. But now, (7) is known as a Bellman dynamic programming equation (see Fleming and Rishel, 1975) and formally, it can be solved as in the following section.

§4. Let $(b_t^1 \ldots b_t^n)$ be the complex brownian motion in $\mathbb{C}^n$ and (Ω, P) its probability space. Let $\mathcal{H}$ be the class of mappings $(s,\omega) \in \mathbb{R}^+ \times \Omega \to (\sigma_{ij}(s,\omega))_{ij}$ with (σ_{ij}) positive hermitian n×n matrix such that $\det \sigma\sigma^* \geqslant 1$, which are not anticipating over the future of brownian path. For every $\sigma \in \mathcal{H}$, let $X^{(\sigma,z)}$ be its response

$$X_s^{(\sigma,z)^i}(\omega) = z^i + \int_0^s \sigma^{ij}(t,\omega)\, db_t^j(\omega) \tag{8}$$

σ is called a kählerian control, (8) is a stochastic integral but not a stochastic differential equation. Let $\zeta_{(\sigma z,)}$ be the first exit time from D of $X^{(\sigma,z)}$ and

$$\begin{cases} w(z,\sigma) = E\left(-\int_0^{\zeta_{(\sigma,z)}} f(X_s^{(\sigma,z)})\,ds + \phi\left(X_{\zeta_{(\sigma,z)}}^{(\sigma,z)}\right)\right) \\ u(z) = \operatorname*{Inf}_{\sigma\in\mathcal{H}} w(z,\sigma) \end{cases} \tag{9}$$

If u were C^2, it would be easy to prove that u, defined by (9), is, indeed, a solution of (7) and then of (5). But it is, in general, very difficult to prove a regularity theorem for u even when one has a strictly elliptic optimal control problem (which is not the case here, because the operators Δ_a are not uniformly elliptic with a lower bound of ellipticity independent of a). So, one uses another device proving directly that the function u is indeed a solution of (5) in the generalized sense of Bedford-Taylor (1976) (without going through the differential form of (7)).

§5. First, suppose that $D = \{z \in \mathbb{C}^n / p(z) < 0\}$ where p is a C^2 strictly plurisubharmonic function, so that D is strictly pseudoconvex domain. Suppose that $f \geqslant 0$, bounded, uniformly continuous on D and call ψ its modulus of continuity. Suppose that ϕ is continuous on ∂D and call Φ its modulus of continuity. Define u by (9). One proves:

(i) there exist some constants C_1, C_2 depending only on D, such that for every $z_0 \in \partial D$, $z \in D$

$$|u(z) - \phi(z_0)| \leqslant C_1 \Phi\left(C_2 |z-z_0|^{1/2}\right) \tag{10}$$

(ii) for every relatively compact domain D' in D, there exist C_3, C_4, C_5 depending only on D', $D \| f \|_\infty$, $\| \phi \|_\infty$ such that for every z, $z' \in D'$

$$|u(z) - u(z')| \leqslant C_3(\psi(C_4|z-z'|) + \Phi(C_5|z-z'|^{1/2}) \tag{11}$$

In particular, u is continuous on $\overline{D}$ and is ϕ on ∂D.

(iii) u satisfies the Bellman principle (integral version of (7)); that is, for every $z \in D$, for every domain $D' \subset \overline{D}' \subset D$ with $z \in D'$, u satisfies

$$u(z) = \operatorname*{Inf}_{\sigma \in \mathfrak{X}} E\left(-\int_0^{\zeta(\sigma,z)} f(X_s^{(\sigma,z)})ds + u\left(X^{(\sigma,z)}_{\zeta(\sigma,z)}\right)\right) \tag{12}$$

where $\tau_{(\sigma,z)}$ is the first exit time of $X^{(\sigma,z)}$ from D'.

(iv) from (12) and the remark of section 2, u is a plurisubharmonic function in D.

(v) u is a solution, in the generalized sense (of Bedford-Taylor, 1976) of the complex Monge Ampère problem (5).

REMARK 1. This is proved in B. Gaveau (1977). Bedford and Taylor prove existence in the case where f and ϕ are Hölder functions.

REMARK 2: All this works with the real Monge Ampere equation, strictly convex domains, and convex functions in exactly the same way. Moreover, it applies to general Monge Ampere equations to define generalized solutions in a new way.

§6. One can also consider the problem

$$\begin{cases} (i\partial\overline{\partial}u)^{\Lambda^n} = f(z,u) & \text{on } D \\ u = \phi & \text{in } \partial D \\ u & \text{plurisubharmonic} \end{cases} \tag{13}$$

where f is continuous, $\geqslant 0$ in $\mathbb{R} \times \overline{D}$, and locally Lipschitz in u.

Call

$$M(D) = \frac{\max_D(-p)}{\min_D\left(\det \frac{\partial^2 p}{\partial z_i \partial \bar{z}_j}\right)^{1/n}}$$

Then if $M(D)$ is sufficiently small, (13) has a continuous solution in $\bar{D}$, plurisubharmonic on D. (See also Bedford-Taylor, preprint, for similar problems and B. Gaveau, 1977, for detailed conditions.)

§7. As an application of (9) to the function algebra of the ball, one can prove that, for any $z \in B$, the convex combinations of kählerian harmonic measures of z on ∂B for all Kähler metrics are dense in the set of all Jensen measures of $A(\bar{B})$ for z. (See Fleming and Rishel, 1975, and B. Gaveau, 1977.)

§8. One can extend the result of section 5 to any bounded symmetric domain D in $\mathbb{C}^n$; but then the boundary value is given only on the Silov boundary of D. The class of controls $\mathcal{K}_z(D)$ is then such that if $\sigma \in \mathcal{K}_z(D)$, the response $X^{(\sigma,z)}$ has to converge towards a point of the Silov boundary at the first exit time of D (all the other conditions being the same). Then $\bigcup_{z \in D} \mathcal{K}_z$ is a kind of infinite dimensional bundle over D which can be trivialized when D is homogeneous; one can also get a regularity result (if f and ϕ are C^2, then u is $L^\infty_{2,loc}$) just as in Bedford-Taylor (1976). One also obtains partial results for some classes of weakly pseudoconvex domains (for example, domains such as $\{q(z) < 0\}$ where q is C^3 in a neighborhood of $\bar{D}$ and plurisubharmonic in D, or for analytic polyhedra). We prove that the function u given by (9) (where the infimum is taken over $\mathcal{K}_z(D)$) is plurisubharmonic and is the supremum of the class $\mathcal{B}(f,\phi)$ where $\mathcal{B}(f,\phi)$ is the class of all plurisubharmonic functions in D such that:

$$\begin{cases} \limsup_{z \to z_0} u(z) \leqslant \phi(z_0) & \forall z_0 \in \text{Silov } D \\ \Phi(u) \geqslant \frac{2f}{n} \end{cases} \tag{14}$$

where $\Phi(u)$ is defined as

$$\left(\det \frac{\partial^2 u}{\partial z_i \partial \bar{z}_j}\right)^{1/n}$$

if u is C^2 and then extended as in Bedford-Taylor to all plurisubharmonic functions.

Also, u is a decreasing limit of continuous plurisubharmonic functions u_n such that $\Phi(u_n) = f$. (See B. Gaveau, 1977, for details.)

9. Further applications to polynomially convex hulls and holomorphically convex hulls will appear in Debiard-Gaveau (1977). The optimal control method can be used to define an extremal capacitary potential in $\mathbb{C}^n$ and a new kind of capacity. Moreover, it is also used to obtain partial results about quasilinear elliptic equations in 3 real variables, which is degenerate (in the sense that the principal symbol is always degenerate). See Debiard-Gaveau (1977).

REFERENCES

Bedford, E., and Taylor, B.A. A Dirichlet problem for a complex Monge Ampère equations. Inventiones Mathematicae (1976).

Bedford, E., and Taylor, B.A. Preprint.

Debiard, A., and Gaveau, B. Méthodes de contrôle optimal en analyse complexe: applications aux envellopes d'holomorphie et à un problème de Dirichlet quasilinéaire dégénéré. Comptes Rendus Acad. Sc. Paris (July, 1977).

Fleming, W., and Rishel, R. Deterministic and stochastic optimal control. Springer-Verlag (1975).

Gaveau, B. Méthodes de contrôle optimal en analyse et en topologie. Comptes Rendus Acad. Sc. Paris (January, 1977).

Gaveau, B. Méthodes de contrôles optimal en analyse complex I. Résolution d'équations de Monge Ampère. J. of Functional Analysis (August, 1977).

II. Cas des domaines faiblement pseudoconvexes (to appear).

(I. has been announced in Comptes Rendus Acad. Sc. Paris, 284, (March, 1977) p. 593).

Malliavin, P. Comportement à la frontière distinguée d'une fonction de plusieurs variables complex. <u>Comptes Rendus Acad. Sc. Paris</u> (Feb., 1969).

Malliavin, P. Fonctions de Green et classe de Nevanlinna d'un ouvert strictement pseudoconvexe. <u>Comptes Rendus Acad. Sc. Paris</u> (Jan., 1974).

Malliavin, P. Diffusions et géométrie différentielle globale. C.I.M.E. (1975).

18, rue Gassendi
59000 Lille
France

THE HOLOMORPHIC TRACE PROBLEM AND THE MINIMAL COMPLEX ENVELOPE

Gary A. Harris

Suppose M is a k-dimensional real-analytic submanifold of C^n and consider the following question: "Which complex valued functions defined on M are locally the restrictions to M of an ambient holomorphic function?" In case M is a C.R. submanifold, it is known that a real-analytic function $f:M \to C$ is locally the trace on M of a holomorphic function if and only if f is a C.R. function. (See Tomassini, 1966.) In case the Tangential Cauchy Riemann equations for M may have singularities (i.e., M is not a C.R. submanifold) the above question is answered in part by the following theorem. For the remainder of this discussion all functions and sets are to be thought of as germs at the origin.

THEOREM A. Suppose $k \geq n$ and M is generic in some neighborhood of each point which is not a singularity of the Tangential Cauchy Riemann equations. There exist holomorphic coordinates $w_1,\ldots,w_n$ for C^n, an open dense subset $N \subset M$ and real-analytic vector fields $Y_1,\ldots,Y_k$ defined on N such that

(i) $Y_iY_j = Y_jY_i$ for each $i,j = 1,2,\ldots,k$.

(ii) For each $q \in N$, $\{Y_{iq},\ldots,Y_{kq}\}$ is a basis for T_qM and $\{Y_{n+1q},\ldots,Y_{kq}\}$ give the Tangential Cauchy Riemann equations for M at q.

(iii) for each $j = 1,2,\ldots n$

$$Y_{jq} = \frac{\partial}{\partial w_j}\Big|_q \bmod T''_q C^n$$

($T''C^n$ denotes the (0,1)-vector fields in TC^n.)

Moreover, for any such $Y_1,\ldots,Y_k$, a given real-analytic function $f:M \to C$ is the trace of a holomorphic function if and only if

a. $Y_1^{\alpha 1}\ldots Y_n^{\alpha n}(f)|_N$ extends real-analytically to all of M for each n-tuple $\alpha = (\alpha_1,\ldots,\alpha_n)$ and

b. $Y_jf|_N \equiv 0$ for each $j = n+1,\ldots,k$.

For example, consider the submanifold $M \equiv \{(z_1,z_1\bar{z}_1)|z_1 \in C\} \subset C^2$ and let $N = M\setminus\{(0,0)\}$,

$$Y_1 = \frac{\partial}{\partial z_1} - \frac{\bar{z}_1}{z_1}\frac{\partial}{\partial \bar{z}_1} \quad \text{and} \quad Y_2 = \frac{\partial}{\partial z_2} + \frac{1}{z_1}\frac{\partial}{\partial \bar{z}_1} + \frac{\partial}{\partial \bar{z}_2}$$

Let f be real-analytic in a neighborhood of (0,0). It suffices to consider f of the form $f = \sum a_{mn} z_1^m \bar{z}_1^n$. For all $\alpha = (\alpha_1,\alpha_2)$, $Y_1^{\alpha_1}Y_2^{\alpha_2}(f)$ is analytically extendable across $(0,0) \Leftrightarrow$ for every $v = 0,1,2,\ldots (1/z_1\ \partial/\partial\bar{z}_1)^v(f)$ is analytically extendable across $(0,0) \Leftrightarrow a_{mn} = 0$ whenever $m < n$.

The development and proof of Theorem A is the content of Sections 1,2, and 3 of Harris (submitted) and can be sketched as follows. By the local nature of the theorem one may assume $0 \in M$, $U \subset R^k$ is an open polydisc about 0 and M has a real-analytic parametric representation $\Phi:U \to M \subset C^n$ with $\Phi(0) = 0$. Let $\tilde{\Phi} \equiv (\phi_1,\ldots,\phi_n)$ denote the unique holomorphic extension of Φ to an open polydisc $V \subset C^k = C \otimes_R R^k$. It now suffices to consider a more general question: "Given a holomorphic mapping $\tilde{\Phi}:C^k \to C^n$ and a holomorphic function $f:C^k \to C$, when does there exist a holomorphic function $F:C^n \to C$ such that $f = F \circ \tilde{\Phi}$?"

This factoring problem is solvable if the generic rank of $\tilde{\Phi}$ is n. This means there is an $n \times n$ minor A of the matrix $(\partial\phi_i/\partial z_j)$ such that $\det A \not\equiv 0$. Notice it is not assumed that rank $(\partial\phi_i/\partial z_j)(0) = n$. (In this case, the problem can be solved by appealing to the Rank Theorem.) The solution proceeds in two stages.

STAGE 1. Necessary conditions on f are found which lead to the existence of $F \in C[[w_1,\ldots,w_n]]$, the ring of formal power series centered at 0, such that $F \circ \tilde{\Phi} \equiv F(\phi_1,\ldots,\phi_n) \equiv f$. ($F(\phi_1,\ldots,\phi_n)$ is well defined because $\phi_j(0) = 0$ for each $j = 1,2,\ldots,n$.) The existence of F is shown by construction as follows. Because the generic rank of $\tilde{\Phi}$ is n, one can construct a set $\{Z_1,\ldots,Z_k\}$ of commuting linearly independent holomorphic vector fields defined on $C^k\backslash E$ where $E \equiv \{z \in C^k \mid \text{rank}(\partial\phi_i/\partial z_j)(z) < n\}$, such that for each $i,j = 1,2,\ldots,n$, $Z_i(\phi_j) = \delta_i^j$ and for each $i = n+1,\ldots,k$, $Z_i(\phi_j) \equiv 0$. (δ_i^j is the Kronecker Delta.) It follows that if $F: C^n \to C$ is holomorphic and $F \circ \tilde{\Phi} = f$, then

(i) For each $\alpha = (\alpha_1,\ldots,\alpha_n)$, $Z_1^{\alpha_1}\ldots Z_n^{\alpha_n}(f)$ extends holomorphically across E, and

(ii) $Z_{n+1}(f) \equiv \ldots \equiv Z_k(f) \equiv 0$.

Moreover, for each $\alpha = (\alpha_1,\ldots,\alpha_n)$

$$Z_1^{\alpha_1}\ldots Z_n^{\alpha_n}(f) = \frac{\partial^{|\alpha|}F}{\partial w_1^{\alpha_1}\ldots\partial w_n^{\alpha_n}}(\phi_1,\ldots,\phi_n)$$

On the other hand, if f satisfies (i) and (ii), the obvious candidate for the solution F is the formal series

$$F \equiv \sum_{|\alpha|=0}^{\infty} \frac{Z_1^{\alpha_1}\ldots Z_n^{\alpha_n}(f)(0)}{\alpha_1!\ldots\alpha_n!} w_1^{\alpha_1}\ldots w_n^{\alpha_n}$$

The first stage of the proof is completed by deriving the formal identity $F \circ \tilde{\Phi} \equiv f$.

STAGE II. F is shown to have positive radius of convergence by appealing to the following algebraic result. (See Paul Eakin and Gary Harris, to appear.)

THEOREM B. Let $\tilde{\Phi}$ be a holomorphic mapping from C^k to C^n with $\tilde{\Phi}(0) = 0$. The generic rank of $\tilde{\Phi}$ is n if and only if for each $F \in C[[w_1,\ldots,w_n]]$, $F \circ \tilde{\Phi}$ convergent implies F is convergent.

Theorem A extends the above mentioned results of Tomassini to certain $(k \geqslant n)$ non C.R. submanifolds M which are generic off the singular set for the Tangential Cauchy Riemann equations. Moreover, the proof of Theorem A recovers Tomassini's results for a general real-analytic C.R. submanifold (the Rank Theorem can be applied).

At this point it appears natural to consider the following question: "Does there exist a complex submanifold $\tilde{M}$ of complex dimension $s \leqslant k$ and containing M such that M is generic in $\tilde{M}$, thus reducing to the hypothesis of Theorem A?" Such a submanifold $\tilde{M}$ must be unique and is called the minimal complex envelope of M. The minimal complex envelope need not always exist. (See Section 6 of Gary Harris, submitted, for examples.) Indeed, such $\tilde{M}$ exists if and only if the generic rank of $\tilde{\Phi}$ is s and there is a biholomorphic mapping $G:C^n \to C^n$ such that $G \circ \tilde{\Phi}$ is of the form $(\gamma_1,\ldots,\gamma_s,0,\ldots,0)$. Without loss of generality, assume the generic rank of $\tilde{\Phi}' \equiv (\phi_1,\ldots,\phi_s)$ is s. It is an easy exercise to see that such G exists if and only if, for every $j = s + 1,\ldots,n$, there is a holomorphic function $F_j:C^s \to C$ such that $\phi_j = F_j \circ \tilde{\Phi}'$. But this is the situation of the factoring problem with $\tilde{\Phi}$ replaced by $\tilde{\Phi}'$. Because $\tilde{\Phi}':C^k \to C^s$ has generic rank s, the factoring problem is solvable as indicated above. In theory this solves the question of existence of the minimal complex envelope; however, the solution is not very pleasant and remains to be simplified and hence clarified.

REFERENCES

Eakin, Paul and Harris, Gary. "When $\Phi(f)$ Convergent Implies f is Convergent." Mathematische Annalen (to appear).

Harris, Gary. "The Traces of Holomorphic Functions on Real Submanifolds." Submitted.

Tomassini, Giuseppe. "Tracce Delle Funzioni Olomorfe Sulle Sottovarieta Analitiche Reali D'una Varieta Complessa." Ann. Scuola Normale Sup., Pisa, 20 (1966), pp. 31-43.

University of Kentucky
Lexington, Kentucky

THE NON-EXISTENCE OF ELLIPTIC COORDINATES IN GENERAL RIEMANNIAN MANIFOLDS

Hans-Christoph ImHof*

Let M denote a complete riemannian manifold of dimension n. For two fixed points $p,q \in M$ we define the functions $e,h:M \to \mathbb{R}$ by $e(m) = d(m,p) + d(m,q)$ and $h(m) = d(m,p) - d(m,q)$, respectively. A local coordinate system $(x_1,\ldots,x_n)$ is called elliptic if $x_1 = e$, $x_2 = h$, and if for all suitable constants $c_1,\ldots,c_n$ the submanifolds $\{m \in M;\ e(m) = c_1,\ h(m) = c_2\}$ are orthogonal to the submanifolds $\{m \in M;\ x_k(m) = c_k$ for $k \geqslant 3\}$.

In euclidean goemetry, elliptic coordinates are a convenient tool for problems involving two distinguished points, but in general riemannian manifolds they do not exist. More precisely, there are elliptic coordinates for each pair of points if and only if either the given manifold is two-dimensional or has constant sectional curvature.

The existence of elliptic coordinates for two given points is equivalent to the integrability of a certain distribution induced by these two points. Using Jacobi field techniques, it is shown that the integrability of this distribution for any pair of points implies constant sectional

*Supported by the Swiss National Science Foundation.

curvature, provided the dimension of the given manifold is at least three. Conversely, if a manifold has dimension two or constant sectional curvature, then the distribution induced by a pair of points is always integrable.

University of California
Berkeley, California

THE POINCARE LEMMA FOR $d\omega = F(x,\omega)$

Howard Jacobowitz

This paper develops an analogue of the Frobenius theorem or, equivalently, Poincaré's lemma for the equation $d\omega - F(x,\omega)$ where F is a map of p-forms to $p + 1$ forms. According to the Frobenius theorem, the system of partial differential equations $\partial f/\partial x_j = F_j(x,f)$ may be solved locally in a region, with f taking on a given value at any point of this region, if and only if in this region $\partial F_j/\partial x_k = \partial F_k/\partial x_j$ when the derivatives $\partial f/\partial x_j$ which occur in the use of the chain rule are replaced by $F_j(x,f)$.

Let us say that $d\omega = F(x,\omega)$ is solvable in a region $D \subset R^n$ if for each $x_0 \in D$ and each exterior p-form ω_0 at x_0 there is some differential p-form ω defined in a neighborhood of x_0 with $d\omega = F(x,\omega)$ in this neighborhood and $\omega = \omega_0$ at x_0.

THEOREM. The equation $d\omega = F(x,\omega)$ is solvable in D if F satisfies the following conditions for each $x_0 \in D$:

This work was supported in part by NSF grants MPS 75-05577 and MCS 76-06978.

$$dF(x,\beta) = 0 \text{ at } x_0 \text{ for each differential } p\text{-form } \beta \text{ which satisfies } d\beta = F(x_0,\beta) \text{ at } x_0 \qquad (*)$$

The theorem also holds for systems of such equations. Further, there is a more restricted definition of solvability which is actually equivalent to (*).

Note two special cases. If ω is a 0-form, i.e., a function, then we have the Frobenius theorem. And if F does not depend on ω then we have the Poincaré lemma: If $dF = 0$ then there is some ω with $d\omega = F$.

The proof starts with an algebraic lemma which allows us to factor $dF(x,\omega)$ in terms of $d\omega - F(x,\omega)$. The rest of the proof is based on the proof of the Frobenius theorem. We integrate the equation in one direction and then use the compatibility conditions (*) to show the equation holds in all directions.

EXAMPLE 1. We give a simple proof for the following result which is the crucial step in establishing a converse to Lie's third fundamental theorem. (Summations are taken over all repeated indices.)

LEMMA. Given n^3 constants c^i_{jk} with $c^i_{jk} = -c^i_{kj}$, there exist n linearly independent one forms ω_i satisfying

$$d\omega_i = \frac{1}{2}\sum c^i_{jk}\omega_j\omega_k$$

if and only if these constants satisfy

$$\sum(c^i_{jk}c^j_{rs} + c^i_{jr}c^j_{sk} + c^i_{js}c^j_{kr}) = 0$$

for all $1 \leqslant i,k,r,s, \leqslant n$.

Proof. Let $F_i(\omega_1,\ldots,\omega_n) = \frac{1}{2}\sum c^i_{jk}\omega_j\omega_k$. If one can solve $d\omega_i = F_i(\omega)$, then $dF_i(\omega) = 0$. That is, $\sum c^i_{jk}c^j_{rs}\omega_r\omega_s\omega_k = 0$. This is equivalent, assuming $\{\omega_i\}$ is linearly independent, to $\sum(c^i_{jk}c^j_{rs} + c^i_{jr}c^j_{sk} + c^i_{js}c^j_{kr}) = 0$. Thus, this condition is necessary. For the converse, assume one has 1-forms $\{\beta_i\}$ which satisfy $d\beta_i = F(\beta)$ just at

*An abstract of a paper to appear in the Journal of Differential Geometry.

some point x_0. It is clear that $\sum(c^i_{jk}c^j_{rs} + c^i_{jr}c^j_{sk} + c^i_{js}c^j_{kr}) = 0$ implies $dF(\beta) = 0$ at x_0. Thus, (*) is satisfied and we may solve $d\omega_i = F_i(\omega)$, specifying initial values at some point. If this set of initial values is linearly independent at the point, then $\{\omega_i\}$ is linearly independent in some neighborhood of this point.

EXAMPLE 2. If one studies $d\omega = F(\omega)$ on R^2, where ω is a 1-form, then (*) is trivially satisfied and the equation is always solvable. We may use this observation to study curvature tensors on two dimensional manifolds.

THEOREM. Let $R(X,Y)$ be smooth $\mathrm{Hom}(TR^2,TR^2)$-valued 2-form defined in a neighborhood of some point $p \in R^2$. There exists some torsion free connection defined near p, which has R as its curvature tensor.

Proof. If ω_1 and ω_2 are linearly independent 1-forms on R^2, then the dual tangent vectors e_1 and e_2 are well defined. Let $\phi_{ij}(X,Y) = \omega_i(R(X,Y)e_j)$ and consider the system

$$d\omega_i = -\sum\omega_{ij}\omega_j$$

$$d\omega_{ij} = -\sum\omega_{ik}\omega_{kj} + \phi_{ij}$$

for $1 \leqslant i,j,k \leqslant 2$. At p choose initial values so that ω_1 and ω_2 are linearly independent. Again, the differential equations may be solved. In terms of the now known vectors e_1 and e_2, we may write $R(X,Y)e_j = \sum_i r_{ij}(X,Y)e_i$ and this defines the 2-forms R_{ij}. Finally, note $\phi_{ij} = R_{ij}$. Thus, $\{\omega_{ij}\}$ are the connection forms of a connection whose curvature tensor is R.

Rice University
Houston, Texas

SOME APPLICATIONS OF GEOMETRIC QUANTIZATION

Jedrzej Sniatycki

Classical dynamical variables can be obtained from the corresponding quantum operators in the classical limit of vanishing Planck's constant, $\hbar \to 0$. The aim of quantization procedures is to reverse this process, that is, to obtain the quantum operators from the corresponding classical functions. A classical system is described by a symplectic manifold representing the phase space of the system together with its Lagrange bracket. This structure is insufficient for the quantization purposes. In the geometric quantization scheme one uses an auxiliary structure on the phase space consisting of prequantization, polarization and a metaplectic structure (Blattner, 1973).

The phase space of a single particle is isomorphic to $\mathbb{R}^6$ with the Lagrange bracket given by

$$\omega = \sum_{i=1}^{3} dp_i \wedge dq_i$$

where p_i and q_i are the momentum and the position coordinates, respectively. The prequantization and the metaplectic structure are essentially unique. Taking the vertical polarization spanned by

$$\left(\frac{\partial}{\partial p_1}, \frac{\partial}{\partial p_2}, \frac{\partial}{\partial p_3}\right)$$

one obtains the Schroedinger representation of quantum mechanics. The geometric quantization of energy and of components of the position, the linear momentum and the angular momentum vectors, yields the standard results. The same method applied to the square of the total angular momentum encounters some difficulty due to possible divergences in the evaluation of the Blattner-Kostant-Sternberg kernels. It seems that if one tried to avoid this difficulty one would get a quantum operator with the spectrum $\hbar^2\ell(\ell + 1)$ modified by a constant correction term depending on the method used (Elhadad, 1977 and Rawnsley, 1977). Blattner-Kostant-Sternberg kernels given by a non-relativistic energy function coincide with the Van Vleck approximation to solutions of the corresponding Schroedinger equation. The iteration of this approximation, if convergent, leads to the Feynman path integral (Blackman, 1976).

REFERENCES

Blackman, J. "Geometric Quantization and the Feynman Path Integral." Preprint, Syracuse University (1976).

Blattner, R. J. "Quantization and Representation Theory" in Harmonic Analysis on Homogeneous Spaces. Proceedings of Symposia in Pure Mathematics, 26, A. M. S., Providence, R. I. (1973), pp. 147-165.

Elhadad, J. "Quantification du flot geodesique de la sphere S^n." Preprint, Université de Provence (1977).

Rawnsley, J. H. "A Non-unitary Pairing of Two Polarizations of the Kepler Manifold." Preprint, Dublin Institute for Advanced Studies (1977).

University of Calgary
Calgary, Alberta
Canada

THE ABSTRACT CAUCHY PROBLEM

Stanly Steinberg

This is a report on some work with R. Hersh that is currently in progress.

To describe the problem we are interested in, we need some notation. Let X be a Banach space, P_i, $0 \leqslant i \leqslant N$, be possibly unbounded operators on X and y_i, $1 \leqslant i \leqslant N$ be elements of X. The Abstract Cauchy

Problem is the problem of finding smooth X-valued function $x(t)$ that is a solution of

$$\sum_{i=0}^{N} P_i \left(\frac{d}{dt}\right)^i x(t) = 0 \tag{1}$$

$$\left(\frac{d}{dt}\right)^j x(t)\Big|_{t=0} = y_j, \quad 0 \leqslant j \leqslant N - 1 \tag{2}$$

We also make the standard assumption that P_N is the identity.

The critical assumptions that we make are that each P_i is polynomial in n operators A_k, $1 \leqslant k \leqslant n$,

$$P_i = P_i(A_1, \ldots, A_n) \tag{3}$$

and that the A_k are a basis of a Lie algebra.

If we also assume that each A_k is a group generator and that the joint analytic vectors of the A_k are dense in X, then we can view (1) as an invariant differential operator on a Lie group G. It is then possible to give existence and uniqueness results for equations on Lie groups that are analogous to the Cauchy-Kowalewski theorem or the results for parabolic and hyperbolic equations that have constant coefficients.

These existence and uniqueness results can be used to give results for certain degenerate hyperbolic or parabolic equations. These results are not easily achieved by any other method.

As examples of Lie algebras of operators we have:

(I) $A_1 = ix$, $A_2 = d/dx$, $A_3 = iI$

(II) $A_1 = ix^2$, $A_2 = i\, d^2/dx^2$, $A_3 = xd/dx + 1/2$

(III) $A_M = \sum_{i,j=1}^{m} x_i M_{ij}\, \partial/\partial x_j$, M a matrix.

Here $i^2 = -1$, I is the identity and X is the square integrable functions on the real line or m dimensional Euclidean space. Clearly I and II have higher dimensional analogs.

Let us consider some operators:

$$\partial/\partial t + \sum_{i=1}^{n} a_i A_i + b \tag{4}$$

$$\partial/\partial t + \sum_{i,j=1}^{n} a_{ij} A_i A_j + \sum_{i=1}^{n} b_i A_i = c \tag{5}$$

$$\partial^2/\partial t^2 + \sum_{i,j=1}^{n} a_{ij} A_i A_j + \sum_{i=1}^{n} b_i A_i + c \tag{6}$$

In (4), if the a_i are real, then the operator will be first order hyperbolic. Such equations have many applications and viewing them as differential equations on a Lie group helps explain many of the Baker-Campbell-Hausdorff identities available for these equations, see Steinberg, to appear in Journal of Differential Equations and In preparation.

In (5), if a_{ij} is a positive definite matrix (or semi-definite), then (5) is parabolic (or degenerate parabolic). There is a growing interest in these operators and their applications, see Jorgensen, 1975. In (6), if a_{ij} is positive definite, then (6) is strictly hyperbolic. The study of these operators and their applications is just beginning, see Hersh and Steinberg, In preparation.

These results can be used to show that the initial value problem for certain unusual initial value problems is well posed, for example,

$$\frac{\partial^2 f}{\partial t^2} = - \frac{\partial^2 f}{\partial x^4} + x^2 \frac{\partial^2}{\partial x^2} + 2x \frac{\partial f}{\partial x} + (\frac{1}{4} - x^4) f \tag{7}$$

$$\frac{\partial^2 f}{\partial t^2} = (x^2 + y^2) \left(\frac{\partial^2}{\partial x^2} + \frac{\partial^2}{\partial y^2} \right) f. \tag{8}$$

Here $f = f(x,t)$ and

$$f(x,0) = f_0(x), \quad \frac{\partial f(x,0)}{\partial t} = f_1(x).$$

Equation (7) is not of standard type, while (8) is hyperbolic with degeneracies at zero and infinity. Both are strictly hyperbolic on the appropriate Lie group.

The Cauchy-Kowalewski results can be applied to equations of the type

$$\sum_{i,j,k=0}^{N} a_{ijk} x^i (\partial/\partial x)^j (\partial/\partial t)^k f(x,t) = 0 \tag{9}$$

If

$$d = \sup\{(i+j)/(N-k); a_{ijk} \neq 0\} \tag{10}$$

and $d \leqslant 2$, then we obtain solvability results for (9).

In addition to existence results, it is possible to give integral representations for the solutions of the parabolic and hyperbolic equations. Let G be the Lie group generated by the A_i, π be the representation of G on X and $d\mu$ be a Haar measure on G. If the equation we are considering is given by

$$\begin{aligned} &\frac{dx(t)}{dt} - P(A_1,\ldots,A_n)x(t) = 0 \\ &x(0) = y \end{aligned} \tag{11}$$

then

$$x(t) = \int_G F(g,t)\pi(g)y\,d\mu(g)$$

where $F(g,t)$ is a fundamental solution of a differential equation on G related to (11) in an elementary way. In the Abelian case, this representation has interesting applications, see Donaldson and Hersh (1970) and Bobisud and Hersh (1972).

REFERENCES

Bobisud, L., and Hersh, Reuben. Perturbation and Approximation Theory for Higher-Order Abstract Cauchy Problems. _Rocky Mountain J. of Math._, 2 (1972), pp. 57-73.

Donaldson, James A., and Hersh, Reuben. A Perturbation Series for Cauchy's Problem for Higher-Order Abstract Parabolic Equations. _Proc. Nat. Acad. Sci._, _67_ (1970), pp. 41-44.

Hersh, Reuben, and Steinberg, Stanly. The Abstract Cauchy Problem and Hyperbolic Equations on a Lie Group. In preparation.

Jorgensen, Palle E. T. Representations of Differential Operators on a Lie Group. J. Functional Analysis, 20 (1975), pp. 105-135.

Steinberg, Stanly. Applications of the Lie Algebraic Formulas of Baker, Campbell, Hausdorff and Zassenhaus to the Calculation of Explicit Solutions of Partial Differential Equations. To appear in J. of Differential Equations.

Steinberg. On Exponential Representations of Solutions of Evolution Equations whose Coefficients Depend on Time. In preparation.

University of New Mexico
Albuquerque, New Mexico

LAGRANGE INTERSECTION THEORY AND ELLIPTIC BOUNDARY VALUE PROBLEMS

R. C. Swanson

The oriented intersection theory, due to V. Arnol'd, of curves of lagrangian subspaces of a symplectic vector space, is extended to an infinite dimensional setting: the author constructs a certain Fredhold manifold of lagrangian Hilbert subspaces, which is shown to possess the stable homotopy type of the direct limit of a family of finite dimensional (lagrange) grassmann manifolds. A formula is obtained for the computation of the intersection number of a continuous arc of lagrangian subspaces with a singular submanifold, the "Maslov cycle", which generalizes previous finite dimensional work of J. Duistermaat and L. Hormander. This material relies on some published work and, as a consequence, is somewhat expository.

The intersection theory is applied to establish a version of the Morse-Smale index theorem for an elliptic boundary value problem with a one-parameter family of Dirichlet boundary conditions arising from a smooth contracting deformation of the underlying compact domain. The proof depends on the interpretation of the Morse index and the number of conjugate points of the deformation as oriented intersection numbers of suitable curves of infinite dimensional largrange subspaces. Finally, a general (non-Dirichlet) variational self-adjoint boundary value problem is considered. A formula is given which expresses the Morse index of the variational problem as a sum of the number of conjugate points (in the associated Dirichlet problem) and a "boundary index" arising from a certain family of mixed boundary conditions.

University of Missouri
Columbia, Missouri

CRYSTALLINE INTEGRANDS AND THE CALCULUS OF VARIATIONS

Jean E. Taylor

Physical crystals provide examples of integrands which are not elliptic: with a fixed orientation for a given crystalline substance, the energy per unit surface area required to expose a given face defines a function $F:G_0(3,2) \to R^+$ (here $G_0(3,2)$ is the Grassmannian of oriented planes through the origin R^3); the total surface energy of a piece of crystal C is then $\int_{x\in\partial C} F(\mathrm{Tan}(\partial C,x))\,d\mathcal{H}^2 x$. Given any continuous $F:G_0(n+1,n) \to R^+$, the problem of finding the shape of the open set in R^{n+1} of volume one, whose center of gravity is at the origin, and whose surface minimizes the integral of F among all such open sets, is solved by the Wulff Construction. Integrands F corresponding to crystalline materials have Wulff shapes which are polyhedral, whereas elliptic integrands correspond to uniformly convex open sets. By analogy, F is called crystalline if and only if the corresponding Wulff shape is polyhedral; this shape is called the crystal of F. (Since the lattice structures of physical crystals impose restrictions on the symmetries of F, there are many crystalline integrands which do not come from physical crystals.)

A much more general problem is the prescribed boundary problem. The existence (as limits of varifolds corresponding to integral currents) of "surfaces" having a given integral cycle of dimension $n - 1$ as boundary and minimizing the integral of F is easily shown; one can define and select a "locally volume maximizing" solution V from the set of all such solutions, and show that at $\|V\|$ almost all points the support of V either has a tangent plane in a certain finite set (determined by the crystal of F alone) or is a cylinder in a neighborhood of that point with tangent plane lying in the dual of the $(n - 1)$-skeleton of the crystal of F. An explicit procedure for calculating these surfaces is being worked out. Such a procedure is of interest not only for its own sake, but also as a means of calculating explicit approximations to surfaces minimizing the integrals of elliptic integrands, since any elliptic integrand can be approximated by crystalline integrands.

Rutgers University
New Brunswick, New Jersey

FUNCTION THEORY ON ALGEBRAIC MANIFOLDS WITH NONNEGATIVE KODAIRA DIMENSION

Marcus Wright

Introduction

Let M be a compact algebraic manifold and K the canonical bundle of M. If $\{\phi_0,\ldots,\phi_N\}$ is a basis of $H^0(M,K^s)$ for s an integer, we can obtain a mapping Φ_s from M into P^N via $M \ni p \mapsto (\phi_0(p),\ldots,\phi_N(p))$.

Definition. The Kodaira dimension $\kappa(M)$ of M is equal to:

$$\begin{cases} -\infty & \text{if } H^0(M,K^s) = 0 \text{ for all } s \geqslant 1 \\ \max\limits_s \dim_C \Phi_s(M) & \text{otherwise.} \end{cases}$$

Thus, $\kappa(M) \in \{-\infty,0,1,2,\ldots,n = \dim_C M\}$ and $\kappa(M) \geqslant 0$ if and only if there is a nontrivial global section α of K^s in $H^0(M,K^s)$ for some positive integer s.

In this announcement are discussed results about those algebraic manifolds for which there exist at least one such global "holomorphic s-form." The case $\kappa(M) = \dim M = n$ has been extensively studied by Kobayashi(1972), Griffiths (1971), Kodaira (1971), Kobayashi and Ochiai (1975), and many others. We consider this case in Section 1; the results concern a kind of generalization of the Bergman kernel and distance for such manifolds; we indicate how these can be used to study the group of biholomorphic self-mappings (henceforth referred to as automorphisms) of such M and to study the surjective holomorphic mappings between two such manifolds. In Section 2, we discuss the case $0 \leqslant \kappa(M) < n$, and present a result which corresponds to work of Kiernan (1977) and Kobayashi and Ochiai (1975) in the case $\kappa(M) = \dim M$. Details of proof for all new results will appear in Wright (to appear).

§1. $\kappa(M) = \dim M$

When $\kappa(M) = \dim M$, M is said to be algebraic of general type. The following fundamental lemma is proved in Kodaira (1971).

Lemma 1.1. Let $\kappa(M) = \dim M$ and L be a line bundle over M such that the global sections of L give an imbedding of M into projective space. Then there is a positive integer m such that $H^0(M,K^mL^{-1}) \neq 0$.

Suppose that $\alpha \in H^0(M,K^mL^{-1})$ is nontrivial. Then the homomorphism $I:H^0(M,L) \to H^0(M,K^m)$ defined by $I(\phi) = \alpha \otimes \alpha$ is injective. If L has the property stated in the lemma (i.e., L is "very ample"), then we may obtain an imbedding Φ of M into projective space using sections of K^m. We define $\Phi(p) = (\alpha(p)\phi_0(p),\ldots,\alpha(p)\phi_N(p))$, where $\{\phi_0,\ldots,\phi_N\}$ is a basis for $H^0(M,L)$. The pseudo-volume form

$$\eta(p) = |\alpha(p)|^{2/m}\left(\sum_{i=1}^{N} \phi_i(p)\overline{\phi_i(p)}\right)^{1/m}$$

will vanish where α does. But $i\partial\bar{\partial}\log \eta$ is an everywhere positive definite (1,1)-form (see Kobayashi, 1976). This fact leads to an equidimensional Schwarz lemma (see Griffiths, 1971, Kiernan, 1977, Kobayashi and Ochiai, 1975). Using this, Kobayashi and Ochiai have shown that the set of meromorphic surjective mappings from M' to M is finite (M' is any Moisezon space). In particular, the group of automorphisms of M is finite, as was earlier proved by Kobayashi (1972). Other results about the function theory of M have been proved by exploiting the existence of a volume form satisfying a Schwarz lemma. (See Griffiths, 1971, Kiernan, 1977).

We intend to sketch a different proof of the finiteness of the automorphism group of M which uses the existence of a "Bergman metric." This kind of metric, in the case of negative first Chern class of M, was introduced by Narasimhan and Simha (1968). Then, we will state some new results which can be obtained using this approach.

Suppose that $\phi \in H^0(M,K^m)$. Define

$$\|\phi\| = \left[\int_M (\zeta)^{1/m}(\phi \otimes \overline{\phi})^{1/m}\right]^{m/2}$$

where $\zeta = \{i^{n^2}(-1)^n\}^{-m}$. Clearly

(i) $\|\phi\| = 0 \Leftrightarrow \phi \equiv 0$ and

(ii) $\|\lambda\phi\| = |\lambda|\,\|\phi\|$ for all $\lambda \in C$.

Let $U(M) = \{\phi \in H^0(M,K^m) \mid \|\phi\| \leqslant 1\}$. A generalization of the Bergman kernel is the following: if $p \in M$, define

$$\tau(p) = \sup_{\phi\in U(M)} (\zeta)^{1/m}(\phi \otimes \overline{\phi})^{1/m}(p)$$

It can, in general, only be shown that τ is a continuous pseudo-volume form, for it may vanish where α does. However, if f is an automorphism of M, then $f^*U(M) = U(M)$ which implies that $f^*\tau(p) = \tau(p)$, a property which τ shares with the Bergman kernel form. Now if $\phi_1,\phi_2 \in IH^0(M,L)$ then

$$(\phi_1,\phi_2) = \int_M \zeta\frac{\phi_1 \otimes \overline{\phi}_2}{\tau^{m-1}}$$

defines a nondegenerate hermitian inner product which has the property that $(f^*\phi_1,f^*\phi_2) = (\phi_1,\phi_2)$ if f is an automorphism of M. This inner product induces a hermitian metric, denoted h, on M, and corresponding to h there is a distance d_h on M which is invariant under automorphisms of M.

THEOREM 1.1. If $\kappa(M) = \dim(M)$, then the group of automorphisms of M is finite. It is a closed subgroup of the group of isometries of M with respect to d_h.

As mentioned before, this result was obtained without introduction of d_h by Kobayashi (1972). The present method of proof leads to the following extension of the previously quoted result of Kobayashi and Ochiai.

THEOREM 1.2. Suppose M and M' are algebraic of general type (i.e., $\kappa(M) = \dim M$, $\kappa(M') = \dim M'$). Then the set of surjective holomorphic mappings from M' to M is compact.

In Kobayashi and Ochiai (1975), it was shown that the set of surjective meromorphic mappings from M' to M is compact.

The pseudo-volume form η is also an analog of the Bergman kernel. Unlike the Bergman kernel, η might not be invariant under automorphisms. But we can show:

THEOREM 1.3. Let

$$\gamma(p) = \left(\sum_{i=0}^{N} \phi_i \overline{\phi}_i (p) \right)^{1/m}$$

where $\{\phi_0, \ldots, \phi_N\}$ is an orthonormal basis of $\mathbb{H}^0(M,L)$ with respect to the inner product defined above. Then γ is invariant under automorphisms of M.

It turns out that the distance d_h is associated to the positive definite (1,1)-form $i\partial\overline{\partial}\log \gamma$.

2. $\dim M > \kappa(M) \geqslant 0$.

We are assuming that M is an algebraic manifold such that for some positive integer s there is a nontrivial section $\alpha \in H^0(M,K^s)$. As before, let L be a very ample line bundle with $\{\phi_0, \ldots, \phi_N\}$ a basis for $H^0(M,L)$. We have a projective imbedding $\Phi: M \to P^N$ defined by $p \mapsto (\alpha(p)\phi_0(p), \ldots, \alpha(p)\phi_N(p))$. There is no obvious way to concoct a pseudo-volume form corresponding to η without knowing something more about L. In fact, a good way to describe the difficulty of studying automorphisms or other function theory of M by using the existence of

Φ is to say that, without further assumptions on M, an automorphism of M doesn't necessarily lift to the P^N in which it is imbedded. (This is what happens when $\kappa(M) = \dim M$.) See Kobayashi (1972), pp. 106-112 for a discussion of this point.

We are able to exploit Φ if we make an assumption about the (measure-) hyperbolicity of M.

Definition 2.1. If M is a complex manifold of dimension n, $p \in M$ and $w = (w_1, \ldots, w_n)$ a coordinate system at p, then the Kobayashi-Eisenman pseudo-volume form μ is expressed in coordinates w as $k_w(q) dw_1 \wedge d\bar{w}_1 \wedge \ldots \wedge dw_n \wedge d\bar{w}_n$ where

$$k_w(q) = \inf \frac{1}{|J_f(0)|^2}$$

and the infimum is taken over all holomorphic mappings f from the Euclidean unit ball in C^n to M with $f(0) = q$ and $J_f(0)$ is the complex Jacobian determinant of f computed with respect to standard coordinates on B and coordinates w on M.

Definition 2.2. M is strongly measure hyperbolic if $\mu \geqslant \nu$ for some continuous volume form ν on M.

THEOREM 2.3. Let M be an algebraic strongly measure hyperbolic manifold with $\dim H^0(M, K^s) \neq 0$ for some positive integer s. Then if $\{f_j : M \to M\}$ is a sequence of automorphisms of M, there is a subsequence of $\{f_j\}$ which converges to a nondegenerate meromorphic mapping $f: M \to M$.

The proof of this theorem proceeds by first noting that if $g: B \to M$ is an imbedding of a ball in M, then there is a commutative diagram of meromorphic mappings:

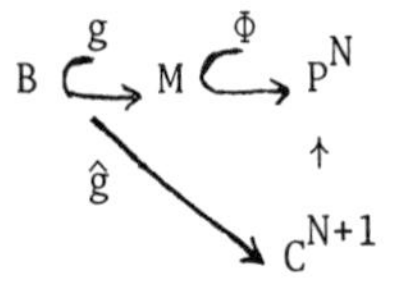

The existence of such a lifting of g follows because $g^* L$ is holomorphically trivial. Without using this, Kiernan (1977) obtains liftings in the case $\kappa(M) = \dim(M)$ which are shown to behave like mappings into bounded domains. In the context of Theorem 2.3, we employ the two hypotheses to show that when g comes from an automorphism, such a "good" $\hat{g}$ exists. This is the key step in the proof.

REFERENCES

Griffiths, P. "Holomorphic mappings into canonical algebraic varieties." Ann. of Math., 93 (1971) pp. 439-458.

Kiernan, P. "Meromorphic mappings into compact complex spaces of general type." Proc. Symp. in Pure Math., Vol. 30, A. M. S. (1977), pp. 239-244.

Kobayashi, S. Tranformation Groups in Differential Geometry. Springer Verlag (Ergebnisse series no. 70) Berlin/New York (1972).

Kobayashi, S. "Intrinsic distances, measures and geometric function theory. B. A. M. S. 82 (1976), pp. 357-416.

Kobayashi, S. and Ochiai, T. "Mappings into compact complex manifolds with negative first Chern class." J. Math. Soc. Japan, 23 (1971), pp. 137-148.

Kobayashi, S. and Ochiai, T. "Meromorphic mappings onto compact complex spaces of general type." Invent. Math., 31 (1975), pp. 7-16.

Kodaira, D. "Holomorphic mappings of polydiscs into compact complex manifolds." J. Diff. Geom., 6 (1971), pp. 33-46.

Narasimhan, M. S. and Simha, R. R. "Manifolds with ample canonical class." Invent. Math., 5 (1968), pp. 120-128.

Wright, M. W. "Automorphisms of compact algebraic manifolds with nonnegative Kodaira dimension." To appear.

University of Kentucky
Lexington, Kentucky

THE DIRICHLET PROBLEM FOR AN EQUATION OF COMPLEX MONGE-AMPERE TYPE

Eric Bedford* and B. A. Taylor†

In Bedford and Taylor (1976), generalized solutions were constructed for the Dirichlet problem for the complex Monge-Ampère operator $(dd^c)^n = dd^c \wedge \dots \wedge dd^c$. If $u = u(z_1,\dots,z_n)$ is a function of class C^2, $d = \partial + \bar{\partial}$, $d^c = i(\bar{\partial} - \partial)$, then $dd^c u = 2i\partial\bar{\partial}u$ and

$$(dd^c u)^n = 4^n n! \det\left[u_{i\bar{j}}\right] dV$$

where

$$u_i = \frac{\partial u}{\partial z_i}, \quad u_{\bar{j}} = \frac{\partial u}{\partial \bar{z}_j}, \quad u_{i\bar{j}} = \frac{\partial^2 u}{\partial z_i \partial \bar{z}_j}$$

and

$$dV = \prod_{j=1}^{n} \frac{i}{2} dz_j \wedge d\bar{z}_j$$

If $\Omega \subset \mathbb{C}^n$ is an open set and if $P(\Omega)$ denotes the cone of plurisubharmonic functions on Ω, then $(dd^c)^n : C^2(\Omega) \cap P(\Omega) \to M(\Omega)$ has a continuous extension to $C(\Omega) \cap P(\Omega)$, where $M(\Omega)$ is the set of Borel measures on Ω with the weak topology. With $(dd^c)^n$ interpreted in this sense, we study the problem

$$\begin{aligned} &u \in P(\Omega) \cap C(\bar{\Omega}) \\ &u = \Phi \quad \text{on} \quad \partial\Omega \\ &(dd^c u)^n = F(u,z)dV, \quad z \in \Omega \end{aligned} \qquad (1)$$

*Research supported in part by a National Science Foundation Project at Princeton University.
†Research supported in part by a National Science Foundation Grant at the University of Michigan

where Ω is a bounded, strictly pseudoconvex domain in $\mathbb{C}^n$ with C^2 boundary. We will show how the proofs of Bedford and Taylor (1976) may be modified suitably to establish the following existence theorem, (see Gaveau (1977) for a related result).

THEOREM A. If $\Phi \in C(\partial\Omega)$ is given, $F(u,z) \in C(R \times \bar{\Omega})$, $F \geq 0$, and $F^{1/n}$ is convex and nondecreasing in u, then there exists a unique solution to the Dirichlet problem (1). If, in addition, $F^{1/n} \in \mathrm{Lip}(\mathbb{R} \times \bar{\Omega})$ and $\Phi \in C^2(\partial\Omega)$, then $u \in \mathrm{Lip}(\bar{\Omega})$.

The best regularity we can obtain for a solution is when Ω is the unit ball in $\mathbb{C}^n$.

THEOREM B. Suppose that $\Omega = B^n$ is the unit ball in $\mathbb{C}^n$. If $\Phi \in C^2(\partial\Omega)$ and $F^{1/n} \in C^2(\mathbb{R} \times \bar{\Omega})$ is nonnegative, convex, and increasing in u, then the solution u of (1) has locally bounded second partial derivatives.

Now we consider the transformation $v = e^{-u}$ or $-\log v = u$, which yields the identity

$$e^{-(n+1)u}(dd^c u)^n = J(v) \tag{2}$$

where J is defined as the (n,n) current

$$J(v) = (-1)^n[v(dd^c v)^n - n\, dv \wedge d^c v \wedge (dd^c v)^{n-1}] \tag{3}$$

It is easily seen that

$$J(v) = n!\ (-4)^n \det \begin{bmatrix} v & v_{\bar{k}} \\ v_j & v_{j\bar{k}} \end{bmatrix} dV$$

The determinant is related to the asymptotic behavior of the Bergman Kernel function (see Fefferman, 1976). By (2), the equation

$$\det \begin{bmatrix} u_{i\bar{j}} \end{bmatrix} = e^{ku} \tag{4}$$

is transformed to

$$J(v) = \frac{v^{n+1-k}}{4^n n!}\, dV$$

With the function $F(u,z) = g(z)e^{(\alpha+n+1)u}$, the problem (1) is transformed to

$$\begin{aligned} &v \in S(\Omega) \cap C(\overline{\Omega}) \\ &v = \psi \quad \text{on} \quad \partial\Omega \\ &J(v) = v^{-\alpha} g(z)\, dV \end{aligned} \tag{5}$$

where $\psi > 0$ on $\partial\Omega$ and $S(\Omega) = \exp(-P(\Omega))$. This exponential mapping allows us to define $J(v)$ as a measure for $v \in S(\Omega) \cap C(\Omega)$, via equation (2). It is also possible to define J directly on $\pm P(\Omega) \cap C(\Omega)$ via (3), since $dv \wedge d^c v \wedge (dd^c v)^j$ is a positive $(j+1,j+1)$ current if $v \in P(\Omega) \cap C(\Omega)$. (See Bedford and Taylor, Proposition 3.2.) Thus, we can deduce the following result from Theorem A.

THEOREM A'. If $\psi \in C(\partial\Omega)$, $\psi > 0$, and $g \in C(\overline{\Omega})$, $g \geqslant 0$, $\alpha \in R$, $\alpha \geqslant -(n+1)$, then there is a unique solution v to (5).

The equation (4) arises naturally in a geometric context (see Calabi, 1975 and Nirenberg, 1974). A Kähler metric $g = \Sigma g_{ij} dz_i d\overline{z}_j$ may be written locally as

$$g_{ij} = \frac{\partial^2 u}{\partial z_i \partial \overline{z}_j}$$

for some function u. The Ricci form for this metric is given by

$$R = -dd^c \log\left(\det u_{i\overline{j}}\right)$$

Thus, $R = -kdd^c u$, i.e., the metric is Kähler-Einstein, if u satisfies (4). Such metrics were constructed on tube domains by Calabi, Cheng, Nirenberg, and Yau (1977). They show that for a solution of (4) on a tube domain Ω, the condition that $u \to +\infty$ at $\partial\Omega$ is equivalent to the completeness of the metric g. For such a function u, $v = e^{-u}$ solves (5) with $\psi = 0$, $g \equiv 1$, and $\alpha = k - n - 1$. The transformation between u and v becomes singular for $\psi = 0$ so that the existence and uniqueness of (1) does not apply to this case. Kähler-Einstein metrics have also been constructed on compact manifolds, see Yau (1977); the analysis there is different because there is no boundary.

The operator $(dd^c)^2$ has a continuous extension from $C^2(\Omega) \cap P(\Omega)$ to the Sobolev space $H^1(\Omega,\text{loc}) \cap P(\Omega) \supset C(\Omega) \cap P(\Omega)$. If solutions are considered in this larger class, then it is possible to obtain solutions that are singular at the boundary.

THEOREM C. Let $\Omega \subset \mathbb{C}^2$ be a bounded, strictly pseudoconvex domain with C^2 boundary. There exists $u \in H^1(\Omega,\text{loc}) \cap P(\Omega)$ such that (4) is satisfied with $n = 2$, and $\lim_{z\to\partial\Omega} u(x) = +\infty$.

Regularity of Envelopes of Subsolutions

The solution u in Theorem A will be obtained via the Perron method. Let Φ be the operator

$$\Phi(u) - 4(n!)^{1/n}\left(\det\left[u_{j\bar{k}}\right]\right)^{1/n}$$

which is defined as a measure for all $u \in P(\Omega)$. Some basic properties of Φ are given in Bedford and Taylor (1976), Section 5. If $\phi \in C(\partial\Omega)$ and $F(u,z) \in C(\mathbb{R}\times\bar{\Omega})$ are given, then we define the following families of subsolutions

$$\mathcal{F}(\phi,F) = \{v \in p(\Omega) \cap C(\bar{\Omega}) : v(z) = \phi(z),\ z \in \partial\Omega, \text{ and } (dd^c v)^n \geqslant F(v,z)dV\}$$

$$\mathcal{B}(\phi,F) = \{v \in P(\Omega) : \limsup_{\zeta\to z} v(\zeta) \leqslant \phi(z),\ z \in \partial\Omega, \text{ and } \Phi(v) \geqslant F(v,z)^{1/n}dV\}$$

$$C\mathcal{B}(\phi,F) = \mathcal{B}(\phi,F) \cap C(\bar{\Omega})$$

Proposition 1. Let Ω be a bounded, strictly pseudoconvex domain in $\mathbb{C}^n$ with C^2 boundary. Let $\phi \in C^2(\partial\Omega)$, $F \geqslant 0$, $F^{1/n} \in \text{Lip}(\mathbb{R}\times\bar{\Omega})$, and let $F(u,z)$ be nondecreasing in u. Then if

$$u(z) = \sup\{v(z) : v \in \mathcal{B}(\phi,F)\},\quad z \in \bar{\Omega}$$

then $u \in \text{Lip}(\bar{\Omega})$, $u(z) = \phi(z)$ for $z \in \partial\Omega$, and $u \in C\mathcal{B}(\phi,F)$.

Proof. The proof requires only simple modifications in the proof of Theorem 6.2, p. 28 of Bedford and Taylor (1976), so we omit it.

The next regularity result we need is the key step in showing that the upper envelope of $\mathcal{B}(\phi,F)$ actually solves (1). The proof is a modification of Theorem 6.7, p. 34 of Bedford and Taylor (1976).

Proposition 2. Let $\Omega = B^n$ be the unit ball in $\mathbb{C}^n$, Let $G(u,z) = [F(u,z)]^{1/n} \in C^2(\mathbb{R} \times \overline{B}^n)$ be a convex nondecreasing function of $u \in \mathbb{R}$, and let $\phi \in C^2(\partial\Omega)$. Then

$$u(z) = \sup\{v(z) : v \in \mathcal{B}(\phi,F)\}$$

has locally bounded second order partial derivate in B^n.

Proof. We will show that

$$\frac{1}{2}[u(z+h) + u(z-h)] \leq u(z) + A(\eta)\ |h|^2,\ |z| \leq 1 - \eta,\ |h| \leq \eta/2 \qquad (1.1)$$

where $A(\eta)$ is a constant independent of h and z. It was proved in Theorem 6.7, p. 34 of Bedford and Taylor (1976) that the proposition follows from (1.1). For $a \in B^n$, let $T_a(z) = \Gamma(a)(z - a)(1 - \overline{a}^t z)^{-1}$ where $\Gamma(a) = a^t\overline{a}(1 - v(a))^{-1} - v(a)I$ and

$$v(a) = \sqrt{1 - |a|^2}$$

Here we are following the notation of Eisenman (1970), p. 6, and points of $\mathbb{C}^n$ are thought of as $n \times 1$ column vectors so that $\Gamma(a)$ is an $n \times n$ matrix. Note that T_a is an analytic automorphism of B^n, $T_a(a) = 0$, $T_{-a} = T_a^{-1}$, and $T_a(z)$ is a smooth function of a. If $|a| \leq 1 - \eta$, $|h| < \eta$, then let $L(a,h,z) = T^{-1}_{a+h}(T_z(z))$ and

$$U = u \circ L,\quad \psi = \phi \circ L,\quad G = G(a,h,z) = G(U,L)$$

(Recall that $G(u,z) = [F(u,z)]^{1/n}$.) Then $U = \psi$ for $z \in \partial B^n$, and ψ is a smooth function of a and h so, for a suitable constant $K_1 = K_1|\eta|$,

$$\frac{1}{2}[U(a,h,z) + U(a,-h,z)] = K_1|h|^2 \leq \phi(z),\quad z \in \partial B^n,\quad |a| \leq 1 - \eta,\quad |h| \leq \eta/2 \qquad (1.2)$$

Define

$$v(z) = v(a,h,z) = \frac{1}{2}[U(a,h,z) + U(a,-h,z)] - K_1|h|^2 + K_2(|z|^2 - 1)|h|^2 \qquad (1.3)$$

where K_2 is a constant to be chosen later. If we can prove $v(z) \in \mathcal{B}(\phi,F)$, (1.1) follows by setting $z = a$ (with $A(\eta) = K_1 + K_2$). Because of (1.2), we already know $v(z) \leqslant \phi(z)$, $z \in \partial B^n$ so it remains only to prove that $\Phi(v) \geqslant G(v(z),z)$.

By the chain rule, $\Phi(U) = \Phi(u)j(a,h,z)$ where $j^{n/2}$ is the absolute value of the Jacobian determinant of the map $z \to L(a,h,z)$. By Theorem 5.7, p. 24 of Bedford and Taylor (1976), Φ is superadditive so from (1.3),

$$\phi(v) \geqslant \frac{1}{2}\,\Phi(u)j(a,h,z) + \frac{1}{2}\,\Phi(u)j(a,-h,z) + 4(n!)^{1/n}K_2|h|^2$$

Since $u \in \mathcal{B}(\phi,F)$, it follows that

$$\Phi(v) \geqslant \frac{1}{2}\,H(a,h,z) + \frac{1}{2}\,H(a,-h,z) + K_3|h|^2$$

where

$$H(a,h,z) = G(U(a,h,z),L(a,h,z))j(a,h,z)$$

Since the function $G(w,L(a,h,z)$ is C^2 in h, there exists a constant K_4 such that

$$|G(w,L(a,h,z)j(a,h,z) + G(w,L(a,-h,z)j(a,-h,z) - 2G(w,L(a,0,z))j(a,0,z)| \leqslant K_4|h|^2 \tag{1.5}$$

for $|w| \leqslant \sup |u|$, $|z| \leqslant 1$, $|h| \leqslant \eta/2$. Consequently, if we apply (1.5) (twice) with $w = U(a.\pm h,z)$ to give a lower bound for the terms on the right hand side of (1.4), then the following inequality if obtained:

$$\begin{aligned}\Phi(v) \geqslant\ & G(U(a,h,z),L(a,0,z))j(a,0,z) \\ & - \frac{1}{2}\,G(U(a,h,z),L(a,-h,z))j(a,-h,z) \\ & + G(U(a,-h,z),L(a,0,z))j(a,0,z) - \frac{1}{2}\,G(U(a,-h,z),L(a,h,z))j(a,h,z) \\ & + (K_3 - K_4)\ |h|^2\end{aligned} \tag{1.6}$$

Write $w(z) = w(a,h,z) = \frac{1}{2}\,[U(a,h,z) + U(a,-h,z)]$. Using the convexity of $u \to G(u,z)$ and noting that $L(a,0,z) \equiv z$, $j(a,0,z) \equiv 1$, we see that the right hand side of (1.6) is at least as large as

$$G(w(z),z) + (K_3 - K_4)\ |h|^2 + I$$

where

$$I = G(w(z),z) - \frac{1}{2} G(U(a,h,z),L(a,-h,z))j(a,-h,z)$$
$$- \frac{1}{2} G(U(a,h,z),L(a,h,z))j(a,h,z)$$

Since $w(z) = v(z) + O(|h|^2)$, we have

$$G(w(z),z) \geqslant G(v(z),z) - K_5|h|^2$$

To estimate I, note that I is (the negative of) the second difference of the smooth function $(w,h) \to G(w,L(a,-h,z))j(a,-h,z)$ at the points $(w(z),0)$, $(w(z) + \Delta w,-h)$, $(w(z) - \Delta w,+h)$, where $\Delta w = \frac{1}{2}[U(a,h,z) - U(a,-h,z)]$. Thus, we have

$$|I| \leqslant K_6(|h|^2 + |\Delta w|^2)$$

However, $u(z)$ is a Lipschitz function by Proposition 1. Thus $|\Delta w| \leqslant K_7|h|$, so that

$$|I| \leqslant K_8|h|^2$$

Combining all these inequalities yields

$$\Phi(v) \geqslant G(v(z),z) + (K_3 - K_4 - K_5 - K_8)\ |h|^2$$

Since K_2 (and therefore K_3) may be chosen as large as we like, we see that for an appropriate choice,

$$\Phi(v) \geqslant G(v(z),z)$$

Hence, $v \in \mathcal{B}(\phi,F)$, which completes the proof.

Existence of Generalized Solutions

Let us begin by showing that the elements of $\mathcal{F}(\phi,F)$ are in fact subsolutions of the problem (1) if $F(u,z)$ is increasing in U. We note that the following proposition also establishes the uniqueness part of Theorem A.

Proposition 3. Let Ω be a bounded domain in $\mathbb{C}^n$. Let $\phi_1, \phi_2 \in C(\partial\Omega)$, $\phi_1 \geqslant \phi_2$, and $F \in C(\mathbb{R} \times \overline{\Omega})$ with $F \geqslant 0$ and $u \to F(u,z)$ increasing. If $u_1,u_2 \in C(\overline{\Omega}) \cap P(\Omega)$ solve

$$u_j = \phi_j \quad \text{on } \partial\Omega$$
$$(dd^c u_j)^n = F(u_j,z) \quad \text{in } \Omega$$

for $j = 1,2$, then $u_1 \geq u_2$ in Ω.

Proof. The proof is similar to that of Theorem A, p. 3, of Bedford and Taylor (1976), and is omitted.

COROLLARY. If we drop the hypothesis $\phi_2 \leq \phi_1$ in Proposition 3, then we may still conclude

$$\max\{|u_1(z) - u_2(z)|: z \in \overline{\Omega} = \max\{|u_1(z) - u_2(z)|: z \in \partial\Omega\}$$

Proof. Let $\eta = \max\{|u_1(z) = u_2(z)|: z \in \partial\Omega\}$. Then $u_2(z) \leq u_1(z) + \eta$ for all $z \in \partial\Omega$. We consider $\omega = \{z \in \Omega: u_2(z) > u_1(z) + \eta\}$. We have $(dd^c u_2)^n = F(u_2,z) \geq F(u_1 + \eta, z) \geq F(u_1,z) = (dd^c u_1)^n = [dd^c(u_1 + \eta)]^n$ on ω and $u_2 = u_1 + \eta$ on $\partial\Omega$, in contradiction to Theorem A of Bedford and Taylor (1976). Thus, $u_2 \leq u_1 + \eta$ in Ω. A similar argument shows $u_2 + \eta \geq u_1$ in Ω, which proves the corollary.

Outline of Proof of Theorem A. The first step is to prove the theorem in the special case where Ω is the unit ball in $\mathbb{C}^n$, $\phi \in C^2(\partial\Omega)$, and $F^{1/n} \in C^2(\mathbb{R} \times \overline{\Omega})$. This follows from Propositions 1 and 2 by the same argument used to prove Theorem 8.1, p. 39, of Bedford and Taylor (1976). The general case then follows by the argument of Theorem 8.3, p. 42, of Bedford and Taylor (1976).

Theorem A may easily be extended to treat the case where F becomes singular. An example of this is the problem

$$\det\left[u_{i\bar{j}}\right] = |u|^{-\alpha} \quad \text{on} \quad \Omega$$

$$u = 0 \quad \text{on} \quad \partial\Omega$$

The real analogue of this arises in certain problems in geometry, see Nirenberg (1974), p. 276.

THEOREM A''. Let F and Ω satisfy the hypotheses of Theorem A except that $F(u,z)$ is defined only for $u \leq 0$, ($F(0,z)$ is possibly infinite.) If $\phi \leq 0$, then there exists a solution to (1).

Proof. By Theorem A, there exists a solution u_ε to (1) with boundary values $\phi_\varepsilon = \min(\phi, -\varepsilon)$. If the sequence $\{u_\varepsilon\}$ converges uniformly to

$u < 0$, then u solves (1). If $f(z) = F(-\|\phi\|, z) \geq 0$ is not identically zero ($\|\phi\| = \sup_{\partial\Omega}|\phi|$), then let v be the solution of (1) with $F = f(z)$. It follows that $u_\varepsilon(z) < v(z) < 0$ for $z \in \Omega$. Finally, if $\varepsilon_1 > \varepsilon_2 > 0$, then

$$\phi_{\varepsilon_1} \leq \phi_{\varepsilon_2} \leq \phi_{\varepsilon_1} + (\varepsilon_1 - \varepsilon_2),$$

and thus by Proposition 3,

$$u_{\varepsilon_1} \leq u_{\varepsilon_2} \leq u_{\varepsilon_1} + (\varepsilon_1 - \varepsilon_2)$$

It follows that the limit solves (1).

Existence for a Singular Dirichlet Problem

We would like to extend Theorem A' to the case where $\psi = 0$. One approach would be to solve (5) for $\psi = \varepsilon$. The solutions v_ε are then monotone decreasing. If a good "maximum principle" were known, then the v_ε would converge uniformly to a solution of (5). We are only able to obtain an existence theorem in $\mathbb{C}^2$ by extending to a still larger class of generalized solutions. It is shown in Bedford and Taylor (Section 5) that $(dd^c)^2$ has a continuous extension to $\{u \in P(\Omega): \nabla u \in L^2(\Omega, \mathrm{loc})\}$. In particular, this set contains $P(\Omega) \cap L^\infty(\Omega, \mathrm{loc})$ (see Proposition 5.2 of Bedford and Taylor). Further, $(dd^c)^2$ may be compared with Φ,

$$(dd^c u)^2 \geq \Phi(u)^2$$

if $u \in P(\Omega) \cap H^1(\Omega, \mathrm{loc})$. As in Bedford and Taylor (1976), Theorem 8.1, we wish to obtain a solution via the Perron method. Let $F(u,z) \in C(\mathbb{R} \times \overline{\Omega})$ be nonnegative and increasing in u. We will say that a function $b(z)$ on Ω <u>is an upper barrier for $F(u,z)$</u> if $b(z) \geq v(z)$, $z \in \Omega$, for all $v \in P(\Omega) \cap L^\infty(\Omega, \mathrm{loc})$ such that

$$\Phi(v) \geq [F(v,z)]^{1/2}$$

Proposition 4. If $k > 0$, then $F(u) = e^{ku}$ has an upper barrier on any open subset of $\mathbb{C}^2$.

Proof. A calculation shows that on the ball $B(\varepsilon) = \{z \in \mathbb{C}^n : |z| < \varepsilon\}$ in $\mathbb{C}^n$

$$\tilde{u}_\varepsilon(z) = -\frac{n+1}{k}\left(1 - \left(\frac{|z|}{\varepsilon}\right)^2\right) + \frac{n}{k}\log\frac{n+1}{k\varepsilon^2}$$

solves (4), and $\tilde{u}(z) \to \infty$ as $z \to \partial B(\varepsilon)$. We claim that if Ω is an open subset of $\mathbb{C}^2$ containing $\overline{B(\varepsilon)}$ and if $v \in P(\Omega) \cap L^\infty(\Omega, \text{loc})$ satisfies $\Phi(v) \geq e^{kv/2}$ then $v \leq u_\varepsilon$, where $u_\varepsilon = \tilde{u}_\varepsilon + \frac{1}{k}\log(4^2 2!)$ (the extra term is needed because

$$(dd^c u)^n = 4^n n! \det\left[u_{j\bar{k}}\right] dV$$

and $n = 2$). Once this is proved, the proposition follows by covering Ω with small balls.

To prove $v \leq u_\varepsilon$, let

$$\chi_\delta = \frac{1}{\delta^{2n}} \chi\left(\frac{z}{\delta}\right)$$

be a family of radial smoothing kernels and $v_\delta = v * \chi_\delta$. Then v_δ decreases to v as $\delta \to 0$. Further, for small $\delta > 0$, the v_δ are plurisubharmonic and smooth on $\overline{B(\varepsilon)}$. Note also that

$$\Phi(v_\delta) \geq e^{kv/2}$$

since, by Theorem 5.7, (3), p. 24 of Bedford and Taylor (1976), we have $\Phi(v_\delta) \geq \Phi(v) * \chi_\delta \geq e^{kv/2} * \chi_\delta \geq e^{kv/2}$, where the last inequality follows from the fact that $e^{kv/2}$ is a plurisubharmonic function. Now set

$$f_\delta = \left[e^{kv_\delta/2} - e^{kv/2}\right]^2$$

Note that $f_\delta \in L^\infty(B(\varepsilon)) \subset L^1(B(\varepsilon))$ and $f_\delta \to 0$ in $L^1(B(\varepsilon))$. Therefore, by Theorem 6.2, Corollary 6.3, and Theorem 6.11 of Bedford and Taylor, there exists $w_\delta \in P(B(\varepsilon)) \cap H^1(B(\varepsilon))$ such that $\Phi(w_\delta) = \sqrt{f_\delta}$, $w_\delta \leq 0$ and $w_\delta \to 0$ in $H^1(B(\varepsilon))$. For small $\eta > 0$, let $w_{\delta,\eta} = w_\delta * \chi_\eta$. Then as before $\Phi(w_{\delta,\eta}) \geq \Phi(w_\delta) * \chi_\eta \geq \sqrt{f_\delta} * \chi_\eta$, so that

$$\Phi(w_{\delta,\eta}) \geq e^{kv_\delta/2} * \chi_\eta - e^{kv/2} * \chi_\eta \geq e^{kv_\delta/2} - e^{kv/2} * \chi_\eta$$

Thus,

$$\Phi(v_\delta + w_{\delta,\eta}) \geq \Phi(v_\delta) + \Phi(w_{\delta,\eta}) \geq e^{-kv/2} * \chi_\delta + e^{kv_\delta/2} - e^{-kv/2} * \chi_\eta$$

so that with $\eta = \delta$,

$$\Phi(v_\delta + w_{\delta,\delta}) \geq e^{+kv_\delta/2} \tag{3.1}$$

Now let $\omega_\delta = \{z \in B(\varepsilon):u_\varepsilon(z) < v_\delta(z)\}$. Since $u_\varepsilon \to +\infty$ at $\partial B(\varepsilon)$ and the v_δ are uniformly bounded on $B(\varepsilon)$, we see that ω_δ is an open subset with $\omega_\delta \subset B(\varepsilon')$, some $\varepsilon' < \varepsilon$, for all $\delta > 0$. From (3.1) and $(dd^c w)^2 \geqslant [\Phi(w)]^2$, we see that $[dd^c(v_\delta + w_{\delta,\delta})]^2 \geqslant (dd^c u_\varepsilon)^2$ on ω_δ while $v_\delta + w_{\delta,\delta} \leqslant u_\varepsilon$ on $\partial\omega_\delta$ since $w_{\delta,\delta} = w_\delta * \chi_\delta \leqslant 0$ and $v_\delta = u_\varepsilon$ on $\partial\omega_\delta$. Thus, by Theorem A of Bedford and Taylor (1976), we have $v_\delta \leqslant u_\varepsilon - w_{\delta,\delta}$ on ω_δ. On $B(\varepsilon)\backslash\omega_\delta$, we clearly have $v_\delta \leqslant u_\varepsilon \leqslant u_\varepsilon - w_{\delta,\delta}$. Thus, since $w_\delta \to 0$ in $H^1(B_\varepsilon)$, we also have that $w_{\delta,\delta} \to 0$ in $H^1(B(\varepsilon),\text{loc})$, so letting $\delta \to 0$ yields $v \leqslant u_\varepsilon$. This completes the proof.

THEOREM 5. Let Ω be a bounded, strictly pseudoconvex set in $\mathbb{C}^2$ with smooth boundary. Let $F \in C(\mathbb{R}\times\overline{\Omega})$, $F \geqslant 0$, be given with $u \to F(u,z)$ increasing in u and $u \to [F(u,z)]^{1/2}$ a convex function of u. If there exists an upper barrier $b(z)$ for F on Ω, then

$$u(z) = \sup\{v(z):v \in P(\Omega) \cap L^\infty(\Omega,\text{loc}),\ \Phi(v) \geqslant [F(v,z)]^{1/2}\}$$

satisfies

$$u \in P(\Omega) \cap L^\infty(\Omega,\text{loc})$$

$$(dd^c u)^2 = F(u,z)$$

$$\Phi(u) = [F(u,z)]^{1/2}$$

$$\lim_{z\to\partial\Omega} u(z) = +\infty$$

Proof. The proof follows closely arguments given in Section 5 of Bedford and Taylor, so we will only outline the essential points. It is clear that $\lim_{z\to\partial\Omega} u(z) = +\infty$ since for any large constant K, we can solve (1) with boundary values K, by Theorem A. Since $b(z)$ is an upper barrier, $u(z)$ is locally bounded and $u^*(z) = \lim_{\zeta\to z} \sup u(\zeta)$, the upper regularization of u, is plurisubharmonic. Further, Φ is uppersemicontinuous (Theorem 5.7, p. 24 of Bedford and Taylor (1976)), so $\Phi(u^*) \geqslant [F(u^*,z)]^{1/2} = [F(u,z)]^{1/2}$ almost everywhere. Thus $u^* = u$ is plurisubharmonic.

We claim that $(dd^c u)^2 = [\Phi(u)]^2 = F(u,z)$. To see this, let $B \subset \Omega$ be a relatively compact ball and let $\phi = u|_{\partial B}$. It follows that $\overline{\partial}_b\phi \in L^2(\partial B)$ (see Corollary 5.6 of Bedford and Taylor). It also follows by directly repeating the arguments used to prove Theorem 6.11 of Bedford and Taylor that there exists $w \in P(B) \cap H^1(B)$ such that $w = \phi$ on ∂B, $(dd^c w)^2 = F(w,z)$ and $\Phi(w) = [F(w,z)]^{1/2}$. Further $w \geqslant u$ on B and the function

$$\tilde{u} = \begin{cases} w(z) & z \in B \\ u(z) & z \in \Omega \setminus B \end{cases}$$

is plurisubharmonic, locally bounded and $\Phi(\tilde{u}) \geqslant [F(\tilde{u},z)]^{1/2}$. Thus, $\tilde{u} \leqslant u$, so on B, $u \leqslant w \leqslant u$; i.e. $u = w$. Thus, $(dd^c u)^2 + [\Phi(u)]^2 = F(u,z)$ on B, and so u has the desired properties. This completes the proof.

REFERENCES

Bedford, E., and Taylor, B. A. The Dirichlet problem for a complex Monge-Ampère Equation. Inv. Math., 37 (1976) pp. 1-44.

Bedford, E., and Taylor, B. A. Variational properties of the complex Monge-Ampère Equation. I. Dirichlet principle.

Calabi, E. A construction of nonhomogeneous Einstein metrics. Proc. Symp. Pure Math., 27 (1975) pp. 17-24.

Calabi, E., Cheng, S.-Y., Nirenberg, L., and Yau, S.-T. In preparation.

Eisenman, D. A. Intrinsic measures on complex manifolds and holomorphic mappings. A. M. S. Memoirs, 96 (1970).

Fefferman, C. Monge-Ampère equations, the Bergman kernel, and geometry of pseudoconvex domains. Ann. of Math., 103 (1976) pp. 395-416.

Gaveau, B. Méthodes de contrôle optimal en analyse complexe. I. Resolution d'equations de Monge-Ampère. J. Funct. Anal., 25 (1977) pp. 391-411.

Nirenberg, L. Monge-Ampère equations and some associated problems in geometry. Proc. Int. Cong. Math., Vancouver, 2 (1974) pp. 275-279.

Yau, S.-T. Calabi's conjecture and some new results in algebraic geometry. Proc. Natl. Acad. Sci., 74 (1977) pp. 1798-1799.

Princeton University
Princeton, New Jersey

University of Michigan
Ann Arbor, Michigan

GLOBAL BEHAVIOR OF SOME TANGENTIAL CAUCHY-RIEMANN EQUATIONS

Daniel M. Burns, Jr.*

To Professor R. C. Taliaferro,
on his seventieth birthday.

§1. This note deals with a small problem in which there is a very intimate relation between a partial differential equation and geometry: the global solvability of H. Lewy-type operators. On the one hand, Lewy showed that the equation $\overline{Z}_o(u) = f$ was locally unsolvable for certain f, where $\overline{Z}_o$ is a complex vector field (cf. §2) which we will think of as giving the tangential Cauchy-Riemann equation on the hypersurface $S^3 = \{(z,w) \in \mathbb{C}^2 \mid |z|^2 + |w|^2 = 1\}$. On the other hand, Kohn's work on the $\overline{\partial}_b$-complex (Kohn, 1965) showed that the analogous (overdetermined) operator on a compact strictly pseudoconvex hypersurface M in a complex manifold X of dimension $n \geqslant 3$ is globally "solvable" in the sense that $\overline{\partial}_b$ has a closed range determined by the null-space of the compatability operator, and orthogonality to a finite-dimensional space of harmonic forms. This breaks down for $\dim X = 2$, but Boutet de Monvel and Sjöstrand have shown recently that if M bounds a relatively compact set D in X of dim 2, then $\overline{\partial}_b$ has closed range. Under this geometric assumption, Boutet de Monvel and Sjöstrand show that the Szegö projector is a pseudolocal Fourier integral operator, hence maps $C^\infty(M) \to C^\infty(M)$ continuously in the C^∞-topology. Our point here is to show that the geometric assumption of Boutet de Monvel and Sjöstrand is also necessary for the closedness of the range of $\overline{\partial}_b$ (as an operator on $C^\infty(M)$) together with the property $S: C^\infty(M) \to C^\infty(M)$ continuously, and thus to show that some examples of Rossi must have $\overline{\partial}_b$-operators which do not have closed range globally, either in the C^∞ or

*Partially supported by the N. S. F. The author would also like to thank the I. H. E. S. for support during a stay there.

L^2 sense.

§2. Let M be a strictly pseudoconvex CR-manifold of real dimension 3, and let $\bar{\partial}_b:\Lambda^{0,0} \to \Lambda^{0,1}$ denote the tangential Cauchy-Riemann operator. ($\Lambda^{0,i}$ = space of $C^\infty(0,i)$-forms on M, thus $\Lambda^{0,0} = C^\infty(M)$. For a general reference on such things we use Folland and Kohn (1972), where $\Lambda^{0,i}$ is denoted $\mathcal{B}^{0,i}$.) If we fix a metric on M, hermitian on the complex subspaces of $T(M)$, we can form the formal adjoint of $\bar{\partial}_b$ denoted $\bar{\partial}_b^*:\Lambda^{0,1} \to \Lambda^{0,0}$, and the "Laplacian" $\Box_b = \bar{\partial}_b^* \bar{\partial}_b:\Lambda^{0,0} \to \Lambda^{0,0}$. We let $S:L^2(M) \to L^2(M)$ denote the orthogonal projection onto the subspace of all $u \in L^2(M)$ which satisfy $\bar{\partial}_b(u) = 0$ in the distributional sense. By restriction, we get an operator $S:C^\infty(M) \to L^2(M)$, which is a priori only continuous in the L^2 norm. The following lemma is modelled on Boutet de Monvel (1974-1975).

Lemma 1. Suppose

1) $\bar{\partial}_b:\Lambda^{0,0} \to \Lambda^{0,1}$ has closed range in the C^∞-topology,

2) $S:C^\infty(M) \to C^\infty(M) \subset L^2(M)$, and is continuous in the C^∞-topology.

Then there is a smooth imbedding $F:M \to \mathbb{C}^N$, where $F = (f_1,\ldots,f_N)$ with $\bar{\partial}_b(f_i) = 0$ for each $i = 1,\ldots,N$. In particular, solutions of $\bar{\partial}_b(u) = 0$ separate points on M.

Remark: By the techniques of Rossi (1965), the conclusion of the preceeding lemma implies that M bounds a relatively compact domain in the complex manifold. The converse of lemma 1 is proven in Boutet de Monvel and Sjöstrand (to appear).

Proof. We show how to separate points in M by solutions, the rest proceeding as in Boutet de Monvel (1974-1975).

Assumption 2) implies we have an orthogonal, topological direct-sum decomposition

$$C^\infty(M) = [\ker(S) \cap C^\infty(M)] \oplus [\operatorname{Im}(S) \cap C^\infty(M)]$$

By definition, $\bar{\partial}_b(\operatorname{Im}(S)) = 0$, and for $u \in \ker(S)$, $\bar{\partial}_b(u) = 0$ implies $u = S(u) = 0$. Letting $R \subset \Lambda^{0,1}$ denote the range of $\bar{\partial}_b$, closed in $\Lambda^{0,1}$ by assumption 1), we have $\bar{\partial}_b:\ker(S) \cap C^\infty(M) \xrightarrow{\sim} R$, the isomorphism being topological, by the open mapping theorem.

Given distinct $p,q \in M$, we find a function $\phi \in C^\infty(M)$ satisfying:

A) $\phi(p) = 0$ and $\bar{\partial}_b(\phi)$ vanishes to infinite order at p.

B) Near p, $\mathrm{Re}(\phi) \geqslant \mathrm{const.}\ |x|^2$, for some coordinate system x centered at p.

C) Outside a smaller neighborhood of p, $\mathrm{Re}(\phi) \geqslant 1$. In particular, $\mathrm{Re}(\phi(q)) \geqslant 1$.

Consider $u = e^{-t\phi}$, for $t > 0$. $u \in C^\infty(M)$, and we have $u = S(u) + (I - S)(u)$. For t large, $u(p) = 1$, $u(q) \sim 0$. We show that $(I - S)(u) \to 0$ in the C^∞-topology, as $t \to +\infty$, and therefore $S(u)$ is a solution separating p and q. By the above, we have only to show $\bar{\partial}_b(u) \to 0$ in R. But $\bar{\partial}_b(e^{-t\phi}) = -te^{-t\phi}\bar{\partial}_b(\phi)$, and A), B), C) guarantee the desired convergence to 0.

§3. On $S^3 \subset \mathbb{C}^2$ as in Section 1, we define a 1-parameter family of CR-structures as follows.

Set

$$\theta = i(\bar{z}\,dz + \bar{w}\,dw)$$

$$\theta^1 = w\,dz - z\,dw$$

$$\bar{Z}_o = w\frac{\partial}{\partial \bar{z}} - z\frac{\partial}{\partial \bar{w}}$$

$$\theta^1_t = (1/\sqrt{1-t^2})(\theta^1 - t\,\bar{\theta}^1), \quad -1 < t < 1$$

$$\bar{Z}_t = (1/\sqrt{1-t^2})(\bar{Z}_o + t\,Z_o), \quad -1 < t < 1$$

Since $d\theta = i\theta^1 \wedge \bar{\theta}^1$, and $\theta(\bar{Z}_t) = 0$, $\bar{\theta}^1_t(\bar{Z}_t) = 1$, setting $\bar{\partial}_{b,t}(u) = \bar{Z}_t(u) \cdot \bar{\theta}^1_t$, we define a strictly pseudoconvex CR-structure on S^3. Denoting SU(2) by G, we have $\theta, \theta^1, \bar{Z}_o$, etc., are all G-invariant. Since $\bar{Z}_t$ is a real analytic, there are many local solutions of $\bar{Z}_t(u) = 0$. The following lemma appears in Rossi (1965), but we give an amusing proof here based on direct analysis of the Caucy-Riemann operator itself.

Lemma 2. (Rossi) For $0 < |t| < 1$, the only solutions of $\bar{\partial}_{b,t}(u) = 0$ are even functions, i.e. $u(z,w) = u(-z,-w)$.

Proof. Decompose $L^2(S^3)$ according to spherical harmonics, $L^2(S^3) = \sum_{p,q\geqslant 0} H^{p,q}$, where $H^{p,q}$ = {restrictions to S^3 of harmonic polynomials on $\mathbb{C}^2$ of type (p,q)}. Each $H^{p,q}$ is invariant and irreducible for G, and it is easy to see that $\bar{Z}_o(H^{p,q}) \subset H^{p+1,q-1}$, $Z_o(H^{p,q}) \subset H^{p-1,q+1}$. One has $z^p \in H^{p,0}$, and $Z_o^p(z^p) = p!\,\bar{w}^p \neq 0$ in $H^{0,p}$, so by Schur's lemma, $Z_o : H^{p,q} \to H^{p-1,q+1}$ is an isomorphism, provided $p - 1 \geqslant 0$, $q \geqslant 0$, and similarly for $\bar{Z}_o$. Setting $H^k = \sum_{p+q=k} H^{p,q}$, each solution u of

$\overline{Z}_t(u) = 0$ may be written $u = u^k$, $u^k \quad H^k$, and $\overline{Z}_t(u^k) = 0$. For k odd, $t \neq 0$, we want to show $d^k = 0$. Assume $u = u^k$, k odd, and write out $\overline{Z}_t(u) = 0$ in $H^{p,q}$-components, where $u = \sum_{p+q=k} u^{p,q}$:

$$
\begin{array}{lrl}
(k,0) & 0 = & \overline{Z}_o(u^{k-1,1}) \\
(k-1,1) & -tZ_o(u^{k,0}) = & \overline{Z}_o(u^{k-2,2}) \\
(k-2,2) & -tZ_o(u^{k-1,1}) = & \overline{Z}_o(u^{k-3,3}) \\
 & \cdots & \\
(1,k-1) & -tZ_o(u^{2,k-2}) = & \overline{Z}_o(u^{0,k}) \\
(0,k) & -tZ_o(u^{1,k-1}) = & 0
\end{array}
$$

Equation $(k,0) \neq 0$, and the injectivity of various Z_o and $\overline{Z}_o$ operators inductively imply $0 = u^{k-1,1} = u^{k-3,3} = \ldots = u^{0,k}$. Similarly, starting from $(0,k)$ one gets $0 = u^{1,k-1} = \ldots = u^{k,0}$. Hence $u = 0$.

Remark: The same sort of reasoning shows that for k even, $u \in H^k$, $\overline{Z}_t(u) = 0$ and $u^{k,0} = 0$ implies $u = 0$.

We note that the space of solutions of $\overline{Z}_t(u) = 0$ in H^2 is 3-dimensional, spanned by

$$X = \frac{\sqrt{2}}{2i}\left[z^2 + w^2 + t(\overline{z}^2 + \overline{w}^2)\right]$$

$$Y = \frac{\sqrt{2}}{2}\left[z^2 - w^2 + t(\overline{z}^2 - \overline{w}^2)\right]$$

$$Z = \sqrt{2}\,[zw - t\overline{zw}]$$

We have

$$X^2 + Y^2 + Z^2 = 2t \tag{*}$$

$$|X|^2 + |Y|^2 + |Z|^2 = 1 + t^2 \tag{**}$$

By the remark after the lemma, the algebra of G-finite (polynomial) solutions of $\overline{Z}_t(u) = 0$ is spanned by polynomials in X, Y, Z, and a dimension count shows that (*) is the only relation. Hence, (X,Y,Z) are the components of a two-to-one CR-immersion π of $M_t = S^3$ with CR-structure given by $\bar{\partial}_{b,t}$, into $\mathbb{C}^3$, and $\pi(M_t) = M_t'$ is defined by equations (*) and (**) above. In particular, M_t' bounds a relatively compact domain D_t in X_t, the affine quadric defined by (*). Giving M_t' the metric making π a local isometry, we denote by S_t' the corresponding Szegö projector, $S_t':L^2(M_t') \to L^2(M_t')$. By the theorem of Boutet de Monvel and Sjöstrand, S_t' is a pseudolocal Fourier integral operator, i.e., its Schwartz kernel $S_t'(x,y)$, $(x,y) \in M_t' \times M_t'$, has singular support only on the diagonal $x = y$. In particular, $S_t':C^\infty(M_t') \to C^\infty(M_t')$, continuously in the C^∞-topology.

On the other hand, decompose $L^2(M_t) = L_+^2 \oplus L_-^2$, $L_\pm^2$ = even, resp. odd, functions in $L^2(M_t)$. If S_t is the Szegö projector for M_t, then $S_t(L_-^2) = 0$, for $t \neq 0$. Let P_+ be the projector onto L_+^2. For each $f \in L_+^2$, there is a unique $\tilde{f} \in L^2(M_t')$ such that $\pi^*(\tilde{f}) = f$, and $\sqrt{2}\|\tilde{f}\| = \|f\|$. From these remarks it follows that, for $f \in L^2(M_t)$, $t \neq 0$,

$$S_t(f) = \sqrt{2}\ \pi^* S_t'(\tilde{f})$$

where $P_+(f) = \pi^*(\tilde{f})$. Hence S_t maps $C^\infty(M_t) \to C^\infty(M_t)$, continuously. We also note that the Schwartz kernel of S_t is given by

$$S_t(x,y) = \frac{1}{2} \cdot S_t'(\pi(x),\pi(y))\ldots \qquad (\#)$$

Combining lemmas 1 and 2, and the properties of S_t, $t \neq 0$, just deduced proves most of the following theorem.

THEOREM. Let M_t be the CR-structure on S^3 given by the vector field $\bar{Z}_t$ as above. Then

1) $\bar{\partial}_{b,t}:\Lambda^{0,0} \to \Lambda^{0,1}$ does not have closed range.
2) (the maximal L^2-extension of) $\bar{\partial}_{b,t}:L^2(M_t) \to L^2_{(0,1)}(M_t)$ does not have closed range. ($L^2_{(0,1)}(M_t) = \{f \cdot \bar{\theta}_t'\ f \in L^2(M_t)\}$.)
3) the Szegö projector $S_t:L^2(M_t) \to L^2(M_t)$ is not a pseudolocal operator.

Proof. 1) follows directly from lemma 1 since the second alternative in the hyposthesis holds for the M_t, $t \neq 0$. 3) follows from (#), which shows that the singular support of $S_t(x,y)$ is the set $\{(x,y)\,|\,y = \pm x\}$, and not just the diagonal.

For the proof of 2), we consider (equivalently) the operator $\bar{Z}_t$. We abuse notation, and use $\bar{Z}_t$ or Z_t to denote the maximal L^2-extension of the differential operator, and denote domains by, e.g., $\mathcal{D}(\bar{Z}_t)$. Note that $H^k = H^k \cap \ker(\bar{Z}_t) \oplus H^k \cap \ker(\bar{Z}_t)^{\perp}$, and that $\bar{Z}_t : H^k \cap \ker(\bar{Z}_t)^{\perp} \to H^k \cap \ker(Z_t)^{\perp}$ is an isomorphism for each k (dimension count). Hence, range $(\bar{Z}_t)$ is dense in $\ker(Z_t)^{\perp}$, and is closed iff there is a bounded operator $R_t : \ker(Z_t)^{\perp} \to \mathcal{D}(\bar{Z}_t) \cap \mathrm{Ker}(\bar{Z}_t)^{\perp}$ satisfying $\bar{Z}_t(R_t(f)) = f$, for all $f \in \ker(Z_t)^{\perp}$.

By what was said earlier about S_t, the space $\ker(\bar{Z}_t)^{\perp} \cap C^{\infty}(M_t) = (I - S_t)(C^{\infty}(M_t))$ is closed in $C^{\infty}(M_t)$, and so is $\ker(Z_t)^{\perp} \cap C^{\infty}(M_t)$. For $f \in L^2(M_t)$, $f = \sum f^k$, $f^k \in H^k$, one has $f \in C^{\infty}(M_t)$ iff $\|f^k\| = O(k^{-N})$, for $N \geq 0$ arbitrary. Given R_t as above, and $f \in \ker(Z_t)^{\perp} \cap C^{\infty}(M_t)$, we have $R_t(f) = g = \sum g^k$, with $R_t(f^k) = g^k \in H^k$. Hence $\|g^k\| = O(k^{-N})$, $N \geq 0$ arbitrary, $g \in C^{\infty}(M_t)$ and $\bar{Z}_t(g) = f$. That is, $\bar{Z}_t(C^{\infty}(M_t)) = \ker(Z_t)^{\perp} \cap C(M_t)$ is closed, contradicting 1). This proves 2).

Remark: A result like 1) and 2) of the theorem above also holds with $\bar{\partial}_{b,t}$ replaced by $\square_{b,t} = -Z_t \cdot \bar{Z}_t$.

The proof above does not really say very clearly why these operators are not globally solvable. Can one associate some invariant to a smooth CR-structure on S^3 which, say, vanishes if that structure has a globally solvable $\bar{\partial}_b$-operator? More concretely, it follows from a theorem of Gray (1959) that a smooth strictly pseudoconvex CR-structure on S^3 close to the standard one is diffeomorphic to one with $\bar{\partial}_b$-operator equivalent to the vector field $\bar{Z}_f = \bar{Z}_0 + fZ_0$, where $f \in C^{\infty}(S^3)$ and $|f| < 1$ throughout S^3. For what f does $\bar{Z}_f$ have closed range? Is this a closed set of f?

REFERENCES

Boutet de Monvel, L. Integration des équations de Cauchy-Riemann induites formelles. Seminaire Goulaouic-Lions-Schwartz, Exposé IX, (1974-1975).

Boutet de Monvel, L., et Sjöstrand, J. Sur la singularité des noyaux de Bergman et de Szegö. To appear.

Folland, G., and Kohn, J. J. The Neumann Problem for the Cauchy-Riemann Complex. Princeton University Press, 1972.

Gray, J. Some global properties of contact manifolds. Ann. Math., 69 (1959).

Kohn, J. J. Boundaries of complex manifolds. Proc. Conf. on Complex Manifolds (Minneapolis), Springer-Verlag, New York (1965).

Rossi, H. Attaching analytic spaces to an analytic space along a pseudoconcave boundary. Proc. Conf. on Complex Manifolds (Minneapolis), Springer-verlag, New York (1965).

Princeton University
Princeton, New Jersey

THE HEAT EQUATION, THE KAC FORMULA AND SOME INDEX FORMULAS*

T. E. Duncan

Introduction

Local formulae for some index theorems such as the Euler-Poincaré characteristic, the Riemann-Roch-Hirzebruch Theorem for Kähler manifolds and the generalized Lefschetz fixed point formula will be computed by the asymptotic properties of the fundamental solutions of the associated heat equations. While all these local formulae have been previously obtained by heat equation methods, the techniques that are used here to obtain these local formulae from the asymptotics of the heat equations are different and seem simpler and perhaps more elementary. Gilkey (1972) introduced an effective use of invariance theory methods of H. Weyl (1946) to determine local formulae for index theorems from the asymptotics of the solutions of the heat equations and these methods have been successfully applied by him and others to compute these local formulae. The techniques in this paper employ Brownian motion and a formula of Kac and its generalizations. In a naive sense it should not be surprising that Brownian motion can be used in these computations because it is intimately associated with the Laplacian. Moreover, Brownian motion has a natural appeal in differential geometry because some of the geometric techniques that are used for smooth curves can be applied to the paths of Brownian motion.

Euler-Poincaré Characteristic

Let V be a finite dimensional vector space over the reals, V^* the dual of V and $A \in \mathrm{Hom}(V,V)$. For $1 \leqslant p \leqslant d$ two elements of $\mathrm{Hom}(\Lambda^p V, \Lambda^p V)$ are $\Lambda^p A$, the pth exterior power of A, and $D^p A$, the derivation extension of A, defined as

*Research supported by N. S. F. Grants ENG 75-06562 and MCS 76-01695 and AFOSR Grant 77-3177.

$$(\Lambda^p A)(v_1 \wedge \dots \wedge v_p) = (Av_1) \wedge \dots \wedge (Av_p)$$

$$(D^p A)(v_1 \wedge \dots \wedge v_p) = \sum_{r=1}^{p} v_1 \wedge \dots \wedge v_{r-1} \wedge Av_r \wedge v_{r+1} \wedge \dots \wedge v_p$$

where $v_1,\dots,v_p \in V$. For convenience define $\Lambda^0 A = I$ and $D^0 A = 0$ as elements of $\mathrm{Hom}(\mathbb{R},\mathbb{R})$. If $B \in \mathrm{Hom}(V,V)$ then $\mathrm{Tr}B$ denotes the trace of B.

The proofs of the following two elementary lemmas can be found in Patodi (1971).

Lemma 1. Let $A_1, A_2, \dots, A_k \in \mathrm{Hom}(V,V)$ where $0 \leqslant k \leqslant d$ and $d = \dim V$. Then

$$\sum_{p=0}^{d} (-1)^p \mathrm{Tr}(D^p A_1 \circ \dots \circ D^p A_k)$$

$$= \begin{cases} 0 & \text{if } k < d \\ (-1)^d K & \text{if } k = d \\ \text{where } K \text{ is the coefficient of } x_1 \dots x_k \text{ in} \\ \det(x_1 A_1 + \dots + x_k A_k) \end{cases}$$

Lemma 2. Suppose that $\dim V = 2n$. Let $A \in V^* \otimes V \otimes V^* \otimes V$, $e_1,\dots,e_{2n}$ be a basis for V and $e_1^*,\dots,e_{2n}^*$ be the dual basis for V^*. Suppose that

$$A = \sum a_{ijk\ell} e_i^* \otimes e_j \otimes e_k^* \otimes e_\ell$$

Then

$$\sum_{p=0}^{2n} (-1)^p \mathrm{Tr}((D^p A)^n)$$

$$= \sum \varepsilon_\sigma \varepsilon_\delta a_{\sigma(1)\delta(1)\sigma(2)\delta(2)} \cdots a_{\sigma(2n-1)\delta(2n-1)\sigma(2n)\delta(2n)}$$

where $\sigma, \delta \in S_{2n}$ (the permutation group of $2n$ letters) and ε_τ is the signature of $\tau \in S_{2n}$.

Let $(\partial/\partial t) - L - B$ be a differential operator where L is a second order elliptic operator and B is a linear transformation. This differential operator will act on a trivial vector bundle over $\mathbb{R}^n$. It is assumed that L and B are smooth. The simplest nontrivial example occurs when L is the sum of the second derivatives in the canonical basis for $\mathbb{R}^n$. The solution of an initial value problem with initial data ϕ at $t = 0$ is $u(t,x) = E_x[Y(t)\phi(X_t)]$ where $(X(t))_{t \in \mathbb{R}_+}$ is the diffusion process with infinitesimal generator L and $(Y(t))_{t \in \mathbb{R}_+}$ satisfies the stochastic matrix differential equation

$$\frac{dY(t)}{dt} = Y(t)B(t,X_t)$$

$$Y(0) = I$$

For $N = 1$ this result was obtained by Kac (1951) and for $N > 1$ the result was obtained by Babbit (1970) and Stroock (1970). This result will play a crucial role in the asymptotic expansion of the heat equation on forms.

Since the smooth manifold M is Riemannian the curvature R can be considered as a tensor field of type $(2,2)$ and therefore D^pR is locally a linear operator from $C^\infty(U,\Lambda^pT^*M)$ into itself where U is a patch on M. The linear operator D^pR and the covariant derivative can be used to describe the Laplacian. Specifically, the Laplace operator Δ on Λ^pT^*M is given by

$$\Delta = \sum_{i,j} g^{ij}\nabla_j \circ \nabla_i + D^pR$$

where (g^{ij}) is the inverse of the matrix of the Riemannian metric g and ∇ is the covariant derivative obtained from the Riemannian connection.

The Chern polynomial C is given by

$$C = \frac{1}{2^{3n}\pi^n n!} \sum \varepsilon_\sigma \varepsilon_\delta R_{\sigma(1)\sigma(2)\delta(1)\delta(2)} \cdots R_{\sigma(2n-1)\sigma(2n)\delta(2n-1)\delta(2n)}$$

where $\sigma,\delta \in S_{2n}$ (the permutation group of $2n$ letters), ε_τ is the signature of $\tau \in S_{2n}$ and $R_{ijk\ell}$ are the components of the curvature tensor.

The following result has be obtained by Patodi (1971), Gilkey (1973), and Atiyah-Bott-Patodi (1973) and McKean and Singer (1967) have obtained a special case of it.

THEOREM 1. Let M be a compact, oriented, smooth, Riemannian manifold of dimension $2n$. Let $e^p(t,x,y)$ be the fundamental solution of the heat operator $(\partial/\partial t) - \Delta$ acting on the p-forms for $0 \leqslant p \leqslant 2n$ and let

$$\mathrm{Tr}e = \sum_{p=0}^{2n} (-1)^p \mathrm{Tr}e^p$$

Then

$$\lim_{t \downarrow 0} \mathrm{Tr}e = C$$

where C is the Chern polynomial.

Proof. A sequence of differential operators on forms will be constructed that converge to the Laplacian. The asymptotic behavior of a trace of the fundamental solutions to this sequence of differential operators will be obtained from which the desired local trace formula for the Laplacian will easily follow.

Let g denote the Riemannian metric. Fix $x_0 \in M$ and choose $r > 0$ so that in $N(x_0;r) = \{X \in T_{x_0}M : g^{1/2}(X,X) < r\}$ the exponential map is a diffeomorphism. Let (r_n) be a sequence of strictly decreasing real numbers such that $r_n \downarrow 0$ and $0 < r_n < r$ for all $n \in \mathbb{N}$. For $n \in \mathbb{N}$ define a sequence (A_n) of local tensor fields as follows. For $x \in \{\exp[\overline{N}(x_0;r_n)]\}$, where $\exp$ is the exponential map at x_0, define $A_n(x) = R(x_0)$, for

$$x \in \left\{\exp\left[N\left(x_0;\frac{r_n+r}{2}\right)\right]\right\}^c$$

define $A_n(x) = R(x)$ and for

$$x \in \left\{\exp\left[N\left(x_0;\frac{r_n+r}{2}\right)\right]\setminus \exp[\overline{N}(x_0;r_n)]\right\}$$

define A_n so that it is C^∞ on M and bounded in norm by the norm of R.

A sequence of differential operators on forms are obtained locally by replacing D^pR by D^pA_n in the Laplacian.

It will be briefly described why an M-valued Brownian motion is the diffusion process whose infinitesimal generator is $\sum g^{ij}\nabla_j \circ \nabla_i$. Probably the easiest way to verify this at a point x_0 in the manifold is to form normal coordinates in a convex neighborhood of x_0. Then, using Taylor's formula and the change of variables formula of K. Itô (1951), the desired result is obtained. A local coordinate description of this operator will follow by using normal coordinates at each point in the convex neighborhood of x_0. This differential operator is smoothly extended to the appropriate Euclidean space and we can construct a fundamental solution to the differential operator $(\partial/\partial t) - \sum g^{ij}\nabla_j \circ \nabla_i$ by constructing the suitable diffusion process.

Let $(Y_t)_{t\in\mathbb{R}_+}$ be the M-valued Brownian motion with $Y_0 \equiv x_0$ whose infinitesimal generator is determined by the Laplacian on functions. Since only the asymptotic behavior of the Laplacian as $t \downarrow 0$ is of interest, the Brownian motion can be restricted to $W = N(x_0;r)$ and the error in the fundamental solution is $o(t^N)$ where N is an arbitrary positive integer (e.g., see McKean and Singer, 1967). For the differential operator

$$\frac{\partial}{\partial t} - \sum_{i,j} g^{ij}\nabla_j \circ \nabla_i$$

on p-forms, it suffices to restrict the construction of the fundamental solution to W as $t \downarrow 0$ from the theory of elliptic differential equations or by the estimates that can be obtained from the probabilistic method of solution. This approximation to the fundamental solution will differ in trace norm from the correct solution by $o(t^N)$ where N is an arbitrary positive integer.

Let $f \in C^\infty(W,\Lambda^p T^*M)$ be some initial data. Using the Kac formula (see Babbitt, 1970, Kac, 1951, and Stroock, 1970), we have a sequence of solutions (U_n) where

$$U_n(t,x) = E_x[X_n^p(t)f(Y_t)]$$

$$\frac{d}{dt}X_n^p(t) = X_n^p(t)D^pA_n(Y_t)$$

$$X_n^p(0) = I$$

It will be shown that for each p and for $t > 0$ sufficiently small, the sequence $(X_n^p(s))_{s\in[0,t]}$ converges uniformly almost surely to $(X^p(s))_{s\in[0,t]}$ where

$$\frac{d}{dt}X^p(t) = X^p(t)D^pR(Y_t)$$

$$X^p(0) = I$$

To verify this uniform convergence, some terminology will be introduced. Let $A,B \in \mathrm{Hom}(V,V)$ where V is a real vector space of dimension k. Define

$$|A| = \max_j \sum_{i=1}^{k} |a_{ij}|$$

Then $|A+B| \leqslant |A| + |B|$ and $|AB| \leqslant |A||B|$. Let

$$\|A\| = \sum_{i,j} |a_{ij}|$$

The following inequalities are satisfied

$$|X_n^p(s) - X^p(s)| \leqslant \|X_n^p(s) - X^p(s) + I - I\| \leqslant \sup_{0\leqslant\tau\leqslant s} \|\dot{X}_n^p(\tau) - \dot{X}^p(\tau)\| s$$

$$= \sup_{0\leqslant\tau\leqslant s} \|X_n^p(\tau)D^pA_n - X^p(\tau)D^pR\| s$$

$$\leqslant k \sup_{0\leqslant\tau\leqslant s} |X_n^p(\tau)D^pA_n - X^p(\tau)D^pR| s$$

$$= ks \sup_{0\leqslant\tau\leqslant s} |X_n^p(\tau)D^pA_n - X^p(\tau)D^pA_n + X^p(\tau)D^pA_n - X^p(\tau)D^pR|$$

$$\leqslant ks \Big[\sup_{0\leqslant\tau\leqslant s} |D^pA_n||X_n^p(\tau) - X^p(\tau)| + \sup_{0\leqslant\tau\leqslant s} |X^p(\tau)||D^pA_n - D^pR| \Big]$$

Since the solutions have been restricted to the convex neighborhood W, it follows from the construction of (A_n) that

$$\sup_\tau \Big[|D^pA_n| + |X^p(\tau)| \Big] \leqslant K$$

Choose t such that $kt\,K < \frac{1}{2}$. Then

$$\sup_{0\leqslant\tau\leqslant t} |X_n^p(\tau) - X^p(\tau)| \leqslant \sup_{0\leqslant\tau\leqslant t} |D^pA_n(Y_t) - D^pR(Y_t)|$$

As $t \downarrow 0$ (with probability arbitrarily near 1), $X_n^p(t)$ can be expressed by Taylor's formula as

$$X_n^p(t) = I + X_n^p(o)t + \ldots + \frac{1}{N!} X_n^{p(N)}(o)t^N + o(t^N) \tag{1}$$

where the remainder formula for Taylor's formula and the assumptions on the sequence (A_n) imply that $|o(t^N)| \leqslant \overline{K}\, t^{N+1}$ and $\overline{K}$ does not depend on n. The differential equation for X_n^p shows that $X_n^{p(k)} = [D_n^p R(x_0)]^k$.

It is known from the theory of elliptic operators on a compact manifold that the index of an elliptic operator is well defined and it is known from the Hodge decomposition that the index can be computed from the dimension of the spaces of harmonic forms. From the asymptotic heat expansion, the index can be obtained from a measure on the manifold canonically determined from the coefficients of the differential operator (e.g., see Atiyah and Bott, 1968, and Gilkey, 1973). In particular, for the Euler-Poincaré characteristic

$$\lim_{t \downarrow 0} \sum_{p=0}^{2n} (-1)^p \mathrm{Tr} e^p(t,x,x)$$

is a continuous measure that can be integrated to give the Euler-Poincaré characteristic.

Since, in the construction of the fundamental solution for the differential operator $(\partial/\partial t) - \sum g^{ij} \nabla_j \circ \nabla_i$, the first term in the infinite series solution is the solution from Euclidean space, the singularity is $(4\pi t)^{-n}$ where $\dim M = 2n$. The other terms in the series can be neglected as $t \downarrow 0$.

From the Taylor formula expressions (1) for the sequence of approximations to the solution of the stochastic ordinary differential equations in the Kac formula and lemma 1 and lemma 2, we have that

$$\lim_{t \downarrow 0} \sum_{p=0}^{2n} (-1)^p \mathrm{Tr} e_k^p(t,x,x) = C(x) \tag{2}$$

where (e_k^p) is the sequence of fundamental solutions on p-forms and C is the Chern polynomial. For fixed $t > 0$ sufficiently small, we have that

$$\mathrm{Tr} e_k^p(t,x,x) \to \mathrm{Tr} e^p(t,x,x)$$

as $k \to \infty$. Since the convergence in (2) is uniform in k by (1) we have

$$\lim_{t\downarrow 0} \sum_{p=0}^{2n} (-1)^p \mathrm{Tr}e^p(t,x,x) = C(x)$$

Riemann-Roch-Hirzebruch Theorem

Let M be a compact Kähler manifold of (complex) dimension n and let ξ be a holomorphic vector bundle over M with an hermitian inner product h. The Laplacian on $C^\infty(M,\xi \otimes \Lambda^q T^{*0,1}M)$ is given by

$$\Delta_{\bar{z}} = \sum g^{\bar{\beta}a}\nabla_\alpha \circ \nabla_{\bar{\beta}} + I_\xi \otimes D^q\tilde{K} - D^q S$$

where I_ξ is the identity operator on ξ

$$\tilde{K} = \sum K^{\bar{\alpha}\bar{\beta}}_{\bar{\beta}\gamma} \frac{\partial}{\partial \bar{z}_\alpha} \otimes d\bar{z}_\gamma$$

$$K = \sum K^{\bar{\alpha}\bar{\gamma}}_{\bar{\beta}\bar{\delta}} \frac{\partial}{\partial \bar{z}_\alpha} \otimes d\bar{z}_\beta \otimes \frac{\partial}{\partial \bar{z}_\gamma} \otimes d\bar{z}_\delta$$

$$S = \sum S^{a\bar{\beta}}_{b\bar{\gamma}}\, s_a \otimes s_b^* \otimes \frac{\partial}{\partial \bar{z}_\beta} \otimes d\bar{z}_\gamma$$

and K and S are curvature fields for the manifold and the bundle respectively.

Let V and W be vector spaces and let $D^qA \in \mathrm{Hom}(W \otimes \Lambda^q V, W \otimes \Lambda^q V)$ be the linear extension of $B \otimes D^qC$ where $B \in \mathrm{Hom}(W,W)$ and $C \in \mathrm{Hom}(V,V)$ and D^qC is the derivation extension of C. By abuse of terminology, D^qA will be called the derivation extension of A.

The proofs of the following three elementary lemmas are given in Patodi (1971).

Lemma 3. Let $A_1,\ldots,A_k \in W \otimes W^* \otimes V^* \otimes V$ where $k < n$ and n is the dimension of the vector space V. Let $D^qA_i \in \mathrm{Hom}(W \otimes \Lambda^q V, W \otimes \Lambda^q V)$ be the derivation extension of A_i, $i = 1,\ldots,k$. Then

$$\sum_{q=0}^{n} (-1)^q \mathrm{Tr}(D^qA_1 \circ \ldots \circ D^qA_k) = 0$$

Lemma 4. Let $A_1,\ldots,A_n \in V^{0,1} \otimes V^{*0,1}$ where $V^{*0,1}$ is the dual of $V^{0,1}$. Let $\phi: V^c \otimes V^{*c} \to V^{*c} \wedge V^{*c}$ where V^c is a complex vector space of dimension n and ϕ is the composition of $\psi \otimes I_d$ and the wedge product where $\psi : V^c \to V^{*c}$ is the canonical map that is obtained from the pairing in V^c induced from the inner product and I_d is the identity on V^{*c}. Then

$$\left(\sum_{q=0}^{n} (-1)^q \mathrm{Tr}(D^q A_1 \circ \ldots \circ D^q A_n)\right) e = (-1)^n \phi(A_1) \wedge \ldots \wedge \phi(A_n)$$

where $e \in \Lambda^{2n} V^*$ is the volume form.

Lemma 5. Let $A_1, \ldots, A_n \in W \otimes W^* \otimes V^{0,1} \otimes V^{*0,1}$. Let $\tilde{\phi} =$ $(\phi_2 \otimes I_d) \circ \phi_1$ where

$$\phi_1 : (W \otimes W^* \otimes V^{*c} \wedge V^{*c})^{\textcircled{n}} \to (W \otimes W^*)^{\textcircled{n}} \otimes \Lambda^{2n} V^{*c}$$

I_d is the identity on $\Lambda^{2n} V^{*c}$ and $\phi_2 : (W \otimes W^*)^{\textcircled{n}} \to \mathbb{C}$ is the trace of the composition of the n maps in $\mathrm{Hom}(W,W)$. Then

$$\left(\sum_{q=0}^{n} (-1)^q \mathrm{Tr}(D^q A_1 \circ \ldots \circ D^q A_n)\right) e = (-1)^n \tilde{\phi}((I_d \otimes \phi) A_1, \ldots, (I_d \otimes \phi) A_n)$$

where I_d is the identity on $W \otimes W^*$ and ϕ and e are defined in Lemma 4.

The following result has been obtained by different methods in Atiyah, Bott and Patodi (1973), Gilkey (1973) and Patodi (1971).

THOEREM 2. Let M be a compact Kähler manifold of (complex) dimension n and let ξ be a holomorphic bundle over M with an hermitian inner product. Let e^q be the fundamental solution of the heat operator $(\partial/\partial t) - \Delta$ acting on $(0,q)$-forms with values in the bundle ξ. Then

$$\lim_{t \downarrow 0} \sum_{q=0}^{n} (-1)^q \mathrm{Tr} e^q * 1 = \sum_{r=0}^{n} ch_{n-r}(\xi) \wedge \tau(C_1(M), \ldots, C_r(M)) \tag{3}$$

where $\sum ch_k(\xi) = ch\xi$ is the Chern character and $\tau(C_1(M), \ldots, C_r(M))$ is the rth Todd polynomial of the first r Chern forms of the manifold M.

Proof. For the infinitesimal generator $\sum g^{\overline{\beta}\alpha} \nabla_\alpha \circ \nabla_{\overline{\beta}}$ a complex Brownian motion is constructed. When the complex structure is used in the computation of Taylor's formula with the change of variables formula of K. Itô

(1951), it is straightforward to verify that the generator of this process is $\sum g^{\beta\alpha}\nabla_\alpha \circ \nabla_{\bar{\beta}}$.

The justification for freezing the derivation extension of curvature at the initial value in the stochastic ordinary differential equation for the fundamental solution of the heat operator on forms is the same as was given in the proof of the Euler-Poincaré characteristic (Theorem 1). Only the verification of the desired formula (3) remains.

Again expand the ordinary differential equations for the fundamental solutions of the heat operators on $(0,q)$-forms $q = 0,1,\ldots,n$ and use lemma 3. Since the manifold is complex, the coordinate system that is used is

$$\frac{\partial}{\partial z} = \frac{1}{2}\left(\frac{\partial}{\partial x} - i\,\frac{\partial}{\partial y}\right)$$

Thus, the variance parameter for the Brownian motion is $t = \frac{1}{4}[2t + 2t]$ and the term from the fundamental solution of the heat operator on functions is $(2\pi t)^{-n}$.

Thus

$$\lim_{t\downarrow 0} \sum_{q=0}^{n} (-1)^q \mathrm{Tr} e^q = \frac{1}{(2\pi)^n n!} \sum_{q=0}^{n} (-1)^q \mathrm{Tr}((D^q A)^n)$$

where $D^q A = D^q S + I_\xi \otimes D^q \tilde{K}$.

$$\sum_{q=0}^{n} (-1)^q \mathrm{Tr}((D^q A)^n) = \sum_{q=0}^{n} (-1)^q \mathrm{Tr}((I_\xi \otimes D^q \tilde{K} - D^q S)^n)$$

$$= \sum_{p,q} (-1)^q \binom{n}{p} \mathrm{Tr}((I_\xi \otimes D^q \tilde{K})^p (D^q S)^{n-p}) (-1)^{n-p}$$

Recall the definition of the Chern character in terms of the curvature form S of ξ, that is,

$$ch_k(\xi) = \frac{(-1)^k}{(2\pi i)^k} \frac{\mathrm{Tr} S^k}{k!}$$

Therefore, it follows, using lemmas 4 and 5, that

$$\lim_{t\downarrow 0} \sum_{q=0}^{n} (-1)^q \mathrm{Tr} e^q * 1$$

$$= \frac{1}{(2\pi)^n n!} \sum_{p,q} (-1)^q (-1)^{n-p} \binom{n}{p} \mathrm{Tr}((I_\xi \otimes D^q\tilde{K})^p (D^q S)^{n-p}) e$$

$$= \frac{(-i)^n (-1)^{n-p}}{(2\pi)^n n!} \sum_p \binom{n}{p} \tilde{\phi}((I_d \otimes \phi)S, \ldots, (I_d \otimes \phi)S) \wedge \phi(\tilde{K}) \wedge \ldots \wedge \phi(\tilde{K})$$

$$= \sum_{p=0}^{n} ch_{n-p}(\xi) \wedge \phi(\tilde{K}) \wedge \ldots \wedge \phi(\tilde{K}) \frac{1}{(2\pi)^p p!} i^p (-1)^p$$

The multiplicative property of the terms after $ch_{n-p}(\xi)$ follows from the wedge product and the binomial theorem. Clearly, these terms can be written in terms of the Chern forms of TM.

It is only necessary to verify that the polynomials in the Chern forms are the Todd polynomials. Since the polynomials have the multiplicative property, they are uniquely determined by a characteristic power series $Q(z)$ (see Hirzebruch, 1966). It suffices to show that

$$Q(z) = \frac{z}{1 - e^{-z}}$$

or, equivalently, that the coefficient of z^n in $(Q(z))^{n+1}$ is 1 (see Hirzebruch, 1966).

Consider the complex projective space $\mathbb{P}_n(\mathbb{C})$. There is a generator $h_n \in H^2(\mathbb{P}(\mathbb{C})\mathbb{Z})$ such that the Chern classes of $\mathbb{P}_n(\mathbb{C})$ are

$$C_i(\mathbb{P}_n(\mathbb{C})) = \binom{n+1}{i} h_n^i$$

The Euler-Poincaré characteric, $\chi(X,\Omega(\xi))$, where $X = \mathbb{P}_n(\mathbb{C})$ and ξ is a trivial bundle over X, is 1. $C_n(\mathbb{P}_n(\mathbb{C}))$, the nth Chern class of $\mathbb{P}_n(\mathbb{C})$, is $n + 1$. Combining these facts, it follows that the coefficient of z^n in $(Q(x))^{n+1}$ is 1.

It is known (see Gilkey, 1972) that when the manifold is not Kähler that the Riemann-Roch invariant is not obtained locally by the asymptotics of the heat equation. The difficulty that arises using the techniques here is that a complex Brownian motion cannot be defined because the Kähler property is lacking.

Euler-Poincaré Characteristic for a Riemannian Manifold with Boundary

Subsequent to his generalization (Chern, 1944) of the Gauss-Bonnet

Theorem, Chern (1945) proved a similar result for manifolds with boundary. Gilkey (1975) has used the asymptotics of the heat equation to obtain this latter formula of Chern.

We shall also obtain this formula here, but the verification will be somewhat less computational than the proofs of Theorems 1 and 2. This approach is more in the spirit of the work of Gilkey and Atiyah, Bott and Patodi.

Let M be a smooth, compact, oriented Riemannian manifold of even dimension with smooth boundary ∂M.

To describe the boundary conditions for the heat operator on p-forms, it is useful to introduce the normal projection on forms. The linear map, N, on forms is said to be the normal projection determined by the normal vector to the boundary if $(x_1,\ldots,x_m)$ is an orthonormal frame on the boundary with x_m normal to the boundary and $(dx_1,\ldots,dx_m)$ is the corresponding family of 1-forms. Then for a p=form

$$\omega = dx_{i_1} \wedge \cdots \wedge dx_{i_p}$$

$N(\omega) = \omega$ if there is one i_j for $1 \leqslant j \leqslant p$ such that $i_j = m$ and $N(\omega) = 0$ if there is no i_j for $1 \leqslant j \leqslant p$ such that $i_j = m$.

Recall that the differential form of Chern (1945) on the boundary is given by

$$B = \frac{1}{\pi^p} \sum_{\lambda=0}^{p-1} (-1)^\lambda \frac{1}{1\cdot 3\cdots(2p-2\lambda-1)^{p+\lambda}\lambda!} \Phi_\lambda \tag{4}$$

where

$$\Phi_k = \sum \varepsilon_{\alpha_1\ldots\alpha_{m-1}} \Omega_{\alpha_1\alpha_2} \wedge \cdots \wedge \Omega_{\alpha_{2k-1}\alpha_{2k}} \wedge \omega_{\alpha_{2k+1}m} \wedge \cdots \wedge \omega_{\alpha_{m-1}m} \tag{5}$$

$m = 2p$,

$$\varepsilon_{\alpha_1\ldots\alpha_{m-1}}$$

is the Kronecker index, Ω is the curvature form, ω is the connection form and the orthnormal frame $(x_1,\ldots,x_m)$ on the boundary has the property that x_m is normal to the boundary.

THEOREM 3. Let e^p be the fundamental solution of the heat operator

$(\partial/\partial t) - \Delta$ acting on p-forms on M where the boundary conditions are $N(\omega) = N(d\omega) = 0$, ω is a p-form and N is the normal projection on forms determined by a normal vector to the boundary. Let

$$\mathrm{Tre} = \sum_{p=0}^{m} (-1)^p \mathrm{Tre}^p$$

and let $\mathrm{Tr}\tilde{e}$ be the restriction of Tre to the boundary of M. Then

$$\mathrm{Tr}\tilde{e} = B$$

where B is the Chern polynomial for the boundary.

Only an outline of the proof of this theorem will be given. The approach will be less naive than the preceding proof for the Euler-Poincaré characteric for a manifold without boundary and the Riemann-Roch-Hirzebruch Theorem for Kähler manifolds. The traces of the fundamental solutions for the heat operators with the (reflecting) boundary conditions given in the statement of the theorem are obtained by forming the double of the manifold and making some natural identifications (e.g., see p. 53 of McKean and Singer, 1967) so that these traces have asymptotic expansions from the results for compact manifolds without boundary. Using these asymptotic expansions and the eigenspaces for these heat operators the first nonzero term in Tre is the constant term whose integral gives the Euler-Poincaré characteristic.

Since the term Tre has already been identified in the interior of the manifold by Theorem 1, it is only necessary to identify this term on the boundary. While the Euler-Poincaré characteristic does not depend on the orientation, an orientation will be chosen for M along with the induced orientation on ∂M. Therefore, the frame bundle is SO(M) and Tre is a form. Since the form is integrated on ∂M, it has to be an $(m-1)$-form, where m is the dimension of M. Consider the heat operators on the boundary described in local coordinates. The fundamental solutions for these operators are defined from the reflected Brownian motion and the Kac formula that is obtained from the terms in the curvature that appear in these operators. This curvature in M is expressed by the equation of Gauss for the curvature of an immersion (see p. 23 of Kobayashi and Nomizu) as a sum of the curvature of the boundary submanifold and some terms from the connection in M that give the normal component. By the uniform convergence almost surely of the sequence of approximations to the fundamental solutions used in the proof of Theorem 1, it follows that we now have that $\mathrm{Tr}\tilde{e}$ is an $(m-1)$-form in the curvature for the boundary submanifold and the terms in the connection that give the normal conponent. Since the Riemannian curvature has the property that $R_{ijk\ell} = R_{k\ell ij}$, the first two indices of the curvature behave as the latter two, that is, as a form. The Riemannian connection has the property that $\Gamma^k_{ij} = \Gamma^k_{ji}$. Thus, the indices α, β and γ in $\Omega_{\alpha\beta}$ and $\omega_{\gamma m}$ in the $(m-1)$-form $\mathrm{Tr}\tilde{e}$ behave as an $(m-1)$-form. Since the indices, excluding m, can be arbitrarily permuted, the terms in $\mathrm{Tr}\tilde{e}$ are only the family $(\Phi_k)_{k=1,\ldots,p-1}$ where Φ_k is given by (5). Specifically, we have

$$\mathrm{Tr}\tilde{e} = \sum \alpha_k \Phi_k$$

and it is only necessary to show that the coefficients (α_k) are the appropriate coefficients in (4). Recall that from the local expressions of the heat operators, $\mathrm{Tr}\tilde{e}$ is an $(m-1)$-form whose coefficients are universal. That is, they depend only on the dimension. Thus, it is only necessary to consider a family of examples that inductively determine the coefficients (α_k).

An elementary family of examples are manifolds that are formed from products of compact manifolds without boundaries and manifolds that are topologically polydiscs. One can compute the coefficients (α_k) for fixed dimension of the manifold by inductively computing with the polydiscs as their dimensions increase.

Generalized Lefschetz Fixed Point Formula

As two other examples of the computation of local formulae for index theorems by the Kac formula, we shall use the de Rham complex and prove the Lefschetz fixed point theorems for isolated fixed points and for fixed points that are smooth submanifolds. A suitable nonsingularity assumption will be made on the map. Lefschetz fixed point theorems for elliptic complexes were introduced by Atiyah and Bott (1967 and 1968).

Let M be a smooth, compact, oriented m-dimensional Riemannian manifold without boundary. Let $T: M \to M$ be a smooth map. The map T induces maps $T_p: H^p(M, \mathbb{C}) \to H^p(M, \mathbb{C})$ in cohomology and the Lefschetz number $L(T)$ is given by

$$L(T) = \sum_p (-1)^p \mathrm{Tr}(T_p)$$

With a nondegeneracy assumption on T, the Lefschetz number has been computed by heat equation methods for the de Rham complex for isolated fixed points by Kotake (1969) and Gilkey (preprint) and for fixed points that are smooth submanifolds by Gilkey (preprint). Donnelly (1976) has obtained some related results. We shall provide the computations of these two cases by the Kac formula.

THEOREM 4. Let $T:M \to M$ be a smooth map with isolated fixed points such that at each fixed point, $\det(I - dT) \neq 0$ where dT is the derivative of T. Let $e^p(t,x,y)$ be the fundamental solution of the heat operator $(\partial/\partial t) - \Delta$ on p-forms. Then the Lefschetz number $L(T)$ is given by

$$L(T) = \lim_{t \downarrow 0} \sum (-1)^p \mathrm{Tr}(T^* e^p) = \sum_{i \in K} \mathrm{sign}(\det(I - dT)(x_i))$$

where $(x_i)_{i \in K}$ are the fixed points of T.

Proof. The map $T:M \to M$ induces the maps $T_p^*:C^\infty(\Lambda^p T^*M) \to C^\infty(\Lambda^p T^*M)$ by the rule $T_p^*(\omega)(x) = \Lambda^p(dT(x))(\omega(Tx))$ where $\Lambda^p(dT)$ is the pth exterior power of the linear map dT. If e^p is the fundamental solution to the heat operator on p-forms, then the map T induces the fundamental solution T^*e^p which can be represented by the spectral representation of e^p as

$$T^*e^p(t,x,y) = \sum_i \exp(-t\lambda_{i,p}) T^*(\phi_{i,p})(x) \otimes \phi_{i,p}(y)$$

where

$$e^p(t,x,y) = \sum_i \exp(-t\lambda_{i,p}) \phi_{i,p}(x) \otimes \phi_{i,p}(y)$$

Since $T^*_{p+1} d = dT_p^*$, the Lefschetz number can be computed as

$$L(T) = \sum_p (-1)^p \mathrm{Tr}(T^* e^p)$$

where Tr is the operator trace.

For small $t > 0$, a fundamental solution for the heat operator in local coordinates globalized to Euclidean space can be constructed by the Kac formula. This fundament solution will differ from the correct fundamental solution in trace norm by $o(t^N)$ for an arbitrary positive integer

N as $t \downarrow 0$.

Let x_0 be an isolated fixed point of T and choose a convex neighborhood W of x_0 that contains only the fixed point x_0. Choose a local coordinate system in this neighborhood so that the corresponding frame is orthonormal at x_0, for example, take an orthonormal frame at x_0 and use the exponential map at x_0 as the local diffeomorphism.

The fundamental solution of the heat operator on functions can be represented as an infinite series by the classical techniques of E. E. Levi (see p. 49 of McKean and Singer, 1967). For small $t > 0$, the dominant term in this series is the first term which is obtained by freezing the metric at the initial point.

Thus, for small $t > 0$, we can use the fundamental solutions for the heat operators in local coordinates from the Kac formula that have been globalized to Euclidean space and, in fact, use the Gaussian density that is the dominant term for the fundamental solution of the heat operator on functions. Then

$$\int_W \mathrm{Tr}(T^*e^P)(t,x,x)dV(x) \approx \int_W \mathrm{Tr}\ f(t,x)p_0(t,x,Tx)\bar{g}(x)dx \tag{6}$$

where $\bar{g}$ is induced from the volume form V, p_0 is the Gaussian density on $\mathbb{R}^m$ and $f(t,x)$ is the conditional expectation of the solution of the stochastic ordinary differential equation conditions on $X_t = x$ where (X_t) is the diffusion process. The difference between the terms in (6) is $o(t)$. Now make the (nonsingular) change of variables $y = Tx - x$ so that the approximation to the trace in (6) becomes

$$\int \frac{\mathrm{Tr}\ f(t,x(y))p_0(t,y,0)}{|\det(I - dT)|}\bar{g}(x(y))dy$$

The change of variables $y \to y\sqrt{t}$ gives

$$\int \frac{\mathrm{Tr}\ f(t,x(y\sqrt{t}))p_0(1,y,0)}{|\det\ (I - dT)|}\bar{g}(x(y\sqrt{t}))dy$$

By elementary convergence properties as $t \downarrow 0$, we have

$$\int_{\mathbb{R}^m} \mathrm{Tr}\ f(0,x_0)\frac{p_0(1,y,0)}{|\det(I - dT)|}dy = \frac{\mathrm{Tr}\ f(0,x_0)}{|\det(I - dT)|}$$

From the Kac formula and the definition of T^*e^P, we have $\mathrm{Tr}\ f(0,x_0) =$

$\mathrm{Tr}\Lambda^p(dT)(x_0)$. By forming the alternating sum over p, we obtain the Lefschetz number for isolated fixed points, recalling that

$$\sum(-1)^p \mathrm{Tr}(A_p) = \det(I - A)$$

where A_p is the pth exterior power of A on p-forms.

For the Lefschetz fixed point theorem for fixed point sets that are smooth submanifolds, we have the following result that has been obtained by Gilkey (preprint) using invariance methods.

THEOREM 5. Let $T:M \to M$ be a smooth map whose fixed point set is the disjoint union of the smooth submanifolds $(N_i)_{i\in K}$ such that $\det(I - dT_\perp) \neq 0$ where $dT_\perp$ is the derivative of T restricted to the normal bundles of the submanifolds $(N_i)_{i\in K}$. Then the Lefschetz number $L(T)$ is given by

$$L(T) = \lim_{t\downarrow 0} \sum_p (-1)^p \mathrm{Tr}\, e^p = \sum_{i\in K} \mathrm{sign}(\det(I - dT\ )) \int_{N_i} C_i$$

where e^p is the fundamental solution of the heat operator on p-forms and C_i is the Chern polynomial for the Euler-Poincaré characteristic for the submanifold N_i.

Proof. Let N be a component of the fixed point set of dimension n and let $TM|N$ be the restriction of TM to N. A decomposition of $TM|N$ into the direct sum $TN \oplus \nu$ will be constructed so that ν is invariant under dT. This decomposition will be uniquely determined. A Riemannian metric on M will be chosen so that the direct sum $TN \oplus \nu$ is an orthogonal direct sum so that ν is the normal bundle $TN^\perp$ of the embedding. This construction by linear algebra has been done by Gilkey (preprint). Let $(\rho_1,\ldots,\rho_n)$ be a local orthonormal frame in TN and choose $(f_1,\ldots,f_{m-n})$ so that $(\rho_1, {}_2,\ldots,\rho_n,f_1,\ldots,f_{m-n})$ is a local frame for $TM|N$. Since TN is an invariant subspace of dT, for a suitable choice of Riemannian metric in TN, $dT(x)$ is represented by the matrix

$$\begin{pmatrix} I_n & A(x) \\ 0 & U(x) \end{pmatrix}$$

where A is an $n \times (m-n)$ matrix and U is an $(m-n) \times (m-n)$ matrix. By the nondegeneracy assumption, $U - I$ is nonsingular. Consider a local change of frame where $(f_1,\ldots,f_{m-n})$ are transformed to $(f'_1,\ldots,f'_{m-n})$ by the equations

$$f'_i = f_i + \sum_{j=1}^{n} E_{ij}\rho_j$$

Since $E = (E_{ij})$ is skew symmetric, dT represented in the basis $(\rho_1,\ldots,\rho_n,f'_1,\ldots,f'_{m-n})$ is

$$\begin{pmatrix} I_n & E \\ 0 & I_{m-n} \end{pmatrix}\begin{pmatrix} I & A \\ 0 & U \end{pmatrix}\begin{pmatrix} I_n & -E \\ 0 & I_{m-n} \end{pmatrix} = \begin{pmatrix} I_n & E(U - I) + A \\ 0 & U \end{pmatrix}$$

By the nondegeneracy assumption $E(U - I) + A = 0$ has a unique solution. The linear span of $(f'_1,\ldots,f'_{m-n})$ determines a unique complementary subbundle of $TM|N$ which is invariant under dT.

Let $x_0 \in N$ be fixed. Choose an orthonormal frame at x_0 so that the local coordinate system is $(a_1,\ldots,a_n,b_1,\ldots,b_{m-n})$ where this coordinate system evaluated at x_0 is the given orthonormal frame and $(a_1,\ldots,a_n)$ are local coordinates in a neighborhood of $x_0 \in N$. Let x_0 be represented as the origin of this coordinate system and $(a_1,\ldots,a_n,0,\ldots,0)$ a local representation of the submanifold.

Recall that the Riemannian metric was chosen so that $TN \oplus TN^{\perp}$ is an orthogonal direct sum. The first term in the infinite series of E. E. Levi that is used to construct the fundamental solution of the heat operator on functions in the coordinate system $(a_1,\ldots,a_n,b_1,\ldots,b_{m-n})$ factors at x_0 as a product of two functions because $TN \oplus TN^{\perp}$ is an orthgonal direct sum. Since dT splits into invariant subspaces of TN and $TN^{\perp}$, we can use the same techniques as was used for isolated fixed points for the integration of the trace formula on $TN^{\perp}$ to show that

$$\lim_{t\downarrow 0} \int Tr(T^*e^p)db = \frac{Tr\Lambda^p(dT_{\perp})(x_0)}{|\det(I - dT_{\perp})(x_0)|}$$

For the integration in the submanifold N there is an n-form in curvature that represents the integrand by the techniques of the proof of Theorem 1. In the Kac formula, the only curvature terms are those of the submanifold so that performing the alternating sum of the traces, we obtain by Theorem 1 the Chern polynomial for the submanifold.

REFERENCES

Atiyah, M. F., and Bott, R. A Lefschetz fixed point formula for elliptic complexes, I. Ann. Math., 86 (1967), pp. 374-407.

Atiyah, M. F., and Bott, R. A Lefschetz fixed point formula for elliptic complexes, II, Applications. Ann. Math., 88 (1968), pp. 451-491.

Atiyah, M., Bott, R, and Patodi, V. K. On the heat equation and the index thoerem. Invent. Math., 19 (1973), pp.279-330.

Babbitt, D. G. Wiener integral representations for certain semigroups which have infinitesimal generators with matrix coefficients. J. Math. Mech., 19 (1970), pp. 1051-1067.

Chern, S. S. A simple intrinsic proof of the Gauss-Bonnet formulas for closed Riemannian manifolds. Ann. Math., 45 (1944), pp. 747-752.

Chern, S. S. On the curvatura integra in a Riemannian manifold. Ann. Math., 46 (1945), pp. 674-684.

Donnelly, H. Spectrum and the fixed point sets of isometries. I. Math. Ann., 224 (1976), pp. 161-170.

Gilkey, P. Curvature and the eigenvalues of the Laplacian for geometrical elliptic complexes. Ph.D. dissertation, Harvard University, June, 1972.

Gilkey, P. B. Curvature and the eigenvalues of the Laplacian for elliptic complexes. Advances in Math., 10 (1973), pp. 344-382.

Gilkey, P. B. Curvature and the eigenvalues of the Dolbeault complex for Kähler manifolds. Advances in Math., 11 (1973), pp. 311-325.

Gilkey, P. B. The boundary integrand in the formula for the Signature and Euler characteristic of a Riemannian manifold with boundary. Advances in Math., 15 (1975), pp. 344-360.

Gilkey, P. B. Spectral geometry and the generalized Lefschetz fixed point formula for the de Rham and signature complexes. Preprint.

Hirzebruch, F. Topological methods in algebraic geometry. Springer, Berlin, 1966.

Itô, K. On a formula concerning stochastic differentials. Nagoya Math. J., 3 (1951), pp. 55-65.

Kac, M. On some connections between probability theory and differential and integral equations. Proc. Second Symp. on Math. Stat. and Prob., Berkeley (1951), pp. 184-215.

Kobayashi, S., and Nomizu, K. Foundations of Differential Geometry, 2. Interscience, New York.

Kotake, T. The fixed point theorem of Atiyah-Bott via parabolic operators. Communications Pure and Appl. Math., 22 (1969), pp. 789-806.

McKean, H. P. and Singer, I. M. Curvature and the eigenvalues of the Laplacian. J. Diff. Geom., 1 (1967), pp. 43-69.

Patodi, V. K. Curvature and the eigenforms of the Laplace operator. J. Diff. Geom., 5 (1971), pp. 233-249.

Patodi, V. K. An analytic proof of Riemann-Roch-Hirzebruch Theorem for Kaehler manifolds. J. Diff. Geom., 5 (1971), pp. 251-283.

Stroock, D. W. On certain systems of parabolic equations. Communications Pure and Appl. Math., 23 (1970), pp. 447-457.

Weyl, H. The Classical Groups. Princeton Univ. Press, Princeton, New Jersey (1946).

University of Kansas
Lawrence, Kansas

CONSTRUCTING BÄCKLUND TRANSFORMATIONS*

Robert B. Gardner

In this paper we give a new approach to a construction of Bäcklund transformations which makes up most of Goursat's memoir (1925). Since the motivation for studying Bäcklund transformations is not well known, we have included a historical survey, in the language of modern differential geometry, of the ideas leading to Bäcklund's original construction. Although there is not agreement on the precise definition of a Bäcklund transformation, the geometric exposition of Bäcklund's original ideas leads to an obvious extension which we suggest should be the definitive definition of Bäcklund transformation in the plane. Finally, as a byproduct of the new formulation and constructions, we give a method for iterating Bäcklund transformations and apply it to the famous Liouville equation.

A Bäcklund transformation in maximal generality is a correspondence between solutions of two systems of p.d.e. The definition is, of course, too general, but there is little agreement on what further structure should be added.

The study of these correspondences has historically been a branch of experimental mathematics. The subject flourished around the turn of the century up until about 1925 (see Lamb, 1976) and then remained dormant until the 1970's (see Muiria, 1976).

In order to get some feeling for the nature of these Bäcklund transformations, let us immediately look at an example. In what follows, we will consistently use the Monge notation for partial derivatives of a function $z(x,y)$

$$p = z_x, \quad q = z_y, \quad r = z_{xx}, \quad s = z_{xy}, \quad t = z_{yy}$$

---

* Elaboration and extension of an invited research announcement at the Park City Conference on Geometry and Partial Differential Equations and of an invited hour address at the Berlin conference on Differential Geometry and Global Analysis in July, 1977.

Example 1. Consider the nonlinear hyperbolic equations

$$s = pq$$

and the Bäcklund transformation

$$z' = \beta(z) = \log|p| - z$$

then

$$q' = \frac{1}{p} s - q = 0$$

and

$$s' = 0$$

How could this be useful? Certainly, we know the most general solution of $s' = 0$ which, in d'Alambert form, is

$$z' = f(x) + g(y)$$

Since we necessarily have

$$q' = g'(y) = 0$$

we may uncover z by solving

$$f(x) = \log|p| - z$$

or

$$p = e^z e^{f(x)}$$

Let $z = \log w$ and $e^{f(x)} = h'(x)$, then

$$\frac{1}{w}\frac{\partial w}{\partial x} = wh'(x) \quad \text{or} \quad \frac{\partial w}{\partial x} = w^2 h'(x)$$

This is solved by

$$w = \frac{-1}{h(x) + g(y)}$$

and hence

$$z = -\log|h(x) + g(y)|$$

a solution of $s = pq$ involving two arbitrary functions.

After meditating on this example, several questions come to mind.

1) Where did this correspondence come from?

2) Why would one want to have a relation which mixes the functions z and its derivatives p?

3) Is the resulting closed form solution expected or does it indicate some special qualitative property of the equation?

Let us begine with question (2). In order to understand this, we need to look into the flow of the consciousness at the turn of the century. The general theory of first order p.d.e. for one unknown

$$F(x^1,\ldots,x^n,z,p^1,\ldots,p^n) = 0 \quad \text{where} \quad p^i = \frac{\partial z}{\partial x^i} \tag{1}$$

and even systems of these was well advanced. There were, in fact, two basic ideas in the wind at this time.

Cauchy Characteristics

Let us take a parametrization of (1), that is, a specific submanifold (Σ_{2n},i) of $J^1(n,1)$ the space of first derivatives of maps from $\mathbb{R}^n$ to $\mathbb{R}$. The initial value problem is classically given by specifying data along a hypersurface in $\mathbb{R}^n$. This leads to maps δ and Δ where the diagram

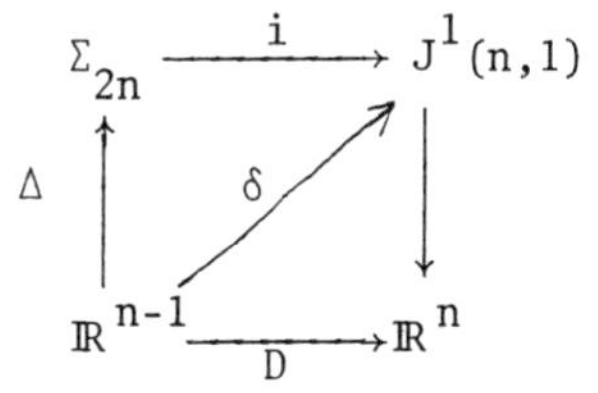

commutes and δ satisfies the integrability condition

$$\delta^* w = 0$$

where $w = dz - \Sigma p_i dx^i$ is the contact 1-form.

Solving the initial value problem is equivalent to constructing a map σ, such that

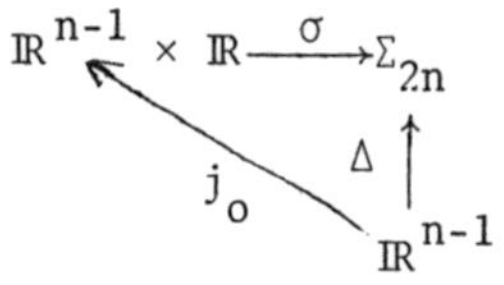 with $j_0(u) = (u,0)$

commutes and satisfies

(i) $* \circ i^*w = 0$

(ii) the image $i \circ \tau$ is transversal to the fibers of $J^1(n,1)$ over $\mathbb{R}^n$ (see Gardner, 1969 and 1977 for more details).

An obvious way to construct σ is via the flow of vector field on Σ_{2n}. In the search for such a vector field, it was found that there was a field

$$i_*x_F = \Sigma F_{p_i} \frac{\partial}{\partial x_i} + \Sigma p_i F_{p_i} \frac{\partial}{\partial z} - \Sigma(F_{x_i} + p_i F_z) \frac{\partial}{\partial p_i}$$

which always works unless x_F was not a transversal to the image of Δ (see Gardner, 1969).

Contact Transformations

A classical result (see Gardner, 1969, for a modern treatment) implies that

$$\mathbb{R}^n \xrightarrow{\sigma} \Sigma_{2n} \xrightarrow{i} J^1(n,1)$$

gives the locus of the 1-graph of a solution if and only if

(i) $(i\circ\sigma)^*(dz - \Sigma_i p_i dx^i) = (i\circ\sigma)^*w = 0$

(ii) $i \circ \sigma$ is transversal to the fibers.

Therefore if there exists a τ

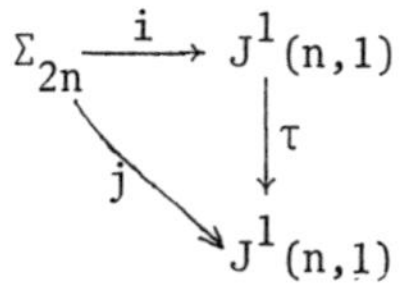

such that $\tau^*w = \lambda w$, then graphs of solutions will be taken into graphs of solutions unless (ii) gets violated. Such a τ is called a contact transformation.

Example w. $pq = x$ (a nonlinear equation). Let $\tau(x,y,z,p,q) = (p,q,z - px - qy, -x, -y)$ then the equation goes into

$$p = xy$$

which is immediately integrable and gives

$$z = \frac{1}{2} x^2 y + g(y)$$

This then allows one to use the inverse mapping to obtain results on the original problem.

Now let us turn to second order p.d.e. in the plane $F(x,y,z,p,q,r,s,t) = 0$ for which we choose a parameterization

$$\Sigma_7 \xrightarrow{i} J^2(2,1)$$

It was known at the turn of the century that

A'. There was no vector field that gave a theory of Cauchy characteristics.

B'. The natural generalization of contact transformation gave no new transformations.

Let us examine B' more carefully. A map

$$\mathbb{R}^2 \xrightarrow{\sigma} \Sigma_7 \xrightarrow{i} J^2(2,1)$$

is known to give the locus of the 2-graph of a solution if and only if

$$(i \circ \sigma)^* \begin{cases} dz - pdx - qdy \\ dp - rdx - sdy \\ dq - sdx - rdy \end{cases} = (i \circ \sigma)^* I = 0 \tag{i}$$

and

$$i \circ \sigma \text{ is transversal to fibers} \tag{ii}$$

Here I is called the second order contact system.

A second order contact transformation is a map

$$\tau : J^2(2,1) \to J^2(2,1)$$

such that

$$\tau^* I \subset I$$

The precise formulation of (B') is

THEOREM. (Bäcklund, 1877) A second order contact transformation in the plane is the prolongation of a contact transformation.

(In fact, the plane may be replaced by $\mathbb{R}^n$ and second order may be replaced by n^{th}-order.)

Proof. The idea is to exploit the first invariant of a nonintegrable Pfaffian system, the first derived system.

Let $I = \{w^1,\ldots,w^q\}$ be a system of 1-forms in the cotangent bundle T^* of some manifold and consider the composite map

$$I \xrightarrow{d} \Lambda^2 T^* \xrightarrow{\Lambda^2\pi} \Lambda^2(T^*/I), \qquad \delta = \Lambda^2\pi \circ d$$

This map is linear over the C^∞-functions since

$$\Lambda^2\pi \circ d(fw) = \Lambda^2\pi(df\wedge w + fdw) = f\Lambda^2\pi \circ dw$$

As such, there is a kernel and short exact sequence

$$0 \to I^{(1)} \to I \to dI \bmod I \to 0$$

This kernel is called the first derived system. We note that

a) If $I = I^{(1)}$ and has constant rank, then the system is completely integrable in the sense of Frobenius.

b) If I = second-order contact system, then

$$I^{(1)} = \{dz - pdx - qdy\}$$

The inclusion from right to left is given by

$$dp \wedge dx + dq \wedge dy \equiv (rdx + sdy) \wedge dx + (sdx + tdy) \wedge dy \equiv 0$$

For the opposite inclusion, see Gardner (1969).

Since maps commute with d,

$$f^* I \subset I \text{ implies } f^* I^{(1)} \subset I^{(1)}$$

which implies Bäcklund's theorem.

Since neither the theory of characteristics nor contact transformations had generalizations, the theory of second order p.d.e. was in rough shape and needed new ideas. The ideas were

A". Monge characteristics

B''. Bäcklund transformations

Examples of Bäcklund transformations predate Bäcklund and go back to very experimental calculations involving solutions of the "Laplace equations" $s = a(x,y)p + b(x,y)q + c(x,y)z$ where it was discovered that the transformation

$$z' = p + bz$$

led to equations of the same type and to certain structure theorems (see Forsythe, 1959).

Bäcklund bacame interested in the idea that bears his name as a result of a very geometric problem. The problem of finding explicit constructions of surfaces of constant negative curvature. The relation of this problem with p.d.e. is the following.

THEOREM. The pieces of surfaces of constant negative curvature -1 are locally parameterized by the solutions $0 < w < \pi$ of

$$\frac{\partial^2 w}{\partial x \partial y} = \operatorname{Sin} w \operatorname{Cos} w \tag{2}$$

Proof. It is classical that every such metric may be written in the form

$$ds^2 = du^2 + 2 \operatorname{Cos} 2w du dv + dv^2$$

where in a realization $2w$ is the angle between asymptotic lines: This metric automatically satisfies the Codazzi equations and the condition (2) is the Gauss equation. Thus, the existence follows from Bonnet's theorem (see Eisenhart, 1960).

During this era, a fashionable question was the investigation of transformations that preserved various geometric quantities and one of those that interested Bäcklund was the study of those transformations which preserved asymptotic lines. An extensive investigation (see Bianchi, 1910 and Eisenhart, 1960) led him to the system of first order p.d.e.

$$p + p' = \operatorname{Sin}(z - z')$$

$$q - q' = \operatorname{Sin}(z + z')$$

which has the property that a solution z of (2) gives rise to a solution z' of (2), and lines of curvature are preserved.

I believe that the key hint here is that the asymptotic lines are the Monge Characteristics of the p.d.e. of the surface and that this is what the classical Bäcklund transformations are preserving. Thus, we propose the definition:

DEFINITION. A Bäcklund transformation between solutions of p.d.e. in the plane is a correspondence taking solutions into solutions, and a family of Monge characteristics into a family of Monge characteristics.

Let us review the notion of Monge characteristics in a differential geometric form. We will restrict to hyperbolic equations, although there are analogous results for parabolic equations. Let

$$\Sigma_7 \xrightarrow{i} J^2(2,1)$$

be a parametrization of a second order p.d.e. and let I now stand for the pulled back system $= \{w^1, w^2, w^3\}$ where $w^1 = i^*(dz - pdx - qdy)$, $w^2 = i^*(dp - rdx - sdy)$, $w^3 = i^*(dq - sdx - tdy)$. There is a canonical conformal C^∞-bilinear form on I defined as follows for $\phi, \psi \in I$. Let

$$d\phi \wedge d\psi \wedge w^1 \wedge w^2 \wedge w^3 = \langle\phi,\psi\rangle d\Sigma_7$$

where $d\Sigma_7$ is a volume element for Σ_7. A p.d.e. is hyperbolic if and only if there is a basis $\{w^1, \pi^2, \pi^3\}$ for I such that the matrix

$$\begin{pmatrix} \langle w^1,w^1\rangle & \langle w^1,\pi^2\rangle & \langle w^1,\pi^3\rangle \\ \langle \pi^2,w^1\rangle & \langle \pi^2,\pi^2\rangle & \langle \pi^2,\pi^3\rangle \\ \langle \pi^3,w^1\rangle & \langle \pi^3,\pi^2\rangle & \langle \pi^3,\pi^3\rangle \end{pmatrix} = \begin{pmatrix} 0 & 0 & 0 \\ 0 & 0 & 1 \\ 0 & 1 & 0 \end{pmatrix}$$

(see Gardner, 1969).

Thus, the isotopic systems of I relative to $\langle , \rangle$

$$M_1 = \{w^1, \pi^2\} \quad \text{and} \quad M_2 = \{w^1, \pi^3\}$$

are intrinsically defined and contact invariant.

In addition there exists an extension of the basis of I to a basis $\{w^1, \pi^2, \pi^3, w^4, w^5, w^6, w^7\}$ of $T^*(\Sigma_7)$ satisfying

$$dw^1 \equiv 0$$

$$d\pi^2 \equiv w^4 \wedge w^5 \text{mod } I$$

$$d\pi^3 \equiv w^6 \wedge w^7$$

and the annhilators

$$V_1 = \{w^1,\pi^2,\pi^3,w^4,w^5\}^{\perp} \quad \text{and} \quad V_2 = \{w^1,\pi^2,\pi^3,w^6,w^7\}^{\perp}$$

are intrinsically defined contact invariant vector field system such that every solution is swept out by a vector in both V_1 and V_2. These are the Monge characteristic systems (see Gardner, 1977).

Now, beginning with second order p.d.e. $F = 0$, we want to construct a correspondence of its solutions with the solution of some second order p.d.e. $G = 0$ so that one of the two sets of Monge characteristics is preserved. As such, it is necessary that an isotropic system M_1^F of $F = 0$ be taken onto an isotropic system M_2^G of $G = 0$. The only reasonable way for this to happen is via a map on the space of variables supporting these systems, schematically

$$F = 0 \;\underset{\text{correspondence}}{\overset{\beta}{\text{- - - - - - - - - - -}}}\; G = 0$$

$$M_1^F \;\underset{\text{map}}{\text{―――――――}}\; M_2^G$$

The problem is that M_2^G is not known.

In order to uncover the natural construction of M_2^G, we must consider additional invariant structure of the isotropic systems.

First, it is a fact that for a hyperbolic p.d.e. the minimal number of variables needed to express an isotropic system M_i (that is, the class of M_i) is six or seven.

If the class of one of them, say M_1, is six, then there exists a C^∞-bilinear form on $M_1 - \{w,\pi^2\}$ defined for ϕ,ψ by $d\phi \wedge d\psi \wedge w^1 \wedge \pi^2 = \langle\phi,\psi\rangle d\Sigma_6$ where $d\Sigma_6$ is a volume on the six variables. An easy calculation shows that there exists a basis $\{w^1,\tilde{\pi}^2\}$ such that

$$\begin{pmatrix} \langle w^1,w^1\rangle & \langle w^1,\tilde{\pi}^2\rangle \\ \langle\tilde{\pi}^2,w^1\rangle & \langle\tilde{\pi}^2,\tilde{\pi}^2\rangle \end{pmatrix} = \begin{pmatrix} 0 & 1 \\ 1 & 0 \end{pmatrix}$$

and hence, such that the equations

$$w^1 = 0 \quad \text{and} \quad \tilde{\pi}^2 = 0$$

are contact invariant.

Example 3. $s = f(x,y,z,p,q)$
Here both isotropic systems have class six and

$$M_1 = \begin{cases} dz - pdx - qdy = w^1 \\ dp - rdx - fdy = \pi^2 \end{cases}$$

The matrix of $<,>$ on M_1 is given by

$$\begin{pmatrix} 0 & 1 \\ 1 & 2f_q \end{pmatrix}$$

hence

$$\tilde{\pi}^2 = \pi^2 - f_q w^1$$

This implies that the Pfaff relative invariants are also contact invariant (see Carton, 1953). Thus, there are two cases

$$(d\tilde{\pi}^2)^2 \wedge \tilde{\pi}^2 \neq 0 \tag{I}$$

and

$$(d\tilde{\pi}^2) \wedge \tilde{\pi}^2 \neq 0 \quad \text{but} \quad (d\tilde{\pi}^2)^2 \wedge \tilde{\pi}^2 = 0 \tag{II}$$

With this preparation, we can describe the algorithm for constructing Bäcklund transformations which Goursat discovered in a different way in 1925.

Construction (I) Find Pfaff coordinates (x,y,z',p',q') so that

$$\tilde{\pi}^2 = dz' - p'dx - q'dy$$

then compute $s' = \partial q'/\partial x$ and express the right hand side as a function of the prime coordinates and x and y by using the formulas for z', p', and q'.

Example 4. (Liouville Equation) $s = e^z$, then

$$M_1^F = \begin{cases} dz - pdx - qdy = w^1 \\ dp - rdx - e^z dy = \pi^2 \end{cases}$$

In this case $\tilde{\pi}^2 = \pi^2$ and we set

$$z' = p, \quad q' = e^z, \quad p' = r$$

Thus

$$z = \beta(z) = \frac{\partial z}{\partial x}$$

is a Bäcklund transformation, written schematically as

$$s = e^z \text{---}\overset{\beta}{\text{---}}\text{---} s' = q'z'$$

Construction (II) Find Pfaff coordinates

$$\tilde{\pi}^2 = dz' - p'dx$$

then $G = s' = 0$.

Example 5. $s = pq$. Hence using example 3, we see

$$\tilde{\pi}^2 = dp - rdx - pqdy - p(dz - pdx - qdy) = dp - pdz - (r - p)^2 dx$$

This is case II since

$$\frac{1}{p}\tilde{\pi}^2 = d(\log|p| - z) - r - p^2)dx$$

Note that the recipe gives $z' = \log|p| - z$ which is the transformation of the first example at the beginning of the paper.

Unfortunately, these two constructions are not exhaustive since $\tilde{\pi}^2$ may not admit a Pfaff normal form involving the differentials dx and dy. This case was ignored by Goursat and needs to be studied.

Example 6. $s = f(x,y,z,p,q)$ with $\partial f/\partial q$ not linear in p or nonconstant in q.

Although it is true that Bäcklund transformations cannot be composed in general, there is an iterative construction which is possible for the hyperbolic equations in the plane. This is achieved by flipping the isotropic systems. In our schematic notation this would lead to the picture

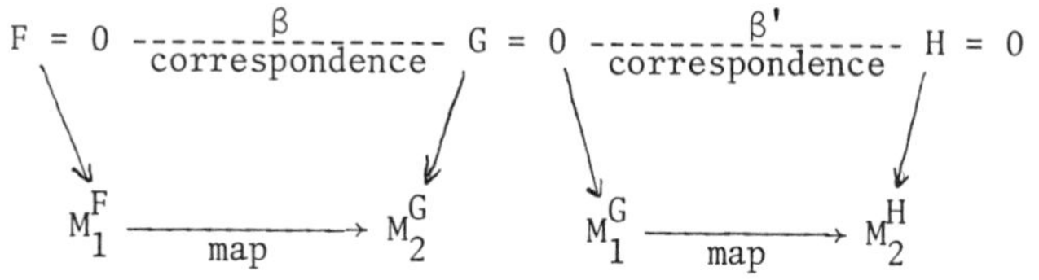
F = 0
β
correspondence
G = 0
β'
correspondence
H = 0
M$_1^F$
map
M$_2^G$
M$_1^G$
map
M$_2^H$

which is inductive as long as both isotropic systems encountered have class six, and until understood, Case III does not appear.

Example 7. (Liouville Equation continued). We take $s = e^z$ for $F = 0$ and, as in example 4, we find a Bäcklund transformation β to $s' = q'z'$. Now

$$M_1^G = \begin{cases} dz' - p'dx - q'dy = w' \\ dp' - r'dx - q'z'dy = \pi^2 \end{cases}$$

and

$$\tilde{\pi}^2 = \pi^2 - z'(dz' - p'dx - q'dy) = d\left(p' - \frac{z'^2}{2}\right) - (r'-p'z')dx$$

Hence we set

$$z'' = p' - \frac{z'^2}{z} \quad \text{and} \quad q' = 0$$

which yields

$$s'' = 0 \quad \text{for} \quad H = 0$$

Thus, we have constructed a composition of correspondences

$$s = e^z \overset{\beta}{-\!-\!-\!-\!-\!-} s' = q'z' \overset{\beta'}{-\!-\!-\!-\!-\!-} s'' = 0$$

which can be used to pull back the general solution $z'' = f(x) + g(x)$ of $s'' = 0$ to the classical closed solution

$$z = 2\,\frac{f'(x)h'(y)}{(f(x) + h(y))^2}$$

of the Liouville equation.

We close by asking two obvious but intriguing questions about this iterative scheme which would hopefully clarify the untouched question 3 mentioned earlier.

4) Classify those $F = 0$ with the above mentioned iterative scheme terminating in $s = 0$ and find an invariant for the number of steps.

5) If a $F = 0$ has the iterative scheme terminating in $s = 0$, is there a solution involving two arbitrary functions and a finite number of their derivatives? If so, find an invariant for the order of the derivatives.

REFERENCES

Bäcklund, A. V. Uber partielle Differential gleichungen hoherer Ordnung. *Math. Ann.*, *13* (1877) pp. 69-108.

Bianchi, L. *Vorlesungen über Differential Geometrie*. B. G. Teubner, Liepzig (1910).

Cartan, E. Les Systèmes de Pfaff a Cinq Variables. *Oeuvres Complètes Gauthier Villars*, Paris (1953) pp. 927-1010.

Cartan, E. Sur les Transformations de Bäcklund. *Oeuvres Complètes Gauthier Villars*, Paris (1953) pp. 1175-1193.

Cartan, E. Sur Certaines Expressions Différentielles et le Problème de Pfaff. *Oeuvres Complètes Gauthier Villars*, Paris (1953) pp. 303-396.

Eisenhart, L. P. *Differential Geometry of Curves and Surfaces*. Dover Publications, New York (1960).

Forsythe, A. R. *Theory of Differential Equations*. Vol VI. Dover Publications, New York (1959).

Gardner, R. A Differential Geometric Approach to Characteristics. *Comm. Pure and Appl. Math.*, *22* (1969) pp. 597-626.

Gardner, R. Differential Geometric Viewpoints on the Development of Shock Waves. In *Conference on Geometric Nonlinear Wave Theory*. Ed., R. Hermann. Math. Sci. Press, Brookline, Massachusetts (1977).

Goursat, E. Le Problème de Bäcklund. *Memor. Sci. Math.*, *6* Gauthier Villars, Paris (1925).

Lamb, G. Bäcklund Transformations at the turn of the Century. In *Lecture notes in Math.*, *Vol. 515*. Springer-Verlag (1976) pp. 69-79.

Muira, R. Bäcklund Transformations. *Lecture notes in Math.*, *Vol. 515*. Springer-Verlag (1976).

University of North Carolina
Chapel Hill, North Carolina

LEFSCHETZ FIXED POINT FORMULAS AND THE HEAT EQUATION

Peter B. Gilkey*

Introduction

The Atiyah-Singer index theorem and the generalization to group actions was first proved by topological methods. It gives a formula for the index of any elliptic complex in terms of characteristic classes. The index theorem contains the Gauss-Bonnet formula for the Euler-Poincaré characteristic, the Hirzebruch formula for the signature, and the Riemann-Roch formula for the arithmetic genus as special cases. We will refer to the generalization of the index theorem to group actions as generalized Lefschetz fixed point formulas. The G-signature theorem and the classical Lefschetz fixed point formula for the DeRham complex are examples of such formulas.

In the past five years, a proof of the Atiyah-Singer index has been given using heat equation methods. This proof substitutes analysis and the techniques of invariant theory for much of the topology of the original proof. More recently, this proof has been generalized to include group actions as well and to give a complete proof of the Atiyah-Singer index theorem in full generality. In this paper, we would like to give a brief historical account of some of this work and to outline the techniques involved.

Let M be a Riemannian manifold of dimension m which is compact and without boundary. If $X = (x_1,\ldots,x_m)$ is a system of local coordinates on M, let $ds^2 = g_{ij}dx^i dx^j$ be the metric tensor; we sum over repeated indices unless otherwise noted. Let g^{ij} be the inverse matrix and let $g = \mathrm{sqrt}(\det(g_{ij}))$. The Riemannian measure is given by $|dvol| = g|dx^1\ldots dx^m|$; we use "$|$" to emphasize that this measure does not depend on the orientation of M.

Let $T:M \to M$ be a smooth map. We assume the fixed point set of T

*Research partially supported by N.S.F. Grant MPS72-04357 and by Sloan Foundation grant BR 1687.

consists of the disjoint union of smooth submanifolds N_i of dimension n_i. Let N be one of the components of the fixed point set of T and let $x_0 \in N$. Let ν be the quotient bundle $\nu = T(M)|_N/T(N)$ over N. Since dT preserves the subbundle $T(N)$, dT induces a map dT_ν on ν. We suppose $\det(I - dT_\nu) \neq 0$ as a nondegeneracy condition. This means T leaves no direction infinitesimally fixed except directions tangent to N. If we put $dT(x_0)$ in Jordan normal form, $T(N)$ is the span of the eigenvectors with eigenvalue 1 and ν is the span of the generalized eigenvectors for the remaining eigenvalues. This induces a natural splitting of $T(M)|_N = T(N) \oplus \nu$. We assume the metric on M is chosen so this splitting is orthogonal. This identifies ν with the normal bundle of N in M.

If T embeds in the action of a compact group on M, we may choose the metric so the group acts by isometries. The conditions above are always satisfied by an isometry, but there are other examples. Let CP_m denote complex projective space of dimension $2m$ and let $A \in GL(m+1,C)$ be a nonsingular complex matrix. Let $T(A)$ be the map induced by A on CP_m. If A is diagonalizable, $T(A)$ is nondegenerate and the fixed points of $T(A)$ are the eigenvectors of A. $T(A)$ is degenerate if A is not diagonalizable. We shall discuss other examples in later sections.

Let (V,d) be an elliptic complex over M. V is a finite collection $\{V_0,V_1,\ldots\}$ of smooth vector bundles over M. d is a finite collection $\{d_i : C^\infty(V_i) \to C^\infty(V_{i+1})\}$ of smooth differential operators. We suppose $d_{i+1}d_i = 0$ and that the complex is exact on the symbol level. If $x \in M$ and if $\xi \in T^*M_x$, the symbol $\sigma(d_i)(x,\xi) : V_i(x) \to V_{i+1}(x)$ is a linear map from the fibre of V_i to the fibre of V_{i+1} over x. The complex is exact on the symbol level if image $\sigma(d_i)(x,\xi) =$ kernel $\sigma(d_{i+1})(x,\xi)$ for all $\xi \neq 0 \in T^*M_x$ and for all $x \in M$. The DeRham complex is the prototypical example of an elliptic complex. Let $V_i = \Lambda^i T^*M$ be the bundle of differential i-forms and let d be ordinary exterior differentiation: $d^2 = 0$. The symbol of d is exterior product $\text{ext}(\xi) : \Lambda^i T^*M \to \Lambda^{i+1} T^*M$ so this is exact on the symbol level.

If (V,d) is an elliptic complex, let $H^i(V,d) = \ker(d_i)/\text{im}(d_{i-1})$ be the finite dimensional cohomology groups. The index $(V,d) = \Sigma_i(-1)^i \dim H^i(V,d)$; the Atiyah-Singer index theorem gives a formula for this integer in terms of characteristic classes and the symbol of d. For example, if (V,d) is the DeRham complex, $H^i(V,d)$ is isomorphic to the ordinary cohomology $H^i(M,C)$ by the DeRham isomorphism. The index of this complex is the Euler-Poincaré characteristic $\chi(M)$.

The g-index is defined as follows: let $T:M \to M$ be a smooth map. A contravariant action $A(T)$ on the elliptic complex (V,d) is a collection of maps $A(T)_i:V_i \to V_i$ which cover the map T. This is equivalent to the following commutative diagram.

$$\begin{array}{ccc} V_i & \xrightarrow{A(T)} & V_i \\ \downarrow & \circ & \downarrow \\ M & \xrightarrow[T]{} & M \end{array}$$

We assume $A(T)$ is smooth and linear on the fibres and that $A(T)$ commutes with the differential operator d, i.e., $dA(T) = A(T)d$ on smooth sections to V. For example, if (V,d) is the DeRham complex, let $A(T) = \Lambda^i(dT)^*:\Lambda^i(T^*M) \to \Lambda^i(T^*M)$. The naturality axiom for exterior differentiation shows $dA(T) = A(T)d$.

Since $A(T)$ commutes with d, it induces maps $A(T)_i^{\#}:H^i(V,d) \to H^i(V,d)$. The g-index or Lefschetz number of this action is defined by $L(T,V,d) = \Sigma_i(-1)^i \mathrm{Tr}\, A(T)_i^{\#}$. If T is the identity map and if $A(T)$ is the trivial action, $L(T,V,d) = \text{index }(V,d)$. Although $L(T,V,d)$ depends on the choice of the action A, we shall supress this dependence for notational simplicity. For suitable actions and elliptic complexes, there are generalized Lefschetz fixed point formulas which express $L(T,V,d)$ in terms of generalized characteristic classes which are evaluated on the fixed point set of a nondegenerate map T.

Let $L(T)_{DER}$ be the Lefschetz number of T relative to the DeRham complex with the action defined above. If we identify the cohomology groups with the ordinary cohomology of M, $A(T)_i^{\#}$ is the usual action of T on cohomology. This implies $L(T)_{DER}$ is the ordinary Lefschetz number of T. The classical Lefschetz fixed point formula expresses $L(T)_{DER}$ in terms of the fixed point set of a nondegenerate map T:

$$L(T)_{DER} = \Sigma_i \mathrm{sign}\ \det(I - dT_\nu)\chi(N_i)$$

where the sum ranges over the components of the fixed point set of T.

We can reexpress this formula in more differential geometric terms as follows: let $R_{ijk\ell}$ denote the curvature tensor of the metric G on M and let $E_m(x,G)$ be the integrand of the Chern-Gauss-Bonnet theorem (see Chern, 1944). $E_m = 0$ for m-odd, while for m even,

$$E_m = c(m)\varepsilon(i_1 \ldots i_m)\varepsilon(j_1 \ldots j_m)R_{i_1 i_2 j_1 j_2} \ldots R_{i_{m-1} i_m j_{m-1} j_m}$$

We evaluate E_m in an orthonormal frame for $T(M)$ and sum over repeated indices. ε is zero if $(i_1 \ldots i_m)$ are not distinct. It is the sign of the permutation if these indices are distinct; $c(m)$ is a universal constant given by Chern. For example, $E_2 = (2\pi)^{-1}R_{1212}$. The Chern-Gauss-Bonnet theorem expresses

$$\chi(M) = \int_M E_m(x,G)|dvol(x)|$$

Let G_N denote the restriction of the metric to N and let $|dvol|_n$ be the Riemannian measure induced by G_N on N. We may use the Chern-Gauss-Bonnet theorem to express

$$L(T)_{DER} = \Sigma_i \text{ sign det}(I - dT_\nu)\int_{N_i} E_{n_i}(x,G_{N_i})|dvol|_{n_i}$$

In this sum, we may ignore components of odd dimension since $E_n = \chi(N) = 0$ if n is odd. E_n is the representative of the Euler class using the Chern-Weyl isomorphism so this equation expresses $L(T)_{DER}$ in terms of generalized characteristic classes. There are other similar formulas for the signature, spin, and Dolbealut complexes which we shall discuss later.

We can use the heat equation to obtain a local formula for the Lefschetz number of any elliptic complex. Let (V,d) be an elliptic complex. Choose a local fibre metric on each V_i. Let d^* be the collection of adjoints and let $D^* = dd^* + d^*d$ be the associated Laplacians. $D_i^* = d_{i-1}d_{i-1}^* + d_i^*d_i$. We shall assume henceforth that d is a first order differential operator. Since (V,d) is elliptic, D^* is an elliptic second order differential operator. It is well known (see Seeley, 1967) that for $t > 0$, $\exp(-tD): L^2(V) \to C^\infty(V)$ is of trace class. The fundamental algebraic lemma is the following.

Lemma 0.1. $L(T,V,d) = \Sigma_i(-1)^i \text{Tr}_{L^2} A(T)_i \exp(-tD_i)$ for any $t > 0$.

Proof. By the Hodge decomposition theorem, the spectrum of D is discrete and pure point. Decompose $L^2(V_i) = \oplus E(\lambda)_i$ into a direct orthogonal sum of the eigenspaces $E(\lambda)_i$ of D_i; each $E(\lambda)_i$ is a finite dimensional space of smooth sections Let $\pi(\lambda)_i$ be orthogonal projection on $E(\lambda)_i$. Since $d^2 = 0$, $dD = Dd$. This implies d preserves the eigenspaces of D and commutes with $\pi(\lambda)$. If $\lambda \neq 0$, the induced complex: $0 \to E(\lambda)_0 \xrightarrow{d} E(\lambda)_1 \to \ldots \to 0$ is a long exact sequence. Let $A(T)_i^\lambda = \pi(\lambda)_i A(T)$; $A(T)^\lambda$ commutes with d and induces a chain map. Consequently $\Sigma_i (-1)^i \mathrm{Tr}\, A(T)_i^\lambda = 0$ for $\lambda \neq 0$. If $\lambda = 0$, we identify $H^i(V,d)$ with $E(0)_i$; under this identification, $A(T)_i^0 = A(T)_i^\#$. Therefore

$$\Sigma_i (-1)^i \mathrm{Tr}_{L^2}\, A(T)_i \exp(-tD_i) = \Sigma_{i,\lambda} (-1)^i \exp(-t\lambda) \mathrm{Tr}\, A(T)_i^\lambda$$

$$= \Sigma_i (-1)^i \mathrm{Tr}\, A(T)_i^0 = \Sigma_i (-1)^i \mathrm{Tr}\, A(T)_i^\# = L(T,V,d)$$

This purely algebraic fact, together with the asymptotic expansion of the following theorem, gives a local formula for the Lefschetz number.

THEOREM 0.2. Let $D: C^\infty(V) \to C^\infty(V)$ be an elliptic, self-adjoint second order differential operator, let $T: M \to M$ be nondegenerate, and let $A(T)$ be an action of V which covers T. We do not need to assume $A(T)D = DA(T)$. Then $\exp(-tD)$ is of trace class and there is an asymptotic formula

$$\mathrm{Tr}_{L^2}\, A(T)\exp(-tD) \sim \Sigma_i (4\pi t)^{-n_i/2} \sum_{k=0}^{\infty} t^k \int_{N_i} a_{k,n_i}(x,T,D,) |dvol_{n_i}|$$

$a_{k,n}(x,T,D)$ is a smooth <u>local</u> invariant of the jets of the differential operator D and of the jets of the map T and action $A(T)$ on the components of the fixed point set.

If we let $a_{k,n}(x,T,V,d) = \Sigma_i (-1)^i a_{k,n}(x,T,D)$, then this yields the formula

$$L(T,V,d) \sim \Sigma_i (4\pi t)^{-n_1/2} \sum_{k=0}^{\infty} t^k \int_{N_i} a_{k,n_i}(a,T,V,d) |dvol(x)|_{n_i}$$

Since the left hand side is independent of t, all the terms on the right hand side must vanish except those for $2k = n_i$. In particular, we may ignore the contribution by the components of the fixed point set of odd dimension. This proves

COROLLARY 0.3. Let T be nondegenerate and act on an elliptic complex (V,d), then

$$L(T,V,d) = \Sigma_i \int_{N_i} a_{k,n_i}(x,T,V,d) |dvol|_{n_i}$$

where we sum over the even dimensional components of the fixed point set of T and over $2k = n_i$. $a_{k,n}$ is a smooth local invariant of (T,V,d), the metric on M, the action A, and the fibre metric on V.

Corollary 0.3 for $T = I$ and (V,d) = the DeRham complex led McKean and Singer (1967) to conjecture that these local invariants could be used to give a proof of the Chern-Gauss-Bonnet theorem. They conjectured that $a_{k,m}(x,I,V,d) = 0$ for $2k < m$ and that $a_{k,m} = E_m$ if $2k = m$. They were able to prove this for $m = 2$ (which proves the Gauss-Bonnet theorem for surfaces), but were unable to prove the general case.

Their conjecture is, in fact, true and there are other similar results for the other classical elliptic complexes. We shall define the twisted signature, spin, and Dolbeault complexes later in this paper. We summarize below the results which generalize McKean and Singer's conjecture.

THEOREM 0.4. Let $T:M \to M$ be nondegenerate.

1) Let (V,d) be the DeRham complex and let $a^{DER}_{k,n}(x,T) = a_{k,n}(x,T,V,d)$. Then

a) $a^{Der}_{k,n}(x,T) = 0$ for $2k < n$

b) $a^{Der}_{k,n}(x,T0 = \text{sign}\det(I - dT_\nu)E_n(x,G_N)$ if $2k = n$

c) $a^{Der}_{k,n}(x,T) \neq 0$ generically if $2k > n$ and if n is even. However, there exists a smooth local 1-form valued invariant $q^{Der}_{k,n}$ so $a^{Der}_{k,n}$ = divergence $q^{Der}_{k,n}$.

2) Let m be even and let M be oriented. We assume T is an orientation preserving isometry. Let $a^{s}_{k,n}(x,T,E) = a_{k,n}(x,T,V,d)$ for (V,d) the twisted signature complex with coefficients in a vector bundle E. We assume T acts on E as well as on the twisted signature complex. The components of the fixed point set of T are all even dimensional. Since the signature complex depends on the orientation of M, we shall regard $a^{s}_{k,n}(x,T,E)$ as an invariant n-form, not a measure. Then

a) $a^{s}_{k,n}(x,T,E) = 0$ for $2k < n$

b) $a^{s}_{k,n}(x,T,E)$ is the formula given by Atiyah and Singer (1968) in the g-signature theorem giving $L(T,E)_s$ in terms of the generalized characteristic classes of $T(N)$, ν, and of E if $2k = n$.

c) $a^{s}_{k,n}(x,T,E) \neq 0$ generically if $2k < n$. However, there exists a smooth n-1 form valued local invariant $q^{s}_{k,n}$ defined on N such that $a^{s}_{k,n} = dq^{s}_{k,n}$.

d) If $a^{sp}_{k,n}(x,T,E)$ is the invariant of the twisted spin complex, a similar result holds true.

3) Let M be an almost complex manifold and let T be an isometry of M preserving the almost complex structure. Let $(\bar{\partial} + \delta): C(\Lambda^{o,even}) \to C(\Lambda^{o,odd})$ be the two term Dolbeault complex. Let $a_{k,n}^{Dol}(x,T,E) = a_{k,n}(x,T,V,d)$ where (V,d) is the Dolbeault complex with coefficients in E. Then

a) $a_{k,n}^{Dol}(x,T,E) = 0$ for $2k < n$ if M is holomorphic and the metric is Kähler. Otherwise, $a_{k,n} \neq 0$ generically if $k > n$. (n is the real dimension of a component of the fixed point set.)

b) If M is holomorphic and if the metric is Kähler, then $a_{k,n}^{Dol}$ is the local formula given by Atiyah and Singer (1968) in terms of generalized characteristic classes. Otherwise, these two local formulas do not agree pointwise generically. The difference of the formulas is the divergence of a functorial 1-form.

c) $a_{k,n}(x,T,E) \neq 0$ for $k > n$ generically. However, if $2k \neq n$, $a_{k,n} = \operatorname{div}(q_{k,n})$ where $q_{k,n}$ is a smooth 1-form valued local invariant.

If $T = I$, then 3) implies the Riemann-Roch formula. If the almost complex manifold saisifies the Nirenberg-Neulander integrability condition (and hence is holomorphic), it is not necessary to impose the additional condition that T is an isometry; it suffices to assume that T is holomorphic in order to define the Lefschetz number relative to the Dolbeault complex in this case. D. Toledo and Y. L. Tong (1975) have given a formula for the Lefschetz number in this case in terms of generalized characteristic classes using other methods.

In view of the local vanishing theorems 0.4 (1a and 2a), it is worth noting

Remark 0.5. Let $m > 0$ and let T^m be the torus of dimension m. There is an elliptic complex (V,d) and a metric G on T^m such that $a_{k,m}(x,I,V,d) \neq 0$ for all $k > 0$.

Remark 0.5 is proved in Gilkey (to appear). The vanishing theorems 1a and 2a of Theorem 0.4 are special consequences of the geometric nature of the DeRham and signature complexes. (There are similar vanishing theorems for the Dolbeault complex if $3k < 2n$ which are proved in Gilkey, 1972.) By 0.4 (3b), the integrand of the heat equation for $2k = n$ need not agree pointwise with the integrand of the index theorem. The nonvanishing results of 0.4 are proved in Gilkey (1972); the results expressing the error term in divergence form follow from the methods of

Gilkey (1976). We shall give a complete proof of Theorem 0.2 and the remainder of 0.4 in this paper.

Atiyah, Bott, and Patodi (1973) noted that any elliptic complex is rationally homotopic to the twisted signature complex. Since the index is a rational invariant of the homotopy class of the leading symbol of the complex, this gives a proof of the Atiyah-Singer index theorem in general, using heat equation methods.

A number of authors have worked on proving Lefschetz formulas using heat equation methods. We would like to give a brief history of some of the efforts in this direction. This will necessarily be incomplete as there were a great many people involved. In 1966, Kac posed the question of whether it was possible to determine the number of holes in a plane domain with boundary from the spectrum of the Laplacian acting on functions. McKean and Singer (1967) solved this question and determined some of the lower order terms in the asymptotic expansion of the heat equation. They gave a proof of the Gauss-Bonnet formula in dimension 2 and conjectured Theorem 0.3 (1), but the general case eluded them.

Also in 1967, Seeley derived the meromorphic extension of the zeta function of a general elliptic differential operator with positive definite leading sympol and for $T = I$. (In fact, Seeley was working with a wider class of differential operators, but we shall not need his results in full generality.) Although he was working with the zeta function, the poles of the zeta function are related to the asymptotics of the heat equation by the Mellin transform, so his results are equivalent to Theorem 0.2 for $T = I$. He also remarked that Bott had mentioned to him that the algebraic fact we have denoted by Lemma 0.1 might be used to give a proof of the index theorem.

In 1969, Seeley and Greiner generalized these results to manifolds with boundary. If $dM \neq 0$, there are additional boundary integrals in formula 0.2 wich arise from the boundary value problem imposed. This yields a generalization of Corollary 0.3 to the index of an elliptic complex on a manifold with boundary. At roughtly the same time, Kotake (1969) derived the leading term in the asymptotic expansion for isolated nondegenerate fixed points. This gave a proof of Theorem 0.2 for the case $k = n = 0$; his analysis was sufficient to derive the Lefschetz formulas of Atiyah and Bott (1967).

In 1975, Lee, a student of Seeley's, generalized Seeley's earlier results on manifolds without boundary to include group actions. This gave a complete proof of Theorem 0.2. In 1976 we derived 0.2 independently, since we were not aware of Lee's work at that time. Donnelly (1976) has also given an independent proof of 0.2 if T is an isometry and derived in the first 3 terms in the expansion of the Laplacian acting on functions. He has generalized 0.2 to manifolds with boundary and actions by isometries in his paper generalizing the eta invariant to group actions (to appear).

In 1971, Patodi constructed a direct proof of the index theorem for the Dolbeault and DeRham complexes using a complicated cancellation lemma. He also indicated that his cancellation method could be applied to the signature complex to prove the Hirzebruch signature theorem by heat equation methods. In 1972, we gave an independent proof of Theorem 0.4 for $T = I$ using invariance theory instead of Patodi's cancellation argument. This gave as a byproduct an abstract characterization of the Pontrjagin, Euler, and Chern classes in terms of their functorial properties as local formulas. Our proof did not use H. Weyl's formula on the invarants of the orthogonal group, so Atiyah, Bott, and Patodi gave another version of our proof, using more classical techniques (1973).

During the period 1970 to 1976, Patodi had been working on generalizing Theorem 0.4 to general nondegenerate maps T and published a research announcement (1971) concerning the holomorphic Lefschetz fixed point formula. Unfortunately, his illness and untimely death in December, 1976 prevented him from publishing any of the details of his work on holomorphic Lefschetz formulas. It also prevented him from publishing the details of his work on the G-signature theorem, and his work on the G-signature theorem is not generally known.

In 1976, Kawasaki gave a proof of the G-signature theorem in his thesis on V-manifolds under slightly restrictive hypothesis regarding the isometries to be considered. Donnelly, in a joint paper with Patodi (1977), has given a proof of the G-signature theorem and Donnelly (preprint) has extended this result to arbitrary coefficient bundles thereby giving a proof of 0.4 (2 and 3). We have given an independent derivation of 0.4 which we will present in this paper.

Although we will not have the space to discuss the corresponding results for manifolds with boundary in this paper, we would like to mention some of the papers for the purposes of reference. In 1973, Atiyah, Patodi and Singer generalized the signature theorem to manifolds with boundary. This yields the formula

$$\text{signature}(M) - \int_M L_k = \eta(dM)$$

L_k is the integrand of the Hirzebruch signature formula, $\eta(dM)$ is a spectral invariant of the boundary. There is an additional integral over the boundary which involves the second fundamental form and curvature if the metric is not product near dM. In 1974, we discussed the additional local boundary integrals which arise in this formula if the boundary is not totally geodesic. We also gave a proof of the Chern-Gauss-Bonnet formula for manifolds with boundary using heat equation methods (1975). In 1977, Donnelly (to appear) generalized the Atiyah, Patodi, Singer theorem to actions by compact groups on maniolfds with boundary; his work has immediate application to computing the eta invariant for lens spaces.

The remainder of the paper is divided into four sections. In the first section, we will give a proof of the analytic results that we shall need (0.2) and discuss some of the functorial properties of the local invariants involved. In the second section, we will prove the Lefschetz formulas for the DeRham complex (Theorem 0.4(1)), in the third section, we will prove the Lefschetz formulas for the twisted signature complex (Theorem 0.4(2)), and in the final section, we will discuss the twisted Dolbeault complex (Theorem 0.4(3)).

§1. Let M be a compact Riemannian manifold without boundary of dimension m and let $T:M \to M$ be nondegenerate. Let V be a vector bundle over M and let $A(T)$ be an action of T on V. Let $D:C^\infty(V) \to C^\infty(V)$ be a second order nonnegative self-adjoint differential operator with leading symbol given by the metric tensor. If X is a system of local coordinates and if s is a local frame for V, express

$$D = -g^{ij}\partial/\partial x_i \partial x_j + b^k \partial/\partial x_k + c$$

The b^k and c are square matrices which are not invariant. The total symbol of D is

$$\sigma(D)(x,\xi) = a^2(x,\xi) + a^1(x,\xi) + a^0(x,\xi) \quad \text{for} \quad \xi \in T^*M_x$$

The a^k are homogeneous polynomials of degree k in ξ given by the formulas

$$a^2(x,\xi) = |\xi|^2 = g^{ij}\xi_i\xi_j, \quad a^1(x,\xi) = ib^k\xi_k, \quad a^0(x,\xi) = c$$

Let $\alpha = (a_1,\ldots,a_m)$ be a multiindex and define

$$|\alpha| = a_1 + \ldots + a_m, \quad \alpha! = a_1!\ldots a_m!, \quad x^\alpha = x_1^{a_1}\ldots x_m^{a_m}$$

$$d_\xi^\alpha = (\partial/\partial\xi_1)^{a_1}\ldots(\partial/\partial\xi_m)^{a_m}, \quad d_x^\alpha = (\partial/\partial x_1)^{a_1}\ldots(\partial/\partial x_m)^{a_m},$$

$$D_x^\alpha = (-i)^{|\alpha|}d_x^\alpha$$

Let λ be a complex number which does not lie on the nonnegative real axis. We construct a pseudodifferential operator which approximates the resolvant $(D - \lambda I)^{-1}$ as follows: let $B(\lambda)$ be a pseudodifferential operator with symbol

$$\sigma(B)(x,\xi,\lambda) = b(x,\xi,\lambda) = b_0(x,\xi,\lambda) + \ldots + b_r(x,\xi,\lambda)$$

for some large integer r. Let the complex parameter λ have degree 2 and let b_i be homogeneous of degree $-2 - i$ in the variables (ξ,λ); i.e., $b_k(x,t\xi,t^{1/2}\lambda) = t^{-2-i}b_k(x,\xi,\lambda)$. Then, the formula for the symbol of the composition of two operators yields

$$\sigma(B(\lambda)(D - \lambda I)) \simeq \textstyle\sum_\alpha d_\xi^\alpha b\cdot D_x^\alpha\sigma(D - \lambda I)/\alpha!$$

If we decompose this sum into terms of degree $-k$ in the variables (ξ,λ), then

$$\sigma(B(\lambda)(D - \lambda I)) \simeq \sum_{k=0}^{\infty}\{b_k\cdot(|\xi|^2 - \lambda) + \sum_{j<k,k=j+|\alpha|+2-i} d_\xi^\alpha b_j\cdot D_x^\xi a^i/\alpha!\}$$

We define b so $\sigma(B(\lambda)(D - \lambda I)) \simeq I$ + terms of arbitrarily high order. This yields

$$b_0 = (|\xi|^2 - \lambda)^{-1}, \quad b_k = -b_0 \sum_{j<k,k=j+|\alpha|+2-i} d_\xi^\alpha b_j\cdot D_x^\alpha a^i/\alpha!$$

Since b_0 is scalar, it commutes with any other matrix. It follows by induction on k that we may express the dependence of b_k on the parameter λ in the form

$$b_k = \textstyle\sum_j b_{k,j}(x,\xi)b_0^{(j+k+2)/2}$$

summed over $k + j$ even. The $b_{k,j}$ are to be homogeneous of degree j in $\xi; b_{k,j} = 0$ if $j + k$ is odd.

We use this approximation of the resolvant to define an approximation to $\exp(-tD)$ using the functional calculus. Let γ be a path in the complex plane which is oriented counterclockwise about the positive real axis and let

$$e_k(x,\xi,t) = \frac{i}{2\pi}\int_\gamma b_k(x,\xi,\lambda)\exp(-t\lambda)d\lambda$$

$$= \frac{i}{2\pi}\int_\gamma \sum_j b_{k,j}(x,\xi)(|\xi|^2 - \lambda)^{-(j+k+2)/2}\exp(-t\lambda)d\lambda$$

$$= \sum_j t^{(j+k)/2}\exp(-t|\xi|^2)b_{k,j}(x,\xi)/((j+k+2)/2)!$$

Let $E(t)$ have symbol $e_0 + \ldots + e_r$ for large r and let $H_s(V)$ be the Sobelev space completion of $C^\infty(V)$ with respect to the s-norm. If A is any operator, let $|A|_{s,s'}$ be the operator norm (possibly infinite) of the map $A:H_s \to H_{s'}$. Seeley (1967) proved

$$|\exp(-tD)-E(t)|_{s,s'} \leqslant C(s,s',k)t^k \quad \text{as} \quad t \to 0 \quad \text{if} \quad r > r(s,s',k)$$

If A is any pseudodifferential operator, the kernel of A is given by

$$K(x,y,A) = \int\exp(i(x - y)\cdot\xi)\sigma(D)(x,\xi)d\xi$$

provided this integral converges absolutely. (We include a normalizing constant of $(4\pi)^{-m/2}$ in the $d\xi$ measure.) Let

$$K_k(t,x,y,D) = \int\exp(i(x - y)\cdot\xi)e_k(x,\xi,t)d\xi$$

$$= \sum_j t^{(k+j)/2}\int\exp(i(x - y)\cdot\xi)b_{k,j}(x,\xi)\exp(-t|\xi|^2)/((j + k + 2)/2)!d\xi$$

Since $b_{k,j}$ is homogeneous of degree j in ξ, we make a change of variables $\xi \to \xi/\sqrt{t}$ to transform this to

$$K_k(t,x,y) = t^{(k-m)/2}\sum_j\int\exp(i(x - y)\cdot\xi/\sqrt{t})b_{k,j}(x,\xi)\cdot$$

$$\cdot\exp(-|\xi|^2)/((j + k + 2)/2)!d\xi \tag{1.1}$$

This integral converges absolutely to define a smooth kernel of (t,x,y).

Since $\sum_k K_k$ is the kernel for $E(t)$, Seeley's estimate implies $\exp(-tD)$ has a smooth kernel $K(t,x,y,D)$ as well. This proves $\exp(-tD)$ is of trace class and

$$|K(t,x,y,D) - \sum_k K_k(t,x,y,D)|_s \leqslant c(j,s)t^j \quad \text{as} \quad t \to 0 \quad \text{for} \quad r > r(j,s)$$

If s is large, this shows we may estimate the sup norm of the difference of the kernels as well as estimate the sup norm of the derivatives of this difference to arbitrarily high order in t as $t \to 0$. Consequently, the asymptotic behavior of $K(t,x,y,D)$ is the same as the asymptotic behavior of $\sum_k K_k$. Therefore,

$$\operatorname{Tr}_{L^2}\exp(-tD) = \int_M \operatorname{Tr}_{V_x} K(t,x,x,D)|dvol(x)| \simeq \int_M \sum_k \operatorname{Tr}_{V_x} K_k(t,x,x,D)|dvol(x)|$$

$$= \sum_k t^{(k-m)/2}\int_M\{\sum_j \int \operatorname{Tr}_{V_x} b_{k,j}(x,\xi)/((j+k+2)/2)!\exp(-|\xi|^2)d\xi\}|dvol(x)|$$

Since $b_{k,j}$ is homogeneous of order j in ξ and vanishes for $k + j$ odd, $b_{k,j}$ is of odd order in ξ if k is odd. This implies the terms vanish for odd k. Let

$$a_{k,m}(x,I,D) = (4\pi)^{m/2}\sum_j \int \operatorname{Tr}_{V_x} b_{2k,j}(x,\xi)/((j + 2k + 2)/2)!\exp(-|\xi|^2)d\xi$$

then

$$\operatorname{Tr}_{L^2}\exp(-tD) \simeq (4\pi t)^{-m/2}\sum_k t^k \int_M a_{k,m}(x,I,D)|dvol(x)|$$

$a_{k,m}$ is a local invariant of the operator D. This proves 0.2 for $T = I$.

If $|x - y| > \varepsilon$, we can integrate equation (1.1) by parts to prove that $|K_k(t,x,y)| \leq C(\varepsilon,j)t^j$ as $t \to 0$. Since K_k vanishes to infinite order off the diagonal, $K(t,x,y)$ also vanishes to infinite order off the diagonal. The operator $A(T)\exp(-tD)$ is given by the kernel function $A(T)K(t,Tx,y,D)$. Let

$$b_k(x,\xi,T) = \sum_j A(T)b_{k,j}(Tx,\xi)/((j + k + 2)/2)!$$

then $b_k(x,-\xi,T) = (-1)^k b_k(x,\xi,T)$.

$$\operatorname{Tr}_{L^2}A(T)\exp(-tD) = \int_M \operatorname{Tr}_{V_x} A(T)K(t,Tx,x,D)|dvol(x)|$$
$$\simeq \sum_k t^{(k-m)/2}\iint \exp(i(Tx - x)\cdot\xi/\sqrt{t})\operatorname{Tr}(b_k(x,\xi,T))\exp(-|\xi|^2)d\xi|dvol(x)|$$

Since $K(t,Tx,xD)$ vanishes to infinite order in t if $|Tx - x| > \varepsilon$, we can restrict this integral to an arbitrarily small neighborhood of the fixed point set by multiplying b_k by a suitable cut-off function. Thus, if T has no fixed points, the integral vanishes to infinite order in t which proves Theorem 0.2 for that case.

If T has fixed points, the analysis is more complex. We first describe the local geometry of the fixed point set. Let $x = (x_1,\dots,x_n)$ be a system of local coordinates on N^n and let $e = (e_1,\dots,e_{m-n})$ be an orthonormal frame for the normal bundle. We express $p \in \nu$ in the form $p = \sum_j y_j e_j(x)$ to introduce coordinates (x,y) on ν. We use the geodesic flow to identify a neighborhood of the zero section of ν with a neighborhood of N^n in M^m. This defines coordinates (x,y) on M^m by identifying (x,y) with $\exp_x(ye(x))$. $N^n = \{z : y = 0\}$.

Let $z = (x,y)$ and decompose $T(z) = (T_1(z), T_2(z))$, then

$$dT(x,0) = \begin{pmatrix} I & 0 \\ 0 & dT_\nu \end{pmatrix}$$

by the definition of ν. Consequently, $T_1(z) - x$ vanishes to second order in y along N^n. We integrate $\mathrm{Tr}\, A(T)K(t,Tz,z,D)$ along the fibres of the normal bundle to reduce the integral to an integral on the fixed point set. Decompose $\xi = (\xi_1,\xi_2)$, then

$$\int \mathrm{Tr}\, A(T)K(t,Tz,z,D)\,|\mathrm{dvol}(z)|$$
$$\simeq \sum_k t^{(k-m)/2} \iiiint \exp(i(T_1(x,y)-x)\cdot\xi_1/\sqrt{t})\exp(i(T_2(x,y)-y)\cdot\xi_2/\sqrt{t})\exp(-|\xi|^2)$$
$$g(x,y)\,\mathrm{Tr}\, b_k(x,y,\xi,T)\,d\xi_1 d\xi_2\, dy\, dx$$

where $\mathrm{dvol}(z) = g(x,y)dydx$. Since T is nondegenerate, we can make the change of variables in the fibre coordinates: $(x,w) = (x, T_2(x,y) - y)$. This transforms the integral into the form

$$\simeq \sum_k t^{(k-m)/2} \iiiint \exp(i(T_1(x,w)-x)\cdot\xi_1/\sqrt{t})\exp(iw\cdot\xi_2/\sqrt{t})\exp(-|\xi|^2)g(x,w)$$
$$\mathrm{Tr}\, b_k(x,w,\xi,T)\,|\det(I - dT_\nu)^{-1}|\, d\xi\, dw\, dx$$

We change variables again and replace w by $\sqrt{t}w$ to transform the integral to

$$\simeq \sum_k t^{(k-m)/2} t^{(m-n)/2} \iiiint \exp(i(T_1(x,\sqrt{t}w)-x)\cdot\xi_1/\sqrt{t})\exp(iw\cdot\xi_2)\exp(-|\xi|^2)$$
$$g(x,\sqrt{t}w)\,\mathrm{Tr}\, b_k(x,\sqrt{t}w,\xi,T)\,|\det(I - dT_\nu)^{-1}|(x,\sqrt{t}w)\,d\xi\, dw\, dx \qquad (1.2)$$

As $t \to 0$, dw ranges over the entire fibre of $\nu_x \simeq R^{m-n}$, $d\xi_1 d\xi_2$ ranges over the fibre of T^*M_x, and dx ranges over the fixed point set of T.

Define

$$c_k(x,w,\xi,t) = \exp(i(T_1(x,\sqrt{t}w)-x)\cdot\xi_1/\sqrt{t})\{g|\det(I - dT_\nu)^{-1}|\mathrm{Tr}\ b_k\}(x,\sqrt{t}w,\xi)$$

Since $T_1(x,y) = y$ vanishes to second order in y, $(T_1(x,\sqrt{t}w) - x)/\sqrt{t}$ is smooth in $\sqrt{t}$ and vanishes to first order in $\sqrt{t}$. If we expand c_k in a Taylor series in $\sqrt{t}$, it has the form

$$c_j \simeq \sum_j t^{j/2} c_{k,j}(x,w,\xi)$$

$c_{k,j}$ is *polynomial* in the (w,ξ) variables with coefficients which depend on the jets of T and b_k at $w = 0$. We substitute this expansion into (1.2) to show

$$\begin{aligned}&\mathrm{Tr}_{L^2} A(T)\exp(-tD)\\ &\simeq \sum_{k,j} t^{(j+k-n)/2}\int_N(\iint \exp(iw\cdot\xi_2)c_{k,j}(x,w,\xi)\exp(-|\xi|^2)d\xi dw)dx\end{aligned}$$

Since $c_{k,j}$ is polynomial in (w,ξ), the $d\xi$ integral yields a polynomial in w multiplied by $\exp(-|w|^2)$. Consequently, this iterated integral is well defined. It is easily checked by induction that $c_{k,j}$ is of odd order in (w,ξ) if $k + J$ is odd and consequently the integrals vanish in this case. Let

$$a_{p,n}(x,T,D) = (4\pi)^{n/2}\sum_{j+k=2p}\iint \exp(iw\cdot\xi_2)c_{k,j}(x,w,\xi)\exp(-|\xi|^2)d\xi dw g_N^{-1}$$

where $g_N dx = |dvol_n|$. Then

$$\mathrm{Tr}_{L^2} A(T)\exp(-tD) \simeq (4\pi t)^{-n/2}\sum_p t^p \int_{N^n} a_{p,n}(x,T,D)|dvol_n|$$

If there is more than one component of the fixed point set, we must sum over the components since each component will make a similar contribution to the asymptotic sum. It is clear from the construction we have given that $a_{p,n}$ only depends on the jets of the metric, map T, action $A(T)$, and symbol of the operator D on the fixed point set and hence are local invariants. This completes the proof of Theorem 0.2.

In fact, $a_{p,n}$ is a homogeneous polynomial of order $2p$ in the sense we define below. Although we can derive this fact directly from the construction given, it is instructive to give a proof using the functorial properties of the $a_{p,n}$.

Introduce the following notation for the jets. If f is a scalar or matrix valued function near N^n and if α is a multiindex, let $f_{/\alpha} = d_z^\alpha f$ represent the jets of f on N^n. These variables are not tensorial, they only make sense relative to a coordinate system. Let $z = (x,y)$ be a coordinate system near N^n in M^m. Normalize z by supposing $N^n = \{z:y(z) = 0\}$ and ν is spanned by the $\{\partial/\partial y_j\}$. Let $\{g_{ij/\alpha}, b^k{}_{/\beta}, c_{/\gamma}\}$ be formal variables for the jets of the symbol of D. Let $T = (T^1,\dots,T^m)$ be the components of the map T and let $T^i_{/\delta}$ be the jets of T on N^n. Let $T^i_j = \partial T^i/\partial z_j = dT^i_j(x,0)$ and let Ψ be the collection of variables: $\{g_{ij/\alpha}, b^k_{/\beta}, c_{/\gamma}, T^i_{/\delta}\}$. If $\psi \in \Psi$, let $\psi(Z,s,T,D,)(x)$ be the natural evaluation of ψ as the appropriate jet relative to the coordinate system Z and the local frame field s.

We introduce a grading on the Ψ variables as follows.

$$\text{ord}(g_{ij/\alpha}) = |\alpha|, \quad \text{ord}(b^k_{/\beta}) = 1 + |\beta|, \quad \text{ord}(c_{/\gamma}) = 2 + |\gamma|,$$

$$\text{ord}(T^i_{/\delta}) = -1 + |\gamma|$$

THEOREM 1.1. $a_{p,n}(T,D)(x) = a_{p,n}(\Psi)(Z,s,T,D)(x)$ is a smooth function of the Ψ variables which is a homogeneous polynomial of order 2p in those variables of order > 0 with coefficients which are smooth functions of the $\{g_{ij}, T^i_j\}$ variables.

We have supressed the dependence on the action A(T) to avoid having to introduce additional formal variables for the jets of the action since in applications, A(T) will be a functorial expression in the T^i_j variables.

Before proving Theorem 1.1, we give an example. Let $T = I$ and let D_0 be the Laplacian acting on $C^\infty(\Lambda T^*M)$. $c(D_0) = 0$ and $b^k(D_0)$ is linear in the first derivatives of the metric. Therefore, $b^k_{/\beta}$ is of order $1 + |\beta|$ in the derivatives of the metric, so $a_{p,n}$ is homogeneous of order 2p in the derivatives of the metric. Let

$$R_{i_1 i_2 i_3 i_4; j_1 \dots j_s}$$

be the components of the covariant derivatives of the curvature tensor. This is of order $2 + s$ in the derivatives of the metric. Relative to normal coordinates, we can compute the jets of the metric tensor in terms of the covariant derivatives of the curvature tensor. Theorem 1.1 implies $a_{1,m}$ is linear in the curvature tensor while $a_{2,m}$ is quadratic in the curvature tensor and linear in the second covariant derivatives. McKean and Singer (1967) and Berger (1963) proved

$$a_{0,m}(x,I,D_0) = I, \quad a_{1,m}(x,(,D_0) = -\frac{1}{6}R_{ijij}$$

$$a_{2,m}(x,I,D_0) = \frac{1}{360}(-12R_{ijij;kk}+5R_{ijij}R_{k\ell k\ell}+2R_{ijk\ell}R_{ijk\ell}-2R_{ijik}R_{\ell j\ell k})$$

There is a similar formula for $a_{3,m}(x,I,D_0)$ given by Sakai (1971). We compute in an orthonormal frame for $T(M)$ and sum over repeated indices. A general formula for the first four terms in the asymptotic expansion for general D and $T - I$ is given in Gilkey (1975). Donnelly (1976) has computed $a_{p,n}(x,T,D_0)$ for $p = 0,1,2$ if T is an isometry.

The polynomial dependence and homogeneity will be of great importance in the proof of Theorem 0.4. We prove Theorem 1.1 by dimensional analysis.

Lemma 1.2. Let $\varepsilon > 0$, then $a_{p,n}(\Psi)(Z,s,T,\varepsilon^2 D)(x) = \varepsilon^{2p}a_{p,n}(\Psi)(Z,s,T,D)(x)$

Proof. $\varepsilon^{-2}G$ is the new metric induced by the leading symbol of $\varepsilon^2 D$. Since $\mathrm{Tr}\, A(T)\exp(-t(\varepsilon^2 D)) = \mathrm{Tr}\, A(T)\exp(-\varepsilon^2 tD)$,

$$\textstyle\sum_i (4\pi t)^{-n_i/2} \sum_p t^p \int_{N_i} a_{p,n_i}(x,T,\ ^2D)\,|\mathrm{dvol}(\varepsilon^{-2}G)|_{n_i}$$

$$\simeq \textstyle\sum_i (4\varepsilon^2 t)^{-n_1/2} \sum_p \varepsilon^{2p} t^p \int_{N_i} a_{p,n_i}(x,T,D)\,|\mathrm{dvol}(G)|_{n_i}$$

Using the naturality of the construction of the $a_{p,n}$, we conclude the local invariants also satisfy this relation. Therefore,

$$a_{p,n}(x,T,\varepsilon^2 D)\,|\mathrm{dvol}(\varepsilon^{-2}G)|_n = \varepsilon^{2p-n} a_{p,n}(x,T,D)\,|\mathrm{dvol}(G)|_n$$

Since $|\mathrm{dvol}(\varepsilon^{-2}G)|_n = \varepsilon^{-n}|\mathrm{dvol}(G)|_n$, this proves the lemma.

Lemma 1.3. Let $W = \varepsilon^{-1}Z$ be new coordinates, then $\psi(W)(\varepsilon^2 D) = \varepsilon^{\mathrm{ord}(\psi)}\psi(Z)(D)$ for $\psi \in \Psi$.

Proof. Since $d_W^\alpha = \varepsilon^{|\alpha|} g_Z^\alpha$,

$$g_{ij/\alpha}(W)(\varepsilon^{-2}G) = \varepsilon^{|\alpha|} g_{ij/\alpha}(Z)(G) = \varepsilon^{\mathrm{ord}(g_{ij/\alpha})} g_{ij/\alpha}(Z)(G)$$

$$T^i_{/\delta}(W) = \varepsilon^{-1} d^\delta_W T^i(Z) = \varepsilon^{|\delta|-1} T^i_{/\delta}(Z) = \varepsilon^{\mathrm{ord}(T^i_{/\delta})} T^i_{/\delta}(Z)$$

The T^i_j variables defined the action of dT on T^*M_x and are independent of changes of scale. Similarly,

$$\varepsilon^2 D = -\varepsilon^2 g^{ij}(Z,D)\partial^2/\partial z_i \partial z_j + \varepsilon^2 b^k(Z,s,D)\partial/\partial z_k + \varepsilon^2 c(Z,s,D)$$

$$= -g^{ij}(Z,D)\partial^2/\partial w_i \partial w_j + \varepsilon b^k(Z,s,D)\partial/\partial w_k + \varepsilon^2 c(Z,s,D)$$

$$b^k_{/\beta}(W,s,\varepsilon^2 D) = \varepsilon^{|\beta|+1}\, b^k_{/\beta}(Z,s,D) = \varepsilon^{\operatorname{ord}(b^k_{/\beta})}\, b^k_{/\beta}(Z,s,D)$$

$$c_{/\gamma}(W,s,\ ^2D) = \varepsilon^{|\gamma|+2} c_{/\gamma}(Z,s,D) = \varepsilon^{\operatorname{ord}(c_{/\gamma})}\, c_{/\gamma}(Z,s,D)$$

This completes the proof of Lemma 1.3.

We use these two lemmas to prove Theorem 1.1: since $a_{p,n}$ is independent of the coordinate system, we may use Z or W to compute $a_{p,n}$. By Lemmas 1.2 and 1.3,

$$\varepsilon^{2p} a_{p,n}(\Psi)(Z,s,T,D) = a_{p,n}(\Psi)(Z,s,T,\varepsilon^2 D) = a_{p,n}(\Psi)(W,s,T,\varepsilon^2 D)$$

$$= a_{p,n}\ \varepsilon^{\operatorname{ord}(\Psi)}\Psi\ (Z,s,T,D)$$

Consequently, $a_{p,n}$ is homogeneous of order 2p in the Ψ variables. Since $a_{p,n}$ is smooth, $a_{p,n}$ must be polynomial in the components of those variables of nonzero order. This proves Theorem 1.1. It is worth noting that if p was a half-integer, then $a_{p,n}$ would be of odd order. In this case, $a_{p,n}(\Psi)(Z,s,T,D) = -a_{p,n}(\Psi)(-Z,s,T,D) = -a_{p,n}(\Psi)(Z,s,T,D)$ since $a_{p,n}$ is coordinate free. This gives an independent argument that the terms of odd order do not appear in the expansion.

We could also have proved Theorem 1.1 by expressing the jets Ψ in terms of tensorial expressions. The order of a variable is the number of covariant indices minus the numer of contravariant indices relative to a GL frame. For example, the curvature tensor in a GL frame is given by R^{ℓ}_{ijk}. This variable is of order $2 = 3 - 1$. The proof of Theorem 1.1 is just dimensional analysis in this notation.

There are other functorial properties of the invariants $a_{p,n}$ which we will need in the proof of Theorem 0.4. Although we have given an explicit construction of $a_{p,n}$, the combinatorial difficulties inherent in its use prevents actual computation except for small values of p. However, if $n = p = 0$, the computation only depends on the leading symbol. This formula for $n = p = 0$ was used by Kotake (1969) to derive the Atiyah-Bott Lefschetz formulas.

THEOREM 1.4. $a_{0,0}(T,D) = \operatorname{Tr} A(T) \det(I - dT)^{-1}$.

Proof. Since x is an isolated fixed point, $dT_\nu = dT$. We use the formulas derived above.

$$b_0(x,\xi,\lambda) = (|\xi|^2 - \lambda)^{-1}, \quad e_0(x,\xi,t) = \exp(-t|\xi|^2)$$

$$c_{0,0} = \operatorname{Tr} A(T)|\det(i - dT)^{-1}|$$

$$a_{0,0} = \operatorname{Tr} A(T)|\det(I - dT)^{-1}|\int\exp(iz\cdot\xi)\exp(-|\xi|^2)d\xi dz$$

Since $d\xi$ has a normalizing constant of $(4\xi)^{m/2}$, the integral $= 1$.

COROLLARY 1.5. If (V,d) is the DeRham complex, $a_{0,0} =$ sign $\det(I - dT)$. Thus, if T has only isolated nondegenerate fixed points, $L(T)_{DER} = \sum_i \text{sign} \det(I - dT)$.

This is the classical Lefschetz fixed point formula. There are other similar formulas for the signature and Dolbeault complexes we will discuss later. We prove Corollary 1.5 by noting $\sum_p (-1)^p \operatorname{Tr}\Lambda^p(dT) = \det(I - dT)$ so $a_{0,0} = \det(I - dT)|\det(I - dT)^{-1}|$.

We end the first section by discussing other functorial properties of the invariants $a_{p,n}$. Let (V^ℓ,d^ℓ) be a finite collection of elliptic complexes over M. Let (V,d) be the direct sum of these elliptic complexes.

$$V = \oplus_\ell V^\ell, \quad d = \oplus_\ell d^\ell : C^\infty(V) \to C^\infty(V)$$

If D_i^ℓ is the associated Laplacian on V_i^ℓ, then $D_i = \oplus_\ell D_i^\ell$. Thus, $K(t,x,y,D_i) = \oplus_\ell D(t,x,y,D_i^\ell)$. We take the direct sum action $A(T) = \oplus_\ell A(T)^\ell$. Since trace is additive on direct sums,

THEOREM 1.6. $a_{p,n}(T,V,d) = \sum_\ell a_{p,n}(T,V^\ell,d^\ell)$ is additive over direct sums.

Another property we shall need is the analogue of the Kunneth formula and expresses the multiplicative property with respect to tensor product. Let $(M^\ell,T^\ell,V^\ell,d^\ell)$ for $\ell = 1,2$ be two elliptic complexes. Let

$M = M^1 \times M^2$ with the product Riemannian metric and let $T = T^1 \times T^2 : M \to M$. T is nondegenerate and the fixed point set of T is the product of the fixed point sets of T^1 and T^2. Let $\pi_\ell : M \to M^\ell$ be projection. Define

$$V_i = \oplus_{i=j+k} \pi_1^*(V_j^1) \otimes \pi_2^*(V_k^2)$$

$$d_i = \oplus_{i=j+k} \pi_1^*(d_j^1) \otimes 1 + (-1)^i \otimes \pi_2^*(d_k^2) : C^\infty(V_i) \to C^\infty(V_{i+1})$$

$$D_i = \oplus_{i=j+k} \pi_1^*(D_j^1) \otimes 1 + 1 \otimes \pi_2^*(D_k^2)$$

$$K(t,x,y,D_i) = \oplus_{i=j+k} K(t,x_1,y_1,D_j^1) \otimes K(t,x_2,y_2,D_k^2)$$

This implies that

$$\oplus(-1)^i \exp(-tD_i) = \oplus_{i=j+k}(-1)^{j+k} \exp(-tD_j^1) \otimes \exp(-tD_k^2)$$

Since trace is additive on direct sums and multiplicative on tensor products, this implies $L(T,V,d) = L(T^1,V^1,d^1)L(T^2,V^2,d^2)$. If we multiply the two asymptotic series of Theorem 0.2 together and use the naturality of the invariants $a_{p,n}$, this proves

THEOREM 1.7. $a_{p,n}(T,V,d) = \sum_{j+k=p,\, n_1+n_2=n} a_{j,n_1}(T^1,V^1,d^1)\, a_{k,n_2}(T^2,V^2,d^2)$ is multiplicative on tensor products.

The final property we shall need is a technical fact.

Lemma 1.8. $a_{p,n}(T,V,d)$ is linear in the components of the action $A(T)$.

This is immediate from the construction that we have given.

§2. In this section, we shall prove the Lefschetz fixed point formulas for the DeRham complex. Let T be nondegenerate and let N^n be a component of the fixed point set of T. Let $z = (x,y)$ be normalized coordinates near N^n in M^m. We adopt the convention that indices i,j,k run from 1 through m and index the coordinate frame $\{\partial/\partial z_i\}$ for $T(M)$. Indices a,b,c run from 1 through n and index the subframe induced on $T(N)$. Indices u,v,w run from $n+1$ through m and index the subframe for ν. We decompose a nultiindex α into tangential $\alpha(t)$ and normal $\alpha(\nu)$ parts. This expresses $f_{/\alpha} = f_{/\alpha(t)\alpha(\nu)}$. We shall also use the symbol $f_{/i_1 \ldots i_s}$

for the jets of f.

Let G be a Riemannian metric on M and let $a_{k,n}^{DER}(T,G)(x)$ be the invariant of the heat equation defined in the first section. By Corollary 0.3,

$$L(T)_{DER} \simeq \sum_i (4\pi t)^{-n_i/2} \sum_k t^k \int_{N_i} a_{k,n_i}^{DER}(T,G)(x)$$

Let $\overline{G}$ be the germ of another metric on M^n near N^n such that $T(M)|_N = T(N)\oplus\nu$ is still an orthogonal decomposition. Using a partition of unity, we can alter G to agree with $\overline{G}$ near N^n without changing the original metric near the other components. $L(T)_{DER}$ is independent of the choice of G, so this implies the individual integrals

$$\int_N a_{k,n}^{DER}(T,G)(x)\,|dvol(x)|_n$$

are invariant separately. This integral is independent of the choice of G for all k and for each component of the fixed point set. It vanishes if $2k \neq n$.

The leading symbol of the Laplacian $D_p: C^\infty(\Lambda^p T^*M) \to C^\infty(\Lambda^p T^*M)$ is given by the metric tensor. $b^k(D_p)$ is linear in the first derivatives of the metric; $c(D_p)$ is linear in the second derivatives of the metric and quadratic in the first derivatives of the metric. Therefore, by Theorem 1.1, $a_{k,n}^{DER}(T,G)$ is a homogeneous polynomial of degree 2k in the derivatives of the metric and of T with coefficients which are smooth functions of the $\{g_{ij}, T_j^i\}$ variables.

The proof of Theorem 0.4 is much simpler for metrics which are chosen so the components of the fixed point set are totally geodesic. Since we are only assuming that $T(M)_N = T(N)\oplus\nu$ is an orthogonal splitting, we will need the following lemma to control the effect of the second fundamental form.

Lemma 2.1. Let $P(T,G)$ be an invariant polynomial in the derivatives of the metric and of the map T. Suppose $\int_N P(T,G)(x)|dvol(x)_n|$ is independent of G and suppose $g_{ij/\alpha(\nu)}$ divides some nomomial of P for some nonzero normal multiindex $\alpha(\nu)$. Then there exists a nonzero tangential multiindex $\alpha(t)$ such that $g_{ij/\alpha(t)\alpha(\nu)}$ divides some monomial of P.

Proof. Suppose the lemma is false. Let $z = (x,y)$ be a local coordinate system and let $f(x)$ be a smooth function with compact support within the domain of definition of x on N^n. Let ε be a parameter and let $d\bar{s}^2 = ds^2 + \varepsilon f(x) y^{\alpha(\nu)} dx^i dx^j$ be the germ of a metric $\bar{G}$. If A is any monomial of P, express $A = g^{\ell}_{ij/\alpha(\nu)} A_0$. By hypothesis, $g_{ij/\alpha(t)\alpha(\nu)}$ does not divide A_0 for any nonzero tangential multiindex $\alpha(t)$. Thus, $A(Z,T,\bar{G}) = A(Z,T,G) + \varepsilon \ell f(x) g^{\ell-1}_{ij/\alpha(\nu)} A_0(Z,T,G) + O(\varepsilon^2)$. In particular, $A(Z,T,\bar{G})$ does not depend on the jets of the function f. Therefore,

$$P(Z,T,\bar{G}) = P(Z,T,G) + \varepsilon f(x) P_0(Z,T,G) + O(\varepsilon^2)$$

Since $g_{ij/\alpha(\nu)}$ divides some monomial of P, P_0 is not the zero polynomial. Thus, for some choice of (Z,T,G), $P_0(Z,T,G)(x_0) \neq 0$. Since $\int P(T,G)$ is independent of G, $\int f(x) P_0(Z,T,G)(x)|dvol_n(x)| = 0$ for any function $f(x)$. This implies $P_0(Z,T,G) = 0$. This contradiction establishes the lemma.

Let (M_i,T_i) for $i = 1,2$ be two Riemannian manifolds. Let $M = M_1 \times M_2$ and $T = T_1 \times T_2$. The DeRham complex for (M,T) is the tensor product of the DeRham complexes for (M_1,T_1) and (M_2,T_2). Consequently, by Theorem 1.7, if G is the product metric

$$a^{DER}_{p,n}(T,G)(x_1,x_2) = \sum_{i+j=p} a^{DER}_{i,n_1}(T_1,G_1)(x_1) a^{DER}_{j,n_2}(T_2,G_2)(x_2)$$

If $(M_1,T_1) = (S^1,I)$ is the unit circle with the identity map, then all the derivatives of the metric vanish and $a^{DER}_{j,1} = 0$ for $j > 0$. If $j = 0$, then the integral of $a^{DER}_{0,1}$ is $\chi(S^1) = 0$. Since $a^{DER}_{0,1}$ is homogeneous of order 0, it is constant on (S^1,I). Since it integrates to zero, it vanishes pointwise. This shows $a^{DER}_{p,1}(S^1,I) = 0$. This proves

Lemma 2.2. If $(M,T) = (S^1,I) \times (M_2,T_2)$, then $a^{DER}_{p,n}(T,G)(x_1,x_2) \equiv 0$.

The coordiante system z was chosen so $N^n = \{z : y(z) = 0\}$ and so that $\{\partial/\partial y_i\}(x,0)$ is an orthonormal frame for $\nu(x)$. Let G_N be the restriction of the metric to N^n and let $x_0 \in N^n$. Further normalize z so $g_{ij}(x_0) = \delta_{ij}$ is the Kronecker index and so the first derivatives of the metric G_N vanish at x_0. The $\{g_{ij/\alpha}, T^i_{/\beta}\}$ variables are not free since we have made various normalizations.

Lemma 2.3. Let $\mathcal{P}_n$ be the polynomial algebra in the $\{g_{ij/\alpha}, T^i_{/\beta}\}$ variables of nonzero order with coefficients which are smooth functions of $dT_\nu = T^u_v$. Introduce the relations $g_{ab/c} \equiv g_{au/\alpha(t)} \equiv g_{uv/\alpha(t)} \equiv T^a_{b/\alpha(t)} \equiv T^a_{u/\alpha(t)} \equiv T^u_{a/\alpha(t)} \equiv 0$ for any nonzero tangential multiindex $\alpha(t)$. Then

$a_{p,n}^{DER}(T,G)$ is invariant and belongs to p_n.

Proof. Since ν is orthogonal to $T(N)$, $g_{au}(x,0) = 0$ and $g_{uv}(x,0) = \delta_{uv}$. Therefore, $g_{au/\alpha(t)} = g_{uv/\alpha(t)} = 0$. Since the coordinate system x on N is geodesic to first order relative to G_N, $g_{ab/c}(x_0) = 0$. $dT(x,0) =$

$$dT(x,0) = \begin{pmatrix} I & 0 \\ 0 & dT_\nu \end{pmatrix}$$

so $T_b^a(x,0) = \delta_{ab}$ and $T_u^a(x,0) = T_a^u(x,0) = 0$. This implies $T_{b/\alpha(t)}^a = T_{u/\alpha(t)}^a = T_{a/}^u{}_{(t)} = 0$. Since $g_{ij}(x_0) = \delta_{ij}$, the coefficients only depend on dT_ν. It is easily verified these are the only algebraic conditions imposed on the algebra by our normalizations. This proves $a_{p,n} \varepsilon p_n$. Since $a_{p,n}$ is independent of z, it is invariant. Exponential geodesic coordinates are examples of such z; we do not restrict to these coordinate systems to avoid introducing additional algebraic relations into p_n.

We use Lemma 2.3 to interpret Lemma 2.2 algebraically. Let $\deg_k(g_{ij/\alpha}) = \delta_{i,k} + \delta_{j,k} + \alpha(k)$ for $|\alpha| > 0$ and let $\deg_k(T_{/\beta}^i) = \delta_{i,k} + \beta(k)$ for $|\beta| > 1$. This count the number of times the index k appears in the variables of nonzero order. If A is a nomomial of p_n, let $\deg_k(A)$ be defined by summing with multiplicity over the variables dividing A. Let $R(A) = 0$ if $\deg_1(A) \neq 0$. If $\deg_1(A) = 0$, the index "1" does not appear in A. By renumbering the indices from 2 through m to run from 1 through $m - 1$, we define $R(A) \varepsilon p_{n-1}$ on a submanifold of dimension $n - 1$. Extend R to an algebra morphism mapping $p_n \rightarrow p_{n-1}$.

The only additional realtion which is imposed on the algebra p by restricting to manifolds of the form $(M,T) = (S^1,I) \times (M_2,T_2)$ is that $\psi = 0$ if $\deg_1(\psi) \neq 0$. R has been defined so that for $P \in p_n$,

$$P((S^1 \times M_2, I \times T_2)) = (RP)(M_2,T_2)$$

We may therefore reformulate Lemma 2.2 to be $R(a_{p,n}^{DER}) = 0$.

We can also define the action of R in classical terms: let $R_{ijk\ell;\ldots}$ be the covariant derivatives of the curvature tensor, let $T_{j;\ldots}^i$ be the covariant derivatives of dT, and let $\omega_{abu;\ldots}$ be the covariant derivatives of the second fundamental form. It is not hard to show that if $P \in p_n$ is invariant, then P can be expressed in terms of these variables. By H. Weyl's theorem (1946) on the invariants of the orthogonal group, P

is a combination of expressions involving contractions of indices. Indices from $T(N)$ run from 1 through n and indices from ν run from $n+1$ through m. $R(P)$ is defined by letting the indices from $T(N)$ run from 2 through n instead.

Let R^N_{abcd} be the curvature tensor of G_N on N^n. Let $E_n = 0$ for n odd and let

$$E_n = \frac{(-1)^p}{(8\pi)^p p!}\varepsilon(a_1\dots a_n)\varepsilon(b_1\dots b_n)R^N_{a_1a_2b_1b_2}\dots R^N_{a_{n-i}a_nb_{n-1}b_n}$$

if $n = 2p$ is even. It is clear

$$E_{n_1+n_2}(N_1 \times N_2) = E_{n_1}(N_1)E_{n_2}(N_2)$$

is multiplicative on product metrics. Therefore, E_n vanishes on product manifolds of the form $S^1 \times N_2$, so $R(E_n) = 0$. The following theorem is a generalization of a result we proved for the Euler form (see Gilkey, Ph.D. dissertation, 1972). It expresses the fact that the Euler form is an unstable characteristic class; it vanishes if the structure group can be reduced to $SO(n-1)$ from $SO(n)$.

THEOREM 2.4. Let $P \in \mathcal{P}_n$ be invariant. Suppose that $R(P) = 0$, that P is homogeneous of order k, and that $\int_N P(T,G)(x)|dvol(x)|_n$ is independent of the choice of G near N^n. Then

a) $P \equiv 0$ if $k < n$ or if k is odd.

b) If $k = n$, $P(T,G) = f(dT_\nu)E_n$. $f(dT_\nu)$ is invariant under the action of $O(m-n)$.

We use Theorem 2.4 to complete the proof of Theorem 0.4 (1a and 1b). Since $a^{DER}_{p,n}(T,G)$ satisfies the conditions of Theorem 2.4 and is homogeneous of order $2p$, $a^{DER}_{p,n} = 0$ for $2p < n$. If $2p = n$, $a^{DER}_{p,n} = E_n f(dT_\nu)$. Let $(M,T) = (M_1,I) \times (M_2,T)$ where T_2 has only isolated fixed points. By formula (2.1)

$$a^{DER}_{p,n}(M,T) = \sum_{i+j=p} a^{DER}_{i,n}(M_1,I)a^{DER}_{j,o}(M_2,T_2) = a^{DER}_{p,n}(M_1,I)a^{DER}_{0,0}(M_2,T_2)$$

This implies $f(dT_\nu) = a^{DER}_{0,0}(M_2,T_2) = \text{sign}\det(I - dT_\nu)$ by Corollary 1.5. This completes the proof.

If $T = I$, $\chi(M) = L(I)_{DER} = \int_M E_m(G)|dvol(x)|_m$. Therefore,

$$L(T)_{DER} = \sum_i \text{sign}\det(I - dT_{\nu_i})\int_{N_i} E_{n_i}(G_N)|dvol(x)|_{n_i} =$$

$= \sum_i \text{sign} \det(I - dT_{\nu_i})\chi(N_i)$

This gives a heat equation proof of the classical Lefschetz formulas.

The remainder of this section is devoted to the proof of Theorem 2.4. Let $P \in \mathcal{P}_n$ satisfy the hypothesis of Theorem 0.4. Since $g_{ab/c} = g_{au/c} = g_{uv/c} = 0$, the only nonzero first derivatives are the $g_{uv/w}$ and $g_{ab/u}$ variables. The $g_{ab/u}$ variables are tensorial in these coordinate systems and represent the second fundamental form. Let $P \not\equiv 0$. There is some monomial A of P of the form

$$A = f(dT_\nu)A_0A_1A_2A_3, \qquad f(dT_\nu) \not\equiv 0$$

$$A_0 = g_{u_1v_1/w_1}\cdots g_{u_gv_g/w_g}$$

$$A_1 = g_{a_1b_1/u_1}\cdots g_{a_rb_r/u_r}$$

$$A_2 = g_{i_1j_1/\alpha_1}\cdots g_{i_sj_s/\alpha_s}, \qquad \text{for } |\alpha_\ell| \geq 2$$

$$A_3 = T^{k_1}_{/\beta_1}\cdots T^{k_t}_{/\beta_t} \qquad \text{for } |\beta_\ell| \geq 2$$

Each of the multiindices β_ℓ contains at least one normal derivative. If $k_\ell \leq n$, then β_ℓ must contain at least two normal derivatives.

We prove Theorem 2.4(a) by counting indices. Let $k = \text{ord}(P) \leq n$ and let $R(P) = 0$. This implies $\deg_1(A) \neq 0$ for every such monomial A of P. Since P is invariant, the form of P is not changed by any coordinate permutation of N^n. Thus, $\deg_a(A) \neq 0$ for $1 \leq a \leq n$ and for every such monomial A of P. The form of P is also unchanged by replacing one of the coordinates by minus that coordinate. If

$$X = (-x_1, x_2, \ldots, x_n)$$

$$Z = (X, y)$$

$A(Z,G,T) = (-1)^{\deg_1(A)} A(Z,G,T)$. Since P is invariant, $P(Z,G,T)$ is independent of A. This implies $\deg_1(A)$ is even and $\deg_a(A)$ is even. Therefore, $\deg_a(A) \geq 2$.

We count tangential indices: $2n \leq \sum_a \deg_a(A)$. It is clear $\sum_a \deg_a(A) = 0$ and $\sum_a \deg_a(A_1) = 2r$. Since $|\alpha_\ell| \leq 2$, $\sum_a \deg_a(A_2) \leq 2s + \sum_\ell |\alpha_\ell| \leq 2s + \text{ord}(A_2) \leq 2\text{ord}(A_2)$. If $k_\ell \leq n$, then β_ℓ involves at least two normal derivatives. Conversely, if $k > n$, then β_ℓ involves at least one normal derivative. Thus, in any event, $\sum_a \deg_a(A_3) \leq \sum_\ell |\beta_\ell| - 1 \leq \mathrm{ord}(A_3)$. This chain of inequalities implies $2n \leq \sum_a \deg_a(A) \leq 0 + 2r + 2\mathrm{ord}(A_2) + \mathrm{ord}(A_3) = 2k - \mathrm{ord}(A_3) \leq 2k$. This is impossible if $k < n$ so $P \equiv 0$ for $k < n$. Since every index appears with even degree, k must be even so $P \equiv 0$ as well if k is odd. This proves Theorem 2.4(a).

If $k = n$, these inequalities must all be equalities. This shows $\mathrm{ord}(Z_0) = \mathrm{ord}(A_3) = 0$, so these variables do not appear in any monomial of P. Since $2s + \mathrm{ord}(A_2) = 2\mathrm{ord}(A_2) = \sum_a \deg_a(A_2)$, $|\alpha_\ell| = 2$ and

$$g_{i_\ell j_\ell/\alpha_\ell}$$

involves only tangential indices. Thus, in particular, $g_{ab/u\alpha(t)}$ divides no monomial of P for any nonzero tangential index $\alpha(t)$. By Lemma 2.1, this implies $g_{ab/u}$ divides no monomial of A and thus A_1 does not appear as well. Consequently, every monomial A of P has the form

$$A = f_A(dT_\nu) g_{a_1 b_1/c_1 d_1} \cdots g_{a_p b_p/c_p d_p} \quad \text{for} \quad 2p = n$$

The domain of $f_A(dT_\nu)$ is the set of all $(m-n) \times (m-n)$ matrices U such that $\det(I - U) \neq 0$. f_A is invariant under the action of $O(m-n)$ on this set. Let S be the finite dimensional vector space of all such smooth functions which is generated by $\{f_A\}$ as A ranges over the monomials of P. Let $\{f^i\}$ be a basis of S. Let $f_A(dT_\nu) = \sum_i c_i(A) f^i(dT_\nu)$ and let $P_i = \sum_i c_i(A) \cdot A$. Then, $P = \sum f^i(dT_\nu) P_i(G)$. P_i is independent of dT_ν and satisfies the same hypothesis as P. We shall show this implies $P_i = c_i E_n$. Thus, $P = (\sum_i c_i f^i(dT_\nu)) E_n$. Therefore, it suffices to prove Theorem 2.4(b) under the additional hypothesis that P is independent of dT_ν and dependes only on the $\{g_{ab/cd}\}$ variables.

We use invariance of P relative to the action of $SO(n)$ to obtain further information about P. Let $P \in \mathcal{P}_n$ and let $c(A,P)$ be the coefficient of the monomial A in the polynomial P. A is a monomial of P if $c(A,P) \neq 0$. For a fixed monomial A, $c(A,P)$ is a linear functional on the algebra $\mathcal{P}_n$. If $\theta \in SO(1)$, let

$$z^\theta = (x_1\cos(\theta) + x_2\sin(\theta), x_2\cos(\theta) - x_2\sin(\theta), x_3, \ldots, x_n, y)$$

be the new coordinate system defined by the action of $SO(1)$ on the first two coordinate axes. $O(n)$ is generated by this action of $SO(1)$ and by the coordinate permutations discussed earlier. If $P \in \mathcal{P}_n$, P is invariant under this action of $SO(1)$ if $P(Z,T,G) = P(Z^\theta,T,G)$ for all

$\theta \in SO(1)$. Such invariance has the following algebraic consequences.

Lemma 2.5. Let $P \in \mathcal{P}_n$ be invariant under the action of SO(1) on the first two coordinate axes. Then

a) Suppose $g_{12/\alpha}$ divides some monomial A of P. By changing only '1' and '2' indices, we can construct a new monomial $\bar{A}$ from A which is a monomial of P and which is divisible by a variable of the form $g_{11/\beta}$ for some multiindex β.

b) Let $g_{ij/\alpha}$ divide some monomial A of P. By changing only '1' and '2' indices, we can construct a new monomial $\bar{A}$ from A which is a monomial of P and which is divisible by a variable of the form $g_{k\ell/\beta}$ where $\beta(2) = 0$.

For example, the polynomial $P' = 2g_{12/12}$ is not SO(1) invariant since it violates Lemma 2.5(a). The polynomial $P'' = g_{11/22} + g_{22/11} - 2g_{12/12}$ is invariant under the linear action of SO(1) and satisfies this lemma. Of course, P'' is not invariant under all coordinate transformations. The use of the indices '1' and '2' is for notational convenience; we can apply this lemma to any pair of coordinate axes. The contraction of indices in this lemma is closely related to H. Weyl's theorem (1940).

Although we proved this lemma in Gilkey, Ph.D. dissertation (1972), we give a quick sketch of the proof here, since it is central to the arguments we shall give. We prove (a) since the proof of (b) is completely analogous. Decompose $P = \sum_k P^{(k)}$ into polynomials $P^{(k)}$ of total degree k in the '1' and '2' indices. Each monomial A of $P^{(k)}$ has $\deg_1(A) = \deg_2(A) = k$. The $P^{(k)}$ are invariant separately under the action of SO(1) so we may suppose, without loss of generality, that $P = P^{(k)}$ for some k. Let $\partial/\partial x_1^\theta = \cos(\theta)\partial/\partial x_1 + \sin(\theta)\partial/\partial x_2$ and $\partial/\partial x_2^\theta = \cos(\theta)\partial/\partial x_2 - \sin(\theta)\partial/\partial x_1$. The remaining partial derivatives are unchanged. We compute $P(Z^\theta,T,G)$ by replacing each '1' index by '$\cos(\theta)1 + \sin(\theta)2$' and each '2' index by '$\cos(\theta)2 - \sin(\theta)1$' and by expanding multilinearly. This decomposes

$$P(Z^\theta,T,G) = \sum_{i=0}^{k} \sin(\theta)^i \cos(\theta)^{k-i} P_i(Z,T,G)$$

Since P is invariant, this sum is independent of θ. We substitute $1 = \sin^2 + \cos^2$ to show k is even, $P_{2i+1} = 0$, $p_{2i} = \binom{k}{i}P$.

We use P_1 to prove the lemma. Suppose the lemma is false. Let $A = (g_{12/\alpha})^\ell A_0$ with $\ell > 0$. By hypothesis, $g_{11/\alpha}$ does not divide A_0. Therefore,

$$A(Z^\theta,T,G) = \cos(\theta)^k A(Z,T,G) - \ell\cos(\theta)^{k-1}\sin(\theta)g_{11/\alpha}(g_{12/\alpha})^{\ell-1}A_0 + \cdots$$

The other terms in the expansion either have a higher power of $\sin(\theta)$ or are not of the form $g_{11/\alpha}(g_{12/\alpha})^{\ell-1}A_0$. Since $P_1 = 0$, there must be some other monomial $\bar{A}$ of P which makes a contribution to P_1 which cancels this term. Since the power of $\sin(\theta)$ is 1, $\bar{A}$ can be constructed from $g_{11/\alpha}(g_{12/\alpha})^{\ell-1}A_0$ by changing a single index $1 \to 2$ or $2 \to 1$. Since $g_{11/\beta}$ does not divide any monomial of P by hypothesis, the index changed must have been to transform $g_{11/\alpha}$ to $g_{12/\alpha}$ so $\bar{A} = A$. This contradiction proves the lemma.

We use Lemma 2.5 to complete the proof of Theorem 2.4 as follows. Let $P \in \mathcal{P}_n$ be invariant, homogeneous of order n, and independent of dT_ν. If $R(P) = 0$, then every monomial of P has the form

$$A = g_{a_1b_1/c_1d_1}\cdots g_{a_pb_p/c_pd_p}$$

for $2p = n$. Sincer there are a total of $2n = 4p$ indices which appear in A and since each index from 1 through n must appear at least twice, every index from 1 through n appears exactly twice. By making a coordinate permutation, we may assume $a_1 = 1$. By using Lemma 2.5(a) if needed, we may choose A so $a_1 = b_1 = 1$. Since the index 1 appears exactly twice, this implies the index '1' appears nowhere else in A. Let $P_1 = \sum_A c(A,P)A$ where we sum over the monomials A of P which are divisible by $g_{11/cd}$ for some (c,d). If $P \neq 0$, then $P_1 \neq 0$. Since we have fixed the index '1', P_1 is invariant under the action of $SO(n-1)$. If

$$A = g_{11/c_1d_1}\cdots g_{a_pb_p/c_pd_p}$$

then the index 1 appears nowhere else in A. We may make a coordinate permutation to assume that $c_1 = 2$ and that $d_1 = 2$ or $d_1 = 3$. By Lemma 2.5(b), we may construct a monomial A of P_1 divisible by $g_{11/22}$. Let $P_2 = \sum_A c(A,P)A$ where we sum over the monomials A of P which are divisible by $g_{11/22}$. $P_2 \neq 0$ is invariant under $SO(n-2)$. We proceed inductively in this fashion to show finally that the monomial

$$A = g_{11/22}\cdots g_{n-1,n-1/nn}$$

is a monomial of P. Thus, if $P \neq 0$ satisfies the hypothesis above,

$c(A,P) \neq 0$. This implies the space of all such P has dimension at most 1. Since E_n is such a P, this proves $P = cE_n$ which completes the proof of Theorem 2.4.

§3. In this section, we shall derive the Lefschetz fixed point formulas for the twisted signature complex. This will give a heat equation proof of the g-signature theorem of Atiyah and Singer (1968). We review the construction of the signature complex: let M^m be an orientable Riemannian manifold of even dimension $m = 2q$ and let ORN be an orientation of M. Let $\mathrm{dvol}(\mathrm{ORN}) \in \Lambda^m T^*M$ be the oriented volume form. The Hodge operator $*(\mathrm{ORN})_p : \Lambda^p T^*M \to \Lambda^{m-p} T^*M$ is defined by $\theta \wedge *(\mathrm{ORN})_p \psi = G(\theta,\psi)\mathrm{dvol}(\mathrm{ORN})$ for $\theta,\psi \in \Lambda^p T^*M$. Let $\Lambda M = \oplus_p \Lambda^p T^*M$ be the complete exterior algebra. Let $\tau(\mathrm{ORN}) : \Lambda M \to \Lambda M$ be $\oplus_p (-1)^{p(p-1)/2} *(\mathrm{ORN})_p$. The $\pm$ signs are chosen so $\tau(\mathrm{ORN})^2 = 1$ if g is even and -1 if g is odd. Decompose $\Lambda M = \Lambda^{\pm} M$ into the eigenspaces of $\tau(\mathrm{ORN})$. The roles of $\Lambda^{\pm} M$ are reversed if we reverse the orientation.

If $\xi \in T^*M$, let $\mathrm{ext}(\xi)$ be exterior multiplication and let $\mathrm{int}(\xi)$ be interior multiplication. Let $\mathrm{cl}(\xi) = \mathrm{ext}(\xi) - \mathrm{int}(\xi) : T^*M \otimes \Lambda M \to \Lambda M$ be Clifford multiplication. If $d : C^\infty \Lambda M \to C^\infty \Lambda M$ is exterior differentiation, let δ be the dual. The leading symbol of d is exterior multiplication, the leading symbol of δ is -interior multiplication, and the leading symbol of $(d + \delta)$ is Clifford multiplication. It is easily verified that $\mathrm{cl}(\xi)\tau(\mathrm{ORN}) = -\tau(\mathrm{ORN})\mathrm{cl}(\xi)$ and $(d + \delta)\tau(\mathrm{ORN}) = -\tau(\mathrm{ORN})(d + \delta)$. This implies that $(d + \delta)$ induces maps $(d + \delta)^{\pm} : C^\infty \Lambda^{\pm} M \to C^\infty \Lambda^{\mp} M$. The index of this elliptic complex is the signature of M.

If g is odd, the signature is alway zero. We can get a nonzero index by twisting the signature complex with a coefficient bundle. Let E be a smooth vector bundle over M with a smooth fibre metric and a Riemannian connection ∇_E. Let ∇ be the Levi-Civita connection on TM. Extend ∇ to act naturally on $\Lambda^{\pm}(M)$. Let ∇ be the tensor product connection on $\Lambda^{\pm} \otimes E$. The twisted signature complex is defined by

$$d^{\pm}(\mathrm{ORN},E) : C^\infty \Lambda^{\pm} M \otimes E \xrightarrow{\nabla} C^\infty T^*M \otimes \Lambda^{\pm} M \otimes E \xrightarrow{\mathrm{cl} \otimes 1} C^\infty \Lambda^{\mp} M \otimes E$$

The formal adjoint of $d^{\pm}(\mathrm{ORN},E)$ is $d^{\mp}(\mathrm{ORN},E)$. If we reverse the orientation, $d^{\pm}(\mathrm{ORN},E) = d^{\mp}(-\mathrm{ORN},E)$. We can recover the signature complex by taking E to be the trivial line bundle with a flat connection.

Let $T : M^m \to M^m$ be an orientation preserving isometry.

Lemma 3.1. T is nondegenerate. The components of the fixed point set of T are totally geodesic submanifolds N^n which are of even dimension.

Proof. Let $T(x_0) = x_0$. Since T is an orientation preserving isometry, $dT(x_0):TM(x_0) \to TM(x_0)$ is an orthogonal matrix of determinant 1 relative to any orthonormal frame for $TM(x_0)$. Decompose $TM(x_0) = A \oplus B$ into the span A of the eigenvectors of eigenvalue 1 and the orthogonal complement B. Relative to this splitting, $dT(x_0)$ has the form

$$\begin{pmatrix} I & 0 \\ 0 & U \end{pmatrix}$$

where $\det(I - U) \neq 0$. Let $\dim(A) = n$ and $\dim(B) = m - n$. Since $U \in SO(m-n)$ and $\det(I - U) \neq 0$, $m - n$ is even so n must be even.

Let

$$f(a) = \exp_{x_0}(a):A \to M^m$$

This is a nonsingular map near x_0. Let $\text{range}(f) = N^n$ near x_0. The tangent space of N^n is A. If t is a real parameter, then $f(ta)$ is a geodesic from x_0 with initial direction $a \in A$. Since T is an isometry, $T(f(ta))$ is another geodesic from x_0 with initial direction $dT(x_0)(a) = a$. This shows these two geodesics are equal, so $T(f(a)) = f(a)$. Consequently, N^n is contained in the fixed point set of T. Conversely, let x be close enough to x_0 so there is a unique shortest geodesic γ from x_0 to x. If $Tx = x$, then $T\gamma$ is another shortest geodesic from x_0 to x. Therefore, $T\gamma = \gamma$ and $dT(\dot{\gamma}(x_0)) = \dot{\gamma}(x_0)$. Consequently, $\dot{\gamma}(x_0) \in A$ so $x \in N$. This implies N is the fixed point set of T near x_0, so the fixed point set is regular. Any geodesic tangent to N^n starting at x_0 has the form $f(ta)$ for $a \in A$. Since this geodesic is T invariant, N^n is totally geodesic. This proves the lemma.

The orientation of M^m does not induce a natural orientation on N^n. We suppose the normal bundles ν are orientable and we choose a fixed orientation $ORN(\nu)$ for ν. This defines a natural orientation $ORN(n)$ on N^n from the orientation on M^m so that $ORN(n)ORN(\nu) = ORN$.

T defines an action $\Lambda^{\pm}(dT)$ on the bundles $\Lambda^{\pm}M$ which commutes with the operators $(d + \partial)$. We suppose there is given an action $A(T)$ on E preserving the fibre metric and the Riemannian connection. Let $\Lambda^{\pm}(dT) \otimes A(T)$ be the action on $\Lambda^{\pm}M \otimes F$. This commutes with the fibre metric and with the operators $d^{\pm}(ORN,E)$. Let $L(T,E,ORN)_s$ be the Lefschetz number of this elliptic complex and let $a^s_{p,n}(T,G,E,ORN)$ be the invariant of the heat equation. If we reverse the orientation of M, the roles of Λ^+M and of Λ^-M are interchanged. This implies that $L(T,E,-ORN)_s = -L(T,E,ORN)_s$ and that $a^s_{p,n}(T,G,E,-ORN) = -a^s_{p,n}(T,G,E,ORN)$. Let $a^s_{p,n}(T,G,E) = a^s_{p,n}(T,G,E,ORN)\ dvol(ORN(n))$. $a^s_{p,n} \in \Lambda^n N$ is an invariantly defined n form which changes sign if the fixed orientation of ν is reversed. This is the basic difference between the signature and the DeRham complexes. The integrand of the DeRham complex is an invariant measure while the integrand of the signature complex is an invariant form.

Let $A(T)$ be the action of T on E. If we restrict E to the submanifold N^n, $A(T)$ covers the identity map and defines an ordinary bundle map on $E|_N$ which commutes with the restriction of the connection ∇_E to N. Similarly, dT_ν is a bundle map on ν over N^n which commutes with the restriction of the Levi-Civita connection to ν. We shall give a brief description of the generalized charactertistic forms which are associated to such bundle maps. The Atiyah-Singer formula (1968) expresses $L(T,E,ORN)_s$ in terms of these gneralized characteristic forms.

We restrict attention for the moment to the submanifold N^n. Let V be a vector bundle over N^n with a smooth fibre metric and a Riemannian connection ∇. Let r be the fibre dimension of V. If V is a complex vector bundle, let the structure group $G = U(r)$. If V is an oriented real vector bundle, let $G = SO(r)$. Otherwise, let $G = O(r)$. Let s be a local G-frame for V. s is an orthonormal frame. If $G = SO(r)$, the orientation of s induces the given orientation of V. Let $F: V \to V$ be a bundle map which commutes with the connection ∇ and which preserves the fibre metric. If $G = SO(r)$, we assume F preserves the orientation as well. Let $F^s \in G$ be the matrix of F, let ω^s be the connection matrix of ∇, and let Ω^s be the curvature of ∇ relative to the frame s. The covariant derivative of F is defined by $\nabla F(s) = \nabla(Fs) - F(\nabla s)$. F commutes with ∇ if and only if $\nabla F \equiv 0$. Fix $x_0 \in N^n$ and normalize the G-frame s so $\nabla s(x_0) = 0$. For such a frame, $dF^s(x_0) = 0$.

If F commutes with ∇, $\nabla F = \omega^s F^s + dF^s - F^s\omega^s \equiv 0$. We differentiate this relation to show $d\omega^s F^s - \omega^s dF^s - dF^s\omega^s - F^s d\omega^s \equiv 0$. At x_0, $dF^s(x_0) = \omega^s(x_0) = d\Omega^s(x_0) = 0$ and $d\omega^s(x_0) = \Omega^s(x_0)$. This shows $\Omega^s F^s(x_0) - F^s\Omega^s(x_0) = 0$. Since both F and Ω are tensorial, $\Omega F \equiv F\Omega$.

Let $P(F,W)$ be a smooth function of the components of two $r \times r$ matrices (F,W). We will assume that P is a homogeneous polynomial of degree k in the components of W. We will also assume that $P(gFg^{-1},gWg^{-1}) = P(F,W)$ for all $g \in G$. We restrict the domain of P to $\{(F,W):F \in G, FW = WF$, and W the Lie algebra of $G\}$. If $G = U(r)$, $W + W^* = 0$ while if $G = SO(r)$ or $O(r)$, $W + W^t = 0$. We will sometimes restrict this domain further by imposing additional G-invariant conditions on the F's to be considered. For example, we might assume that $F \in G$ satisfied $\det(I - F) \neq 0$. If the restriction of P to the appropriate domain is nonzero, we will say P defines a nonzero local formula. Since the curvature is a 2-form valued element of the Lie algebra of G, we can evaluate $P(F^s,\Omega^2) \in \Lambda^{2k}(N)$. Since P is invariant under the action of the structure group G, this is independent of the choice of the frame s. We denote this common value by $P(F,\Omega)$.

THEOREM 3.2. $P(F,\Omega)$ is a closed 2k form. If F commutes with two different connections ∇_0 and ∇_1, then $P(F,\Omega_0)$ and $P(F,\Omega_1)$ represent the same cohomology class of $H^{2k}(N,C)$.

We say such a $P(F,\Omega)$ is a G-characteristic form. The cohomology class represented by $P(F,\Omega)$ is a G-characteristic class. We can recover the Chern and Pontrjagin classes of V by taking F to be the identity map. We prove Theorem 3.2 by generalizing the classical proof for the Chern and Pontrjagin classes. Fix $x_0 \in N^n$ and choose a G-frame s for V near x_0 so $\omega^s(x_0) = 0$. Then $dF^s(x_0) = d\Omega^s(x_0) = 0$ so $dP(F^s,\Omega^s)(x_0) = 0$. Since dP is independent of the choice of the frame s, $P(F,\Omega)$ is a closed 2k form.

Let F commute with two connections ∇_0 and ∇_1. Let $\nabla_t = t\nabla_1 + (1 - t)\nabla_0$, then F commutes with the connection ∇_t for $t \in [0,1]$. Let $\eta = \nabla_1 - \nabla_0$; η is a 1-form valued endomorphism of V which is invariantly defined and which commutes with F. The connection matrix of ∇_t is given by $\omega_t^s = \omega_0^s + t\eta^s$. Since $P(F,W)$ is polynomial in the components of W, we polarize P to express $P(F,W) = P(F,W,\ldots,W)$. $P(F,W_1,\ldots,W_k)$ is a symmetric multilinear form in the components of $(W_1,\ldots,W_k)$ which is G-invariant. Define

$$Q(t) = P(F^s,\eta^s,\Omega_t^s,\ldots,\Omega_t^s) \in \Lambda^{2k-1}(N)$$

$$P(t) = P(F^s,\Omega_t^s,\ldots,\Omega_t^s) \in \Lambda^{2k}(N)$$

Both $P(t)$ and $Q(t)$ are independent of the frame s. We will show $kdQ(t) = \partial P/\partial t$. Thus, $P(F,\nabla_1) - P(F,\nabla_0) = \int_0^1 \partial P/\partial t\, dt = d(\int_0^1 kQ(t)dt)$ which will prove Theorem 3.2.

Fix $\tau \in [0,1]$ and choose a frame s so $\omega_\tau^s(x_0) = 0$. Then $dF^s(x_0) = \omega_\tau^s(x_0) = d\Omega_\tau^s(x_0) = 0$. $\partial\Omega_t^s/\partial t(x_0,\tau) = (d\eta^s + \eta^s\wedge\omega_t^s + \omega_t^s\wedge\eta^s)(x_0,\tau) = d\eta^s(x_0)$. Therefore,

$$dQ(x_0,\tau) = P(F^s,d\eta^s,\Omega_t^s,\ldots,\Omega_t^s)(x_0,\tau) = P(F^s,\partial\Omega_t^s/\partial t,\Omega_t^s,\ldots,\Omega_t^s)(x_0,\tau)$$

$$= \frac{1}{k}\,\partial P/\partial t(x_0,\tau)$$

Let $\mathcal{R}^r_{n,m,*}$ be the ring generated by the Pontrjagin forms of $T(N)$, the generalized $SO(m-n)$ characteristic forms of ν relative to the action dT_ν, and by the generalized characteristic forms of E relative to the action $A(T)$. The $SO(m-n)$ characteristic forms are to be defined relative to the domain $\det(I - dT_\nu) \neq 0$. We shall assume E is a complex vector bundle. $\mathcal{R}^r_{n,m,k}$ will denote the subspace of k-forms. The elements of $\mathcal{R}^r_{n,m,k}$ are homogeneous of order k in the jets involved. If $R\in\mathcal{R}^r_{n,m,k}$, let $R(T,G,E) \in \Lambda^k(N^n)$ be the evaluation of R on this data. Let R be a sequence $R^r_{n,m} \in \mathcal{R}^r_{n,m,n}$, we define

$$R(T,E)[M] = \textstyle\sum_i \int_{N_i} R^r_{n,m}(T,G,E)$$

By Thoerem 3.2, this is independent of the choice of T-invariant metric G and the $A(T)$-invariant fibre metrics and connections on E. We can state the Atiyah-Singer Lefschetz fixed point formulas for the twisted signature complex in terms of these invariants.

THEOREM 3.3. (Atiyah-Singer). There exists a sequence of generalized characteristic forms $R = \{R^r_{n,m}\} \in \mathcal{R}^r_{n,m,n}$ such that $L(T,E,ORN)_s = R(T,E)[M]$.

The exact formula is given in Atiyah and Singer (1968). We shall use the following two theorems to complete the proof of Theorem 0.4(2a and 2b).

THEOREM 3.4.

a) $a^s_{p,n} \equiv 0$ if $2p < n$

b) $a^s_{p,n}$ is a generalized characteristic form of $\mathcal{R}^r_{n,m,n}$ if $2p = n$.

THEOREM 3.5. Let $R = \{R^r_{n,m}\} \in \mathcal{R}^r_{n,m,n}$ be a sequence of generalized characteristic forms such that $R(T,E)[M] = 0$ for all (T,E,M). Then the local invariants $R^r_{n,m} \equiv 0$.

Let R_0 be the sequence of Theorem 3.3 and let R_1 be the sequence of the heat equation. By Theorem 3.4, R_1 is a sequence of generalized characteristic forms. By Corollary 0.3, $R_1(T,E)[M] = L(T,E,ORN)_s$. Therefore, $(R_0 - R_1)(T,E)[M] = 0$ so $R_0 \equiv R_1$ by Theorem 3.5. We shall prove Theorem 3.5 by constructing a sequence of test manifolds that will be a dual basis to $\mathcal{R}^r_{n,m,n}$. It is possible to give an independent proof of the g-signature theorem using heat equation methods by checking the Atiyah-Singer formula directly for these test manifolds and by applying Theorem 3.5. We omit the details of this verification since the formula of Theorem 3.3 is well known. The remainder of this section is devoted to the proof of these two results.

Before beginning the proof of Theorem 3.4, we study the local geometry of the submanifold N^n. We use the same indexing conventions as in the second section.

Lemma 3.6. There exist local coordinates $z = (x,y)$ near $x_0 \in N^n$ on M^m such that

a) $N^n = \{z:y(z) = 0\}$ and $\{\partial/\partial y_u\}(x,0)$ is an oriented orthonormal frame for $\nu(x)$.

b) $T(x,y) = (x,U\cdot y)$ for a constant matrix $U \in SO(m-n)$ with $\det(I - U) \neq 0$.

c) $g_{ij}(x_0) = \delta_{ij}$ and all the first derivatives of the metric vanish at x_0.

d) $g_{au/bc}(x_0) = g_{bc/au}(x_0) = 0$ and $g_{au/bv}(x_0) + g_{av/bu}(x_0) = 0$.

Proof. N^n is totally geodesic by Lemma 3.1. Let x be a system of geodesic polar coordinates on N^n centered at x_0. Let $s(x_0)$ be an oriented orthonormal frame for $\nu(x_0)$. Extend s to an orthonormal frame $s(x)$ for $\nu(x)$ by parallel translation along the geodesic rays from x_0. Let $z = \exp_x(y\cdot s(s))$ be the coordinate system defined by the geodesic flow. This system of tubular geodesic coordinates is unique up to the action of the group $O(n) \times SO(m-n)$. (a) is immediate for this system of coordinates.

Since T is an orientation preserving isometry, $dT(s)$ is another orthonormal oriented frame for ν which is parallel along the geodesic rays from x_0. This implies $dT(s) = U\cdot s$ for a constant matrix $U \in SO(m-n)$; $\det(I - U) \neq 0$ by Lemma 3.1. Since T is an isometry, it maps tubular geodesic coordinates to tubular geodesic coordinates so $T(x,y) = (x,U\cdot y)$ which proves (b). Let

$$\Gamma_{ijk} = G(\nabla_{\partial/\partial z_i}\partial/\partial z_j, \partial/\partial z_k) = \frac{1}{2}(g_{ik/j} + h_{jk/i} - g_{ij/k})$$

be the Christoffel symbols of the Levi-Civita connection. Because the rays $(tx,0)$ are goedesics from x_0, $\Gamma_{abi}(x_0) = 0$. $g_{ab/c}(x_0) = \Gamma_{cab}(x_0) + \Gamma_{cba}(x_0) = 0$. $g_{uv}(x,0) = \delta_{uv}$ and $g_{ua}(x,0) = 0$, so $g_{uv/a}(x_0) = g_{ua/b}(x_0) = 0$. $g_{ab/u}(x_0) = -2\Gamma_{abu}(x_0) + g_{au/b}(x_0) + g_{bu/a}(x_0) = 0$. Because the rays $(0,ty)$ are geodesics from x_0, $\Gamma_{uvi}(x_0) = 0$. $g_{uv/w}(x_0) = \Gamma_{wuv}(x_0) + \Gamma_{wvu}(x_0) = 0$. Because $s(x) = \{\partial/\partial y_u\}$ is parallel transported along the geodesic rays from x_0, $\Gamma_{aui}(x_0) = 0$. $g_{au/v}(x_0) = \Gamma_{vau}(x_0) + \Gamma_{vua}(x_0) = \Gamma_{avu}(x_0) + \Gamma_{vua}(x_0) = 0$ so all the first derivatives of the metric vanish at x_0. Since $g_{ij}(x_0) = \delta_{ij}$, this proves (c).

Since $g_{au/c}(x,0) = 0$, $g_{au/bc}(s_0) = 0$ and the second fundamental form of the embedding of N^n in M^m is given by the $\{g_{ab/u}\}$ variables. Since N^n is totally geodesic, the second fundamental form vanishes so $g_{ab/u}(x,0) = 0$. This implies $g_{ab/uc}(x_0) = 0$. Finally, the curves $f(x,ry)$ are geodesics for any (x,y) so $\Gamma_{uva}(x,0) = 0$. $\Gamma_{uva/b}(x_0) = g_{au/bv}(x_0) + g_{av/ub}(x_0) + g_{uv/ab}(x_0) = g_{au/bv}(x_0) + g_{av/ub}(x_0) = 0$ which completes the proof of the lemma.

Let script letters a,b,c index a local frame for E which is normalized so that $\nabla_E s(x_0) = 0$. Let $\nabla_E s_b = \omega^a_{bi} s_a \otimes dx^i$ be the connection form, and let $\omega^a_{bi/\beta}$ be the jets of the connection form. Since ∇_E is Riemannian, $\omega^a_{bi} \equiv -\omega^b_{ai}$. We could further normalize the choice of s by parallel translating a given $U(r)$ frame $s(x_0)$ for $E(x_0)$ along the geodesic rays from x_0. This would fix the choice of s up to the action of $U(r)$. Since $A(T)$ commutes with ∇_E, $A(T)s$ is another such frame. Therefore, $A(T)s = s$ for some constant matrix $A \in U(r)$. In geodesic tubular coordinates and relative to such a frame, only the 0-jets of dT_ν and of $A(T)$ are nonzero. Therefore, in any coordinate system and relative to any frame, we may express all the jets of dT_ν and of $A(T)$ in terms of the $\{g_{ij/\alpha}, \omega^a_{bi/\beta}\}$ variables and in terms of the 0-jets of dT_ν and of $A(T)$. This implies $a^s_{p,n}$ is a polynomial of order $2p$ in the $\{g_{ij/\alpha}, \omega^a_{bi/\beta}\}$ variables with coefficients which are smooth functions of the components of dT_ν and of $A(T)$. By Lemma 1.8, these functions are not arbitrary, but are linear in the components of the action $A(T)$ on E. We shall not restrict to geodesic tubular coordinates and parallel frames to avoid introducing the Bianchi identities into the algebras we shall consider.

Let N^n be a totally geodesic submanifold of M^m, let $U(x)$ be an $SO(m-n)$ valued endomorphism of ν with $\det(I - U)(x) \neq 0$, and let A be an $U(r)$ valued endomorphism of $E|_N$. We shall not suppose $\nabla U = \nabla A = 0$ nor that there is an isometry T of M such that $U = dT_\nu$. Let $P^r_{n,m}$ be the free n-form valued polynomial algebra in the $\{g_{ij/\alpha}, \omega^a_{bi/\beta}\}$ variables with coefficients that are linear in the $\{A^a_b\}$ variables and smooth in the $\{U^u_v\}$ variables. We impose the additional relations

$$\begin{array}{llll} g_{ij/\alpha} = g_{ji/\alpha} & \omega^a_{bi/\beta} = -\omega^b_{ai/\beta} & g_{ij}(x_0) = \delta_{ij} & \\ g_{ij/k}(x_0) = 0 & \omega^a_{bi}(x_0) = 0 & g_{ab/cu}(x_0) = 0 & (*) \\ g_{au/bc}(x_0) = 0 & g_{au/bv}(x_0) + g_{av/bu}(x_0) = 0 & \det(I - U) \neq 0 & \end{array}$$

We shall only consider coordinates $z = (x,y)$ which satisfy (*). We can always choose such coordinates z by Lemma 3.6. We normalize the $U(r)$ local frame s so $\nabla_E s(x_0) = 0$ as well. These are the only normalizations we will make. If $P(Z,s,U,A,G,E) \in \Lambda^n N$ is independent of the choice of (Z,s), we shall say that $P \in P^r_{n,m}$ is invariant.

$a^s_{p,n}$ is not an element of $P^r_{n,m}$ since $a^s_{p,n}$ is only defined for certain (U,A). Let $\mathcal{P}^r_{n,m}$ be the quotient algebra of $P^r_{n,m}$ which is defined by imposing the additional algebraic conditions which are induce by assuming that $U = dT_\nu$ for some isometry T and that $\nabla_E A \equiv 0$. If Ω_ν is the curvature of ν and if Ω_E is the curvature of E, these two assumptions imply that $U\Omega_\nu = \Omega_\nu U$ and that $A\Omega_E = \Omega_E A$. There are many other relations on the algebra which make it difficult to work with $\mathcal{P}^r_{n,m}$ directly.

Let $\pi: P^r_{n,m} \to \mathcal{P}^r_{n,m}$ be the natural projection by restricting a polynomial of $P^r_{n,m}$ to the appropriate subdomain. It is surjective.

Lemma 3.7. Let $P \in \mathcal{P}^r_{n,m}$ be homogeneous of order k and invariant. Then there exists $P' \in \mathcal{P}^r_{n,m}$ which is homogeneous of order k and invariant such that $\pi P' = P$.

Proof. Let P'' be homogeneous of order k such that $\pi P'' = P$. Of course, P'' need not be invariant; P'' is only invariant when restricted to the appropriate subdomain. If we choose geodesic tubular coordinates Z and parallel frames s, the choice of (Z,s) is unique up to the action of the compact group $G = O(n) \times SO(m-n) \times U(r)$. If $g \in G$, let (Z^g, s^g) be the new system of geodesic tubular coordinates and the new parallel frame defined by the action of g. Let dg be Haar measure on the compact group G. Define

$$P'(U,A,G,E) = \int_G P''(Z^g, d^g, U,A,G,E)\,dg$$

It is clear P' is a smooth polynomial function of the jets involved. It is invariant since we have averaged over the structure group. It is homogeneous of order k. If $U = dT_\nu$ for an isometry T and if $\nabla_E A \equiv 0$, then

$$P'(U,A,G,E) = \int_G P''(A^g, s^g, U,A,G,E)\,dg = \int_G P(Z^g, s^g, U,A,G,E)\,dg$$

$$= \int_G P(U,A,G,E)\,dg = P(U,A,G,E)$$

This implies $\pi P' = P$.

Let $R^r_{n,m,n} = \{R \in P^r_{n,m} : (a)\pi R \in \mathcal{R}^r_{n,m,n},\ (b) R$ is invariant, (c) R is homogeneous of order $n\}$. We can also describe this subset as follows: to define $\mathcal{R}^r_{n,m,n}$, we considered polynomials $P(F,W)$ defined on the domain $\{(F,W): FW = WF, F \in G, W + W^* = 0\}$. We can generalize this construction by considering the larger domain $\{(F,W): F \in G, W + W^* = 0\}$. This defines $P(F,\Omega)$ for a larger class of endomorphisms F. Of course, $P(F,\Omega)$ has no cohomological significance if $\nabla F = 0$ since Theorem 3.2 fails in this case. Let $R^r_{n,m,*}$ be the ring generated by the Pontrjagin forms of TN, the $SO(m-n)$ forms of ν relative to any endomorphism U, and the $U(r)$ forms of E relative to any endomorphism A. We suppose $P(U,\Omega_\nu)$ is smooth in U and $\overline{P}(A,\Omega_E)$ is linear in A. By Lemma 3.7, to prove Theorem 3.4 it suffices to prove

THEOREM 3.8. Let $P \in P^r_{n,m}$ be invariant and homogeneous of order k. Then

a) $p \equiv 0$ if $k < n$
b) $P \in \mathcal{R}^r_{n,m,n}$ if $k = n$.

Proof. We passed from the algebra $\mathcal{R}^r_{n,m}$ to $P^r_{n,m}$ to be able to apply Lemma 2.5 to the free algebra $P^r_{n,m}$. Let $0 \not\equiv P \in P^r_{n,m}$ be invariant and homogeneous of order k. Let B be any monomial of P. B has the form

$$B = f(U^u_v, A^a_b) A_0 A_1 dx^1 \wedge \ldots \wedge dx^n \quad \text{for } f \not\equiv 0$$
$$A_0 = g_{i_1 j_1/\alpha_1} \cdots g_{i_s j_s/\alpha_s} \quad \text{for } |\alpha_i| \geq 2$$

$$A_1 = \omega_{b_1 k_1/\beta_1}^{a_1} \cdots \omega_{b_t k_t/\beta_t}^{a_t} \quad \text{for} \quad |\beta_j| \geq 1$$

$f(U,A)$ is smooth in U and linear in A. $k = \text{ord}(B) = \sum_i |\alpha_i| + \sum_j (|\beta_j| + 1) \geq 2s + 2t$.

Since P is invariant under the action of $O(n)$, $\deg_a B = 1 + \deg_z A_0 + \deg_a A_1$ is even for $1 \leq a \leq n$. Therefore, $\deg_a A_0 + \deg_a A_1 \neq 0$. If one of the indices $\{i_1, j_1\} \leq n$, we may make a coordinate permutation to assume B is chosen so $i_1 = n$. If $j_1 < n$, we apply Lemma 2.5(a) to choose B so $j_1 = n$ as well. This proves we may choose B so

$$\deg_a(g_{i_1 j_1}) = 0$$

for $a < n$. We repeat this argument on the indices $< n$ to choose B so

$$\deg_a(g_{i_2 j_2}) = 0$$

for $a < n-1$. Since the monomials which are constructed in Lemma 2.5 only involve changing indices less than n, the indices $i_1 j_1$ were unchanged so

$$\deg_a(g_{i_1 j_1} g_{i_2 j_2}) = 0$$

for $a < n - 1$. We continue this process to construct B so

$$\deg_a(g_{i_1 i_2} \cdots g_{j_s j_s}) = 0$$

for $a \leq n - s$.

We apply Lemma 2.5(b) a total of $n - s$ times to the indices $a \leq n - s$ and to the multiindex α_1 to construct a monomial B of P such that $\alpha_1(a) = 0$ for $a \leq n - s - 1$. We continue inductively to construct a monomial B of P such that $\deg_a A_0 = 0$ for $a \leq n - 2s$. Since $\deg_a A_0 + \deg_a A_1 \neq 0$, $\deg_a A_1 \neq 0$ for $a \leq n - 2s$. We count indices

$$n - 2s \leq \sum_{a=1}^{n-2s} \deg_a(A_1) \leq \sum_i \deg_i(A_1) = t + \sum_j |\beta_j| = \text{ord}(B) - \sum_i |\alpha_i|$$
$$\leq \text{ord}(B) - 2s$$

This shows $n \leq \text{ord}(B)$ so $P \equiv 0$ if $k - \text{ord}(B) < n$. This proves Theorem 3.8(a).

In the extremal case $n - \text{ord}(P)$, all these inequalities must be equalities. Since the $\{|\alpha_i|, |\beta_j|\}$ are unchanged by an application of

Lemma 2.5, $|\alpha_i| = 2$ and $|\beta_j| = 1$ for every monomial B of P. By the normalization (*), the only nonzero second derivatives of the metric are the $\{g_{ab/cd}, g_{au/bv}, g_{uv/\bar{u}\bar{v}}\}$ variables. Therefore, every monomial B of P has the form

$$B = f(U^u_v, A^a_b) B_0 B_1 B_2 B_3 dx^1 \wedge \dots \wedge dx^n$$

$$B_0 = g_{a_1 b_1/a'_1 b'_1} \cdots g_{a_p b_p/a'_p b'_p}$$

$$B_1 = \omega^{a_1}_{b_1 i_1/j_1} \cdots \omega^{a_g}_{b_q i_q j_q}$$

$$B_2 = g_{c_1 u_1/c'_1 u'_1} \cdots g_{c_s u'_s/c_s u'_s}$$

$$B_3 = g_{u_1 v_1/u'_1 v'_1} \cdots g_{u_t v_t/u'_t v'_t}$$

$$n = 2p + 2q + 2s + 2t$$

By Lemma 2.5, we can construct a monomial B of P by changing only tangential indices so $\deg_a B_0 = 0$ for $a > 2p$. The remaining $n - 2p$ tangential indices must appear at least once in $B_1 B_2 B_3$. There are at most $2q + 2s$ tangential indices which appear in $B_1 B_2 B_3$. This implies $n - 2p \leqslant 2q + 2s$. Thus $n \leqslant 2p + 2q + 2s \leqslant 2p + 2q + 2s + 2t = n$. This proves $t = 0$ and the indices $\{c_i, c'_i, i_j, j_j\}$ are all distinct and tangential. This also shows P is a polynomial in the $\{g_{ab/cd}, \omega^a_{ba/b}, g_{au/bv}\}$ variables.

If B is chosen so $\deg_a B_0 = 0$ for $a > 2p$, then the indices $\{c_i, c'_i, i_j, j_j\}$ are a permutation of the indices from $2p + 1$ through n. By making a coordinate permutation, we may choose B so B_1 involves tangential indices running from $2p + 1$ through $2p + 2q$ and so B_2 involves tangential indices running from $2p + 2q + 1$ through n. This proves

Lemma 3.9. Let $P \in P^r_{n,m}$ be invariant and homogeneous of order n. Then there exist integers p,q,s with $2p + 2q + 2s = n$ and a monomial B of P of the form $B = f(U^u_v, A^a_b) B_0 B_1 B_2 dx^1 \wedge \dots \wedge dx^n$. B_0 is a monomial in the $\{g_{ab/cd}\}$ variables involving indices from 1 through 2p. B_1 is monomial in the $\{\omega^a_{ba/b}\}$ variables involving tangential indices from $2p + 1$ through $2p + 2q$. B_2 is a monomial in the $\{g_{au/bv}\}$ variables involving tangential and normal indices from $2p + 2q + 1$ through m. $f \not\equiv 0$ since B is a monomial of P.

Let $T_{p,q,s}(B) = B$ if B is of the form given in Lemma 3.9 and let $T_{p,q,s}(B) = 0$ otherwise. Extend $T_{p,q,s}$ to be a linear map on $P^r_{n,m}$. $T_{p,q,s}$ is not equivariant under the action of the group $0(n) \times SO(m-n) \times U(r)$, but it is equivariant under the action of the group $0(2p) \times 0(2q)$ $0(2s) \times SO(m-n) \times U(r)$. Let Q_n be the subspace of $P^r_{n,m}$ of polynomials in the $\{g_{ab/cd}\}$ variables. Let Q^r_n be the subspace of polynomials in the $\{\omega^a_{ba/b}\}$ variables with coefficient functions which are linear in $\{A^a_b\}$. Let $Q_{n,m}$ be the subspace of polynomials in the $\{g_{au/bv}\}$ variables with coefficient functions which are smooth in $\{U^u_v\}$. Relative to geodesic tubular coordinates and parallel frames, the structure group of Q_n is $0(n)$, the structure group of $Q^r_{n,m}$ is $0(n) \times U(r)$, and the structure group of $Q_{n,m}$ is $0(n) \times SO(m-n)$.

Renumber the tangential indices of Q^r_{2q} to run from $2p + 1$ through $2p + 2q$ instead of from 1 through $2q$. Renumber the tangential and normal indices of $Q_{2s,m-n+2s}$ to run from $2p + 2s + 1$ through m instead of from 1 through $m - n + 2s$. Using this renumbering, there are natural injections

$$j_{p,q,s}: Q_{2p} \otimes Q^r_{2q} \otimes Q_{2s,m-n+2s} \to P^r_{n,m} \quad (n = 2p + 2q + 2s)$$

Since the coefficient functions $f(U^u_v, A^a_b)$ are linear in A, we may decompose $f(U^u_v, A^a_b) = \sum f^b_a(U^u_v)A^a_b$. Since the coefficient functions split in this way, the image of $T_{p,q,s}$ is the range of $i_{p,q,s}$. If P is homogeneous of order n and invariant, $T_{p,q,s} \neq 0$ for some (p,q,s). Decompose $T_{p,q,s}P = \sum_i P^1_i \otimes P^2_i \otimes P^3_i$. We may choose this decomposition so that $P^1_i \in Q_{2p}$ is homogeneous of order $2p$ and $0(2p)$ invariant, $P^2_i \in Q^r_{2q}$ is homogeneous of order $2q$ and $0(2q) \times U(r)$ invariant, and $P^3_i \in Q_{2s,m-n+2s}$ is homogeneous of order $2s$ and is $0(2s) \times SO(m-n)$ invariant.

Let R_n be the span of the Pontrjagin n-forms of $T(N)$, let R^r_n be the span of the generalized $U(r)$ characteristic n-forms of E, and let $R_{n,m}$ be the span of the generalized $SO(r)$ characteristic n-forms of ν. $\pi R^r_n = \mathcal{R}^r_n$ and $\pi R_{n,m} = \mathcal{R}_{n,m}$ are defined by restricting the domain to $\nabla U = \nabla A = 0$.

Lemma 3.10.

a) If $P \in Q_n$ is homogeneous of order n and invariant, $P \in R_n$.

b) If $P \in Q^r_n$ is homogeneous of order n and invariant, $P \in R^r_n$.

c) If $P \in Q_{n,m}$ is homogeneous of order n and invariant, $P \in R_{n,m}$.

We use Lemma 3.10 to complete the proof of Theorem 3.4(b). There are natural inclusions of the generalized characteristic forms

$$R_{2p} \to R^r_{n,m,2p}, \quad R^r_{2q} \to R^r_{n,m,2q}, \quad R^r_{2s,m-n+2s} \to R^r_{n,m,2s}$$

We take the wedge product of these inclusions to define an injection

$$R_{2p} \otimes R^r_{2q} \otimes R_{2s,m-n+2s} \to R^r_{n,m,n} \qquad (n = 2p + 2q + 2s)$$

The map

$$T_{p,q,s}: R^r_{n,m,n} \to R_{2p} \otimes R^r_{2q} \otimes R_{2s,m-n+2s}$$

is a splitting of this inclusion. $R_{n,m,n}$ is the direct sum of these subspaces. Let $P \in P^r_{n,m}$ be invariant and homogeneous of order n. By Lemma 3.10, $T_{p,q,s}(P) \in R_{2p} \otimes R^r_{2q} \otimes R_{2s,m-n+2s}$. Therefore, $T_{p,q,s}(P)$ is in the range of the projection $T_{p,q,s}(R^r_{m,n})$. Choose $\overline{P} \in R^r_{n,m,n}$ such that $T_{p,q,s}(\overline{P}) = T_{p,q,s}(P)$ for $2p + 2q + 2s = n$. $\overline{P} - P$ is invariant and homogeneous of order n. $T_{p,q,s}(\overline{P} - P) = 0$ so by Lemma 3.9, $\overline{P} - P = 0$. This proves Theorem 3.4. We use the projection maps $T_{p,q,s}$ to break the proof of Theorem 3.4 into the corresponding result for Q_n, Q^r_n, and Q^m_n.

Proof. We proved (a) in Gilkey (1973) to give a proof of the Hirzebruch signature theorem using heat equation methods. We prove (b) as follows. Let $P \in Q^r_n$ be homogeneous of order n. If P is invariant, we showed earlier that every monomial B of P has the form

$$B = f_B(A^a_b)\omega^{a_1}_{b_1 a_1/b_1} \cdots \omega^{a_q}_{b_q a_q/b_q} dx^1 \wedge \cdots \wedge dx^n \qquad 2q = n$$

Since there are $2q = n$ indices in the collection $\{a_1, b_1, \ldots, a_q, b_q\}$ and since every tangential index must appear in this collection, each of the indices from 1 to n appears exactly once; $\{a_1, b_1, \ldots, a_q, b_q\}$ is a permutation σ of the indices from 1 to n. Let S be the collection of indices $\{a_1, b_1, \ldots, a_q, b_q\}$. Since P is $O(n)$ invariant and n-form valued, the coefficient functions have the form $f_B(A^a_b) = \text{sign}(\sigma) f_S(A^a_b)$. We normalized the frame s so $\omega^s(x_0) = 0$. Therefore, $\Omega^s(x_0) = d\omega^s(x_0)$ and

$$\begin{aligned} P &= \textstyle\sum_{S,\sigma} f_S(A^a_b)\,\text{sign}(\sigma)\,\omega^{a_1}_{b_1 \sigma(1)/\sigma(2)} \cdots \omega^{a_q}_{b_q \sigma(n-1)/\sigma(n)} dx^1 \wedge \cdots \wedge dx^n \\ &= 2^{-q} \textstyle\sum_S f_S(A^a_b)\,\Omega^{a_1}_{b_1} \wedge \cdots \wedge \Omega^{a_q}_{b_q} \end{aligned}$$

Let

$$P(A,W) = 2^{-q}\sum_S f_S(A_b^a)A_{b_1}^{a_1}\ldots A_{b_q}^{a_q}$$

then $P(A,W)$ is invariant under $U(r)$ and $P(A,\Omega) = P$. This proves Lemma 3.10(b).

The proof of (c) is analogous. If $P \in Q_{n,m}$ is invariant and homogeneous of order n, then every monomial B of P has the form

$$B = f(U_v^u)g_{a_1u_1/b_1v_1}\ldots g_{a_su_s/b_sv_s}dx^1\wedge\ldots\wedge dx^n \qquad 2s = n$$

The collection of indices $\{a_1,b_1,\ldots,a_s,b_s\}$ is a permutation σ of the indices from 1 to n. If S is the collection of indices $\{u_1,v_1,\ldots,u_s,v_s\}$, then $f_B(U_v^u) = \text{sign}(\sigma)f_S(U_v^u)$. Therefore, P has the form

$$P = \sum_{S,\sigma}\text{sign}(\sigma)f_S(U_v^u)g_{\sigma(1)u_1/\sigma(2)u_2}\ldots g_{\sigma(n-1)u_q/\sigma(n)v_q}dx^1\wedge\ldots\wedge dx^n$$

Let $\bar{\Omega}$ be the curvature of the Levi-Civita connection on the normal bundle, then

$$\bar{\Omega}_v^u = \frac{1}{4}(g_{au/bv} - g_{av/bu} - g_{bu/av} + g_{bv/au})dx^a\wedge dx^b$$

By (*), $g_{au/bv} = -g_{bv/au}$ so we can rewrite P in the form

$$P = 4^{-q}\sum_S f_S(U_v^u)\,\bar{\Omega}_{v_1}^{u_1}\wedge\ldots\wedge\bar{\Omega}_{v_q}^{u_q}$$

If we form the corresponding polynomial $P(U,W)$, then P is invariant under the action of $SO(r)$ and $P(U,\Omega) = P$. This completes the proof of Lemma 3.10 and thereby of Theorem 3.4.

The remainder of this section will be devoted to the proof of Theorem 3.5. We shall derive Theorem 3.5 from the following technical lemma.

Lemma 3.11.

a) Let $P \not\equiv 0 \in \mathcal{R}_n$. We can find N^n such that $P[N] \neq 0$.

b) Let $P \not\equiv 0 \in \mathcal{R}_n^r$. We can find a vector bundle E over some N^n and an action A of E with $\nabla_E A \equiv 0$ wuch that $P(E)[N] \neq 0$.

c) Let $P \not\equiv 0 \in \mathcal{R}_{n,m}$. We can find M^m and an isometry T of M^m such that the fixed point set of T consists of a single submanifold N^n of dimension n together with other components of lower dimension and such that $P(T)[N] \neq 0$.

Proof. (a) is well known. N^n can be taken to be the product of complex projective spaces. We prove (b) as follows. Let $P(A,W) \not\equiv 0$ be $U(r)$ invariant. The domain of P consists of pairs (A,W) with $A \in U(r)$, $W = W^* = 0$, and $AW = WA$. Choose an orthonormal basis $\{e_1,\ldots,e_r\}$ for C^r which diagonalizes A and W simultaneously. Then

$$A(e_k) = \exp(i\theta_k)e_k \quad \text{and} \quad W(e_k) = i\lambda_k e_k$$

for real numbers θ_k and λ_k. Let $r \geqslant 1$ and suppose the lemma has been proved for vector bundles of lower fibre dimension. Let $\theta' = (\theta_2,\ldots,\theta_r)$ and $\lambda' = (\lambda_2,\ldots,\lambda_r)$. Expand

$$P(A,W) = \textstyle\sum_j P_j(\theta_1,\theta',\lambda')\lambda_i^j \qquad (0 \leqslant j \leqslant \tfrac{1}{2}n)$$

P_j is a polynomial of order $\frac{1}{2}n - j$ in λ'. Since $P \not\equiv 0$, we may choose some index j so $P_j \not\equiv 0$. Let c_1 be the first Chern class of a vector bundle. By induction, we may choose (N_2,E_2,A_2) so $P_j(\theta_1)(E_2)[N_2] \neq 0$ for some value of θ_1. Let $N_1 = CP_j$ be complex projective space of real dimension $2j$ and let E_1 be the hyperplane line bundle over M_1. Let $N = N_1 \times N_2$, $E = E_1 \otimes E_2$, and $A = \exp(i\theta_1) \otimes A_2$. Then

$$P(E)[N] = (2\pi i)P_j(\theta_1,E_2)[N_2]c_1(E_1)^j[N_1] \neq 0$$

This proves (b). We note that if $j = 0$, CP_j is just a single point and $E_1 = C$.

Since both m and n are even, $m - n$ is even. The maximal torus of $SO(m-n)$ has dimension $\frac{1}{2}(m-n)$ and is isomorphic to $SO(2)\times\ldots\times SO(2)$. By applying a similar diagonalization argument and by taking appropriate products, we can reduce the proof of (c) to the case $m - n = 2$. Since ν is an oriented 2 dimensional vector bundle, we can use the orientation to regard ν as a complex line bundle; $U(1) = SO(2)$. Let $2p = n$ and suppose that $n > 0$. Let $P(A,W) \not\equiv 0$. $A = \exp(i\theta) \neq 0$ and $W = i\lambda$. $P(A,W) = f(\theta)\lambda^p$. Fix θ so that $f(\theta) \neq 0$. Let $M = CP_{p+1}$ be complex projective space of real dimension $m = 2p + 2$ with the standard metric. Let I be the $(p+1) \times (p+1)$ identity matrix and let

$$S = \begin{pmatrix} I & 0 \\ 0 & \exp(i\theta) \end{pmatrix} \in U(p + 2)$$

S induces an isometry T on CP_{p+1}. The fixed points of T are the eigen-lines of S. Since $\exp(i\theta) \neq 1$, T has one isolated fixed point which corresponds to the eigenspace of $\exp(i\theta)$. The remainder of the fixed point set of T consists of a single component N^n of dimension $n > 0$ which is isomorphic to CP_p. On N^n, $dT_\nu = \exp(i\theta) \neq 1$. The normal bundle of CP_p in CP_{p+1} is isomorphic to the hyperplane line bundle. Therefore, $c_1(\nu)^p[N^n] = 0$. This implies $P(T)[N^n] \neq 0$ and completes the proof of (c) for $n > 0$.

If $n = 0$, the construction given above yields $S^2 = CP_1$ with two isolated fixed points. Since there is no isometry of S^2 having just one isolated fixed point, we must give a different construction for $n = 0$, $m = 2$. $P(T) = f(\theta)$ in this case. Let (p,q) be odd integers which are coprime. We will construct an orientation preserving diffeomorphism T of a Riemann surface of genus $\frac{1}{2}(q - 1)$ which is of order $2q$. Since T is finite order, we can average any metric to construct a T-invariant metric. T^p will have a single isolated fixed point at the origin 0 and $dT^p(0) = \exp(i\pi p/q)$. Since f is continuous and $f \not\equiv 0$, $f(\pi p/q) \neq 0$ for some (p,q) which will complete the proof of (c).

Let M be the Riemann surface which is obtained by identifying opposite sides of a regular polygon with $2q$ sides centered at the origin. We identify opposite sides in such a way to make M_q orientable. The vertices belong to different equivalence classes. Let T be a rotation of the polygon through an angle of π/q. It is possible to choose a polygon so T induces a smooth map of order $2q$ on the quotient manifold M.

T^p maps one equivalence class of vertices to the other equivalence class. T^p does not map any side to itself or to the opposite side. In the interior of the polygon, the only fixed point of T^p is at 0. T^p has a single fixed point at 0 and dT^p is a rotation through an angle of $\pi p/q$. This completes the proof of Lemma 3.11.

We use Lemma 3.11 to complete the proof of Theorem 3.5 by considering product manifolds. Let $R = \{R^r_{n,m}\} \in \mathcal{R}^r_{n,m,n}$ be a sequence of generalized characteristic forms such that $R(T,E)[M] = 0$ for all (T,E,M). Suppose that the local invariants $R^r_{n,m}$ do not vanish pointwise. Fix n minimal so $R^r_{n,m} \not\equiv 0$. We will show there exists (T,E,M) so the fixed point set of T consists of a single submanifold N^n of dimension N^n together with other components of lower dimension and such that $R^r_{n,m}(T,E)[N] \neq 0$. Since $R^r_{j,m} \equiv 0$ for $j < n$, $R(T,E)[M] = R^r_{n,m}(T,E)[N] \neq 0$ which will complete the proof of Theorem 3.5.

Let $\tau = (p,q,s)$ with $n = 2p + 2q + 2s$. Let P^1_p be a Pontrjagin

2p form of TN, let P_q^2 be an U(r) generalized characteristic 2q form of E, and let P_s^3 be an SO(m-n) generalized characteristic 2s form of ν. Let $\mathcal{R}^r_{n,m,n,\tau}$ be the subspace of $\mathcal{R}^r_{n,m,n}$ which is generated by products $P_p^1 P_q^2 P_s^3$. There is a direct sum decomposition

$$\mathcal{R}^r_{n,m,n} = \oplus_\tau \mathcal{R}^r_{n,m,n,\tau}$$

Let $R^r_{n,m} = \sum_\tau R_\tau$ and choose $\tau = (p,q,s)$ with p maximal so $R_\tau \neq 0$. Let $M = N_1^{2p} \times N_2^{2q} \times M_3^{m-n+2s}$. Let $T = I \times I \times T_3$ where T_3 is an isometry of M_3. The fixed point set of T_3 is to consist of a single submanifold N_3^{2s} of dimension 2s together with other components of lower dimension. Let $N = N_1 \times N_2 \times N_3$. Let E_2 be a smooth vector bundle over N_2 with an endomorphism A_2 such that $\nabla A_2 = 0$. Extend A_2 and E_2 to be independent of $(N_1 \times M_3)$ to define (E,A) over M. Since $T = I$ on N_2, T commutes with the connection and action of E. Let $\bar\tau = (\bar p,\bar q,\bar s)$ for $\bar p < p$. We compute that

$$P^1_{2\bar p} P^2_{2\bar q} P^3_{2\bar s}(T,E)[N] = P^1_{2\bar p}(N_1 \times N_2 \times N_3) P^2_{2\bar q}(E_2) P^3_{2\bar s}(T_3)[N_1 \times N_2 \times N_3]$$

If $\bar p < p$, this gives a form of degree at most $2\bar p$ on N_1, so this vanishes. If $\bar p = p$, this becomes $P^1_{2\bar p}(N_1) P^2_{2\bar q}(E_2) P^3_{2\bar s}(T_3)$ which again vanishes unless $\bar q = q$ and $\bar s = s$. Consequently, $R(T,E)[M] = R^r_{n,m}(T,E)[N] = R_\tau(T,E)[N]$ for any such product.

This reduces the proof of Theorem 3.5 to the following final lemma.

Lemma 3.12. Let $0 \neq R_\tau \in \mathcal{R}_{n,m,n}$. There exists (M,T,E) of the product form given above such that $R_\tau(T,E)[N] \neq 0$.

Proof. Decompose $R_\tau = \sum c(i,j,k) P^1_{p,i} P^2_{q,j} P^3_{s,k}$. The $\{P^1_{p,i}\}$ are a finite collection of linearly independent Pontrjagin 2p forms. The $\{P^2_{q,i}\}$ are a finite collection of linearly independent generalized U(r) characteristic 2q forms of E, and the $\{P^3_{s,k}\}$ are a finite collection of linearly independent SO(m-n) characteristic 2s forms of ν. Since $R_\tau \neq 0$, $c(i,j,k) \neq 0$ for some (i,j,k).

By Lemma 3.11, we may choose a dual basis $\{N_1^i\}$, $\{(N_2^j, E_2^j)\}$, $\{(M_3^k, T_3^k)\}$ so that

$$P^1_{p,i}[N_1^i] = a_i^i, \qquad P^2_{q,j}(E_2^j)[N_2^j] = b_2^j, \qquad P^3_{s,k}(T_3^k)[N_3^k] = c_k^k$$

for some nonsingular matrices a_i^i, b_j^j, c_k^k. If M is the product $N_1^i \times N_2^j \times M_3^k$, then

$$R_T(T,E)[N] = \sum c(i,j,k)P^1_{p,i}[N_1^{i}]P^2_{q,j}(E_2^{j})[N_2^{j}]P^3_{s,k}(T_3^{k})[M_3^{k}] = \sum c(i,j,k)a_i^{i}b_j^{j}c_k^{k}$$

Since $c \neq 0$ and since the $a_i^{i}, b_j^{j}, c_k^{k}$ matrices are nonsingular, $R_T(T,E)[N] = 0$ for some choice of (i,j,k). This completes the proof of Theorem 3.5 and thereby completes the proof of Theorem 0.4(2a and 2b).

We note that the examples constructed in the proof of Lemma 3.11 are all holomorphic manifolds. The vector bundles in Lemma 3.11(b) are all holomorphic vector bundles. We will use this observation in the final section in discussing the spin and the Dolbeault complexes.

§4. In this section, we shall discuss the twisted spin and the twisted Dolbeault complexes. We first review briefly the facts we shall need concerning these elliptic complexes. A good treatment of spinors may be found in Atiyah, Bott and Shapiro (1964). A good description of the spin and spin^c complexes may be found in Hitchin (1974).

Let W be an m dimensional real vector space with a positive definite inner product. We avoid identifying W with R^m immediately to emphasize our constructions will be coordinate free. Let $SO(W)$ be the group of linear maps of W which preserve the orientation and inner product. Let $T(W)$ be the complete tensor algebra of W. Let $I(W)$ be the two sided ideal of $T(W)$ generated by all elements $x \in T(W)$ of the form $x = v \otimes v + 1(v,v)$. Let $CL(W) = T(W)/I(W)$ be the Clifford algebra with the multiplication inherited from $T(W)$.

The inclusion $i: W \to T(W)$ induces an inclusion $i: W \to CL(W)$. If $\{e_1, \ldots, e_m\}$ is an orthonormal basis of W, then $CL(W)$ is generated as an algebra by the $\{e_i\}$ subject to the relations $e_i^2 = -1$ and $e_i e_j = -e_j e_i$ for $i \neq j$. If $I = \{1 \leq i_1 < \ldots < i_q \leq m\}$, let

$$e_I = e_{i_1} \ldots e_{i_q}$$

$CL(W)$ inherits a natural inner product from W; the $\{e_I\}$ are an orthonormal basis for $CL(W)$. Left or right multiplication by a unit vector of W is an isometry of $CL(W)$.

If $y = w_1 \otimes \ldots \otimes w_k \in T(W)$, let $y^t = w_k \otimes \ldots \otimes w_1$. The map $y \to y^t$ preserves the ideal $I(W)$ and extends to a map of $CL(W)$. Let $SPIN(W) = \{x \in CL(W) : x = w_1 \ldots w_{2k}$ for $w_i \in W$, $|w_i| = 1\}$. This is a group under Clifford multiplication. If w is a unit vector of W, let $\rho(w)y = wyw$.

W is an invariant subspace of $\rho(w)$. $\rho(w)$ acts on W by reflection in the perpendicular hyperplane defined by w. If $x \in SPIN(W)$, let $\rho(x)y = xyx^t : W \to W$. This automorphism of W preserves the inner product and the orientation since we are reflecting in an even number of hyperplanes. This defines a map $\rho: SPIN(W) \to SO(W)$ which is a nontrivial double covering.

If $m > 2$, the fundamental group of $SO(W)$ is Z_2. $SPIN(W)$ is the universal covering group of $SO(W)$. For example, if $m = 3$, $SPIN(W) = S^3$ and $SO(W) = RP_3$ is real projective space. If $m = 2$, $SO(W) = SPIN(W) = S^1$ and the map ρ is the usual double cover of S^1.

Let $SPIN(W)$ act on $CL(S)$ by Clifford multiplication on the left. Extend this to $CL(W) \otimes C$. This representation is not irreducible, but decomposes into $2^{m/2}$ isomorphic representations Δ. Identify one such representation with its representation space. We now assume that m is even and that W is oriented. Let $\{e_1, \ldots, e_m\}$ be an oriented basis and let $\tau = e_1 \ldots e_m \in CL(W)$ be the orientation. Since $\tau^2 = (-1)^{m/2}$, we may decompose Δ into the eigenspaces $\Delta^{\pm}$ under the action of τ. Since τ is in the center of $SPIN(W)$, these spaces are $SPIN(W)$ invariant. This defines two irreducible representations which are known as the half-spin representations. Multiplication by $w \in W$ anitcommutes with τ and preserves Δ so Clifford multiplication defines a map $cl: W \otimes \Delta^{\pm} \to \Delta^{\mp}$. We now identify $W = R^m$ to define $SO(m)$ and $SPIN(m)$.

If M is an oriented Riemannian manifold of even dimension m, let P_{SO} be the principal $SO(m)$ bundle of oriented frames for the tangent or cotangent bundle. A spin structure on M is a principal $SPIN(m)$ bundle P_{SPIN} together with a map $\rho: P_{SPIN} \to P_{SO}$ which induces ρ on each fibre. Equivalently, choose coordinate charts U_i for M and local oriented orthonormal frames for P_{SO} over each U_i. P_{SO} is given by the transition functions $f_{ij}: U_i \cap U_j \to SO(m)$. A spin structure is a lifting F_{ij} of f_{ij} to $SPIN(m)$ such that $\rho F_{ij} = f_{ij}$ and such that the cocycle condition is preserved (i.e., $F_{ij}F_{jk} = F_{ik}$).

Not every Riemannian manifold admits a spin structure. THe obstruction to a spin structure is the second Stiefel Whitney class in $H^2(M; Z_2)$. For example, S^n is a spin manifold for any n, but CP_n is only a spin manifold if n is odd. A spin structure is not unique. Different spin structures are related by a representation of the fundamental group of M in Z_2. Of course, locally a spin structure always exists and is unique up to the action of Z_2.

Let M be a spin manifold and let $\Delta^{\pm}M = P_{SPIN} \otimes \Delta^{\pm}$ be the associated vector bundles. Clifford multiplication induces a map $cl:\Lambda^1 M \otimes \Delta^{\pm} M \to \Delta^{\mp} M$. Lift the Levi-Civita connection to a SPIN(m) connection on P_{SPIN} to define a Riemannian connection on $\Delta^{\pm} M$. Let E be a coefficient bundle with a Riamannian connection. The spin complex with coefficients in E is defined by the diagram:

$$d^{\pm}_{SPIN}: C^{\infty}(\Delta^{\pm}\otimes E)(M) \xrightarrow{\nabla} C^{\infty}(\Lambda^1 \otimes \Delta^{\pm} \otimes E(M) \xrightarrow{cl\otimes 1} C^{\infty}(\Delta^{\mp}\otimes E)(M)$$

The adjoint of $d^{\pm}_{SPIN}$ is $d^{\mp}_{SPIN}$. For example, if $E = (\Delta^+ + \Delta^-)(M) = \Delta(M)$, this is the signature complex. If E is the virtual bundle $E = (\Delta^+ - \Delta^-)$, this is the DeRham complex. This permits us to express the DeRham complex as $(\Delta^+ - \Delta^-)\otimes(\Delta^+ - \Delta^-)$ so the spin complex is the sqare root of the DeRham complex. The Z_2 indeterminacy associated with choosing a local spin structure is equivalent to the $\pm$ indeterminacy involved in taking a local square root.

Let $T:M \to M$ be an orientation preserving isometry. dT induces an action of T on P_{SO}. The pull-back T^*S_{SPIN} is a second spin structure on M. We assume that there is given a fixed principal bundle isomorphism $H:T^*P_{SPIN} \to P_{SPIN}$ covering T such that $\rho H = dT$. This induces an action of T on $\Delta^{\pm}(M)$. The set $S = \{U \in SO(m-n):\det(I = U) \neq 0\}$ is contractible. Thus, $\rho^{-1}S$ decomposes into two arc components in $SPIN(m-n)$ We interpret the isomorphism H as a choice of the lifting of dT_{ν} to $SPIN(m-n)$ over any component of the fixed point set; this lifting is unique up to a choice of sign.

Let E be a coefficient bundle with a Riemannian connection and an action $A(T)$ covering T commuting with the connection and fibre metric. The isomorphism H and action $A(T)$ defines an action on $\Delta^{\pm}\otimes E$ which commutes with $d^{\pm}_{SPIN}$, Let $L(T,E)_{sp}$ be the Lefschetz number of this elliptic complex and let $a^{sp}_{p,n}$ be the invariant of the heat equation. If we reverse the orientation of M, the roles of $\Delta^+ M$ and $\Delta^- M$ are reversed so $L(T,E)_{sp}$ and $a^{sp}_{p,n}$ change sign. Consequently, $a^{sp}_{p,n}$ can be regarded as an $O(n) \times SO(m-n)$ invariantly defined n-form on N^n which is of order $2p$ in the derivatives of the metric and connection.

Locally, a spin structure always exists and is unique up to the action of Z_2. We may always lift dT_{ν} to $SPIN(m-n)$ locally to define the isomorphism H. Therefore, $a^{sp}_{p,n}$ is a well defined local invariant regardless of whether or not M admits a global spin structure. We therefore apply the analysis of the third section to show $a^{sp}_{p,n} = 0$ for $2p < n$ and that $a^{sp}_{p,n}$ is a generalized characteristic form of the kind given in

Theorem 3.4 if $2p = n$.

Let $a^{sp}(T,E)[M]$ be the integral formula defined by the sequence $\{a^{sp}_{p,n}\}$. This only has cohomological significance if M is a spin manifold. We cannot apply Theorem 3.5 directly to show $a^{sp}_{p,n}$ is the local formula given by Atiyah and Singer (1968) since not all of the manifolds of Lemma 3.11 were spin manifolds. Rather than rederive this lemma using spin manifolds, we shall pass to the spin^c complex. This will enable us to discuss the Dolbeault complex as well.

Let $SPIN^c(m) = SPIN(m) \times U(1)$ with the identification $(g,z) = (-g,-z)$. Since $\rho(g) = \rho(-g)$, we extend $\rho: SPIN^c(m) \to SO(m)$ with fibre $U(1)$. A spin^c structure on M is a principal $SPIN^c(m)$ bundle P_{SPIN^c} together with a projection $\rho: P_{SPIN^c} \to P_{SO}$ which induces ρ on the fibres. Any spin^c structure may be represented locally as the tensor product of a local spin structure with a $U(1)$ bundle. This $U(1)$ bundle is equivalent to a complex line bundle L with a given fibre metric. Let ∇ be a Riemannian connection on L. We define a connection on P_{SPIN^c} by taking the product of the connection on L with the lift of the Levi-Civita connection to the local spin structure. The resulting connection on P_{SPIN^c} is independent of the locally flat Z_2 indeterminacy. Although L may not be well defined globally. $L \otimes L$ is a well defined global line bundle.

Extend $SPIN^c(m)$ to act naturally on $CL(R^m) \otimes C$ and extend $\Delta^{\pm}$ to a representation of $SPIN^c$. If M is a spin^c manifold, we use the connection on P_{SPIN^c} and these representations to define the spin^c complex. Let $A(T)$ be an action on the coefficient bundle which preserves the connection. Let $H: T^*P_{SPIN} \to P_{SPIN}$ be an isomorphism of spin^c bundles preserving the connections such that $\rho H = dT$. This defines an action of T on the spin^c complex. Let $L(T,E)_{spc}$ be the Lefschetz number of this action and let $a^{spc}_{p,n}$ be the invariant of the heat equation.

The spin^c complex is determined locally by a line bundle L. Even though L need not be globally defined, the indeterminacy in the definition of L is a locally flat Z_2 action. We may therefore view the generalized characteristic forms of L as well defined local formulas. Let $\mathcal{S}^r_{n,m,*}$ be the ring of local formulas generated by the Pontrjagin forms of $T(N)$, the $U(r)$ generalized characteristic forms of E, the $U(1)$ generalized characteristic forms of L, and the $SO(m-n)$ generalized characteristic forms of ν. The spin^c complex with coefficients in E is locally isomorphic to the spin complex with coefficients in $E\otimes L$. We may express the generalized $U(r)$ characteristic forms of $E\otimes L$ in terms of the generalized $U(r)$ characteristic forms of E and the $U(1)$ generalized characteristic forms of L. Since $a^{spc}_{(p,n)}(T,E) = a^{sp}_{p,n}(T,E\otimes L)$ and since $a^{sp}_{p,n} \in \mathcal{R}^{r}_{n,m,n}$, this proves

Lemma 4.1. $a^{spc}_{p,n}(T,E) = a^{sp}_{p,n}(T,E\otimes L) \in \mathcal{S}^{r}_{n,m,n}$ is a generalized characteristic form.

If M is a spin manifold, we may define a spin^c structure on M corresponding to the trivial line bundle L. With this spin^c structure, $a^{spc}_{p,n} = a^{sp}_{p,n}$. To complete the proof of Theorem 0.4(2a and 2b) and to show $a^{spc}_{p,n}$ is the integrand of the Atiyah-Singer theorem (1968), it suffices to prove

Lemma 4.2. Let $S = \{S^{r}_{p,n}\} \in \mathcal{S}^{r}_{n,m,n}$ be a sequence such that $S(T,E,L)[M] = 0$ for all (T,E,L,M) such that M is a spin^c manifold with structure bundle L. Then the local invariants $S^{r}_{p,n} \equiv 0$.

We shall prove Lemma 4.2 later in this section. If M is a spin manifold, it admits a trivial spin^c structure, but there are manifolds which are not spin manifolds which admit spin^c structures. Any holomorphic manifold admits a natural spin^c structure which we describe as follows. Let W be a real vector space of even dimension m and let J be an almost complex structure on W. $J:W \to W$ is a real linear map such that $J^2 = -1$. Let $\Lambda_c W$ be the complexified exterior algebra. Decompose $\Lambda_c W = \oplus \Lambda^{p,q} W$ into the forms of bidegree (p,q).

Let ext denote exterior multiplication and let int denote the dual map, interior multiplication. Extend ext and int to be complex linear maps on $(W\otimes C)\otimes\Lambda_c W \to \Lambda_c W$. Let

$$\Lambda^{0,+}W = \oplus_q \Lambda^{0,2q}W, \qquad \Lambda^{0,-}W = \oplus_q \Lambda^{0,2q+1}W$$

$$\Lambda^{0,*}W = \oplus_q \Lambda^{0,q}W = \Lambda^{0,+}W\oplus\Lambda^{0,-}W$$

Let $c(w) = 1/\sqrt{2}\{\text{ext}(w-iJw) - \text{int}(w+iJw)\}:W\otimes\Lambda^{0,*}W \to \Lambda^{0,*}W$. Since $c(w)c(w) = -|w|^2$, c extends to a linear algebra morphism of the real Clifford algebra $CL(W)$ into the complex endomorphisms of the vector space $\Lambda^{0,*}W$. Complexify c and extend c to the complex algebra $CL(W)\otimes C$. $\text{SPIN}^c(W)$ is naturally embedded in $CL(W)\otimes C$. The restriction of c to $\text{SPIN}^c(W)$ defines a representation Δ_c on $\Lambda^{0,*}W$ which is isomorphic to the complexification of the representation Δ. Δ_c preserves the decomposition of $\Lambda^{o,*}W = \Lambda^{o,+}W \oplus \Lambda^{o,-}W$. The restriction of Δ_c to $\Lambda^{o,\pm}W$ defines representations $\Delta_c^{\pm}$ which are isomorphic to the complexification of the representations $\Delta^{\pm}$.

Let $U(W)$ be the subgroup of $SO(W)$ which commutes with the action of J. Let W_c be the complex $m/2$ dimensional vector space obtained by letting multiplication by i correspond to the action of J. If W_c^* is the dual vector space, $W \otimes C = (W_c \oplus W_c^*)$ and $\Lambda^{p,q}(W) = \Lambda^p(W_c) \otimes \Lambda^q(W_c^*)$. $U(W)$ forms a group of complex linear maps of W_c. Let $\det: U(W) \to U(1)$ be the determinant of this complex linear map, let $inc: U(W) \to SO(W)$ be the inclusion map, and let $\gamma: U(1) \to U(1)$ be the usual double cover of the circle. We define the lift $\alpha: U(W) \to SPIN^c(W)$ by the diagram

$$\begin{array}{ccc} & \nearrow^{\alpha} & SPIN^c(W) = SPIN(W) \times U(1)/\{(q,u) = (-q,-u)\} \\ & & \downarrow \rho \times \gamma \\ U(W) & \xrightarrow{\ inc \times \det\ } & SO(W) \times U(1) \end{array}$$

This uniquely defines α as a group homomorphism.

We can describe α explicitly. Let $u \in U(W)$ act on W_c. Let $\{\exp(i\theta_j), v_j\}$ be a complete orthnormal decomposition of W_c into eigenvalues and eigenvectors of u. Let

$$\alpha(u) = \Pi_j \exp(\tfrac{1}{2} i\theta_j)(\cos(\tfrac{1}{2}\theta_j) + \sin(\tfrac{1}{2}\theta_j) v_j J(v_j)) \in CL(W) \otimes C$$

It is easily verified that $\alpha(u) \in SPIN^c(W)$ is the desired lifting and that $c\alpha(u)$ is the usual representation of $U(W)$ on $\Lambda^{o,*}$.

Let M be an almost complex manifold. J is a bundle map of $TM \to TM$ with $J^2 = -1$ which gives the almost complex structure. Let T_cM be the real tangent bundle of M with the complex structure given by J. Choose the metric and orientation so J is an orientation preserving isometry. We reduce the structure group of TM to $U(\frac{m}{2})$. The map α lifts the structure group to $SPIN^c$ and defines a natural $spin^c$ structure on M. The local bundle which gives the $spin^c$ structure is a local square root of the canonical bundle $K(M) = \Lambda^{m,o}(M)$. M admits a global spin structure if and only if $c_1K(M) \in H^2(M,Z)$ is even.

Let $T: M \to M$ be an isometry. We assume $JdT = dTJ$. T induces a map dT on T_cM which is complex linear. T also induces a natural action H on the $spin^c$ structure. Let ∇ be the holomorphic connection on $K(M)$. We use this connection to define a natural connection on the principal

spin^c complex. Let E be a coefficient bundle with a Riemannian connection on which T acts. If N is a component of the fixed point set of T, TN and ν inherit natural almost complex structures. Let $\mathcal{T}^r_{n,m,*}$ be the ring of local formulas in the Chern forms of TN, the generalized U(r) characteristic forms of E, and the generalized $U(\frac{1}{2}(m-n))$ characteristic forms of ν. Since the genralized characteristic forms of K(M) can be computed in terms of the Chern forms of TN and the $U(\frac{1}{2}(m-n))$ characteristic forms of ν, $a^{spc}_{p,n} \in \mathcal{T}^r_{n,m,n}$ for the natural spin structure.

There is a natural map from $\mathcal{S}^r_{n,m,n}$ to $\mathcal{T}^r_{n,m,n}$ defined by restricting to the natural spin^c structure of an almost complex manifold. Since this map is not injective, we cannot prove Lemma 4.2 by proving a uniqueness theorem for the integral formulas defined by elements of $\mathcal{T}^r_{n,m,n}$. However, the natural spin^c structure is not the only spin^c structure an almost complex manifold admits. Let L_0 be a complex line bundle on which T acts. Let $\text{spin}^c L_0$ be the spin^c structure defined by the tensor product of the natural spin^c structure and the U(1) bundle L_0. Let $\mathcal{U}^r_{n,m,*}$ be the ring of local formulas in the Chern forms of TN, the generalized U(r) characteristic forms of E, the generalized U(1) characteristic forms of L_0, and the generalized $U(\frac{1}{2}(m-n))$ characteristic forms of ν. $\mathcal{T}^r_{n,m,*}$ is a subspace of $\mathcal{U}^r_{n,m,*}$. There is a natural injective map from $\mathcal{S}^r_{n,m,*}$ to $\mathcal{U}^r_{n,m,*}$ obtained by restricting to the $\text{spin}^c(L_0)$ structure of an almost complex structure. To complete the proof of Lemma 4.2, it suffices to show

THEOREM 4.3. Let $U = \{U^r_{p,n}\} \in \mathcal{U}^r_{n,m,n}$ be a sequence such that $U(T,E,L_0)[M] = 0$ for all almost complex manifolds M and all (T,E,L_0). Then the local invariants $U^r_{p,n} \equiv 0$.

Proof. By taking appropriate product manifolds and by using the same techniques as were used in the proof of Theorem 3.5, we can reduce the proof of Theorem 4.3 to the following analogue of Lemma 3.11.

Lemma 4.4.

a) Let $P \not\equiv 0$ be a Chern n form of TN. We can find an almost complex manifold N^n so that $P[N] \neq 0$.

b) Let $P \not\equiv 0$ be a generalized U(r) n form of E. We can find E over an almost complex manifold N^n and an action with $\nabla_E A \equiv 0$ such that $P(E)[N] \neq 0$.

c) Let $P \not\equiv 0$ be a generalized $U(1)$ form of L_o. We can find L_o over an almost complex manifold N^n and an action with $\nabla_E A \equiv 0$ such that $P(L_o)[N] \neq 0$.

d) Let $P \not\equiv 0$ be a generalized $U(\frac{1}{2}(m-n))$ form of ν. We can find an almost complex M^m and an isometry T of M^m which preserves J such that the fixed point set of T consists of a single submanifold N^n of dimension n together with other components of lower dimension and such that $P(T)[N] \neq 0$.

Proof. (a) is well known. N^n can be taken to be the product of complex projective spaces. In Lemma 3.11(b), the vector bundles were all holomorphic vector bundles over products of complex projective spaces. Therefore, the examples constructed in the proof of Lemma 3.11(b) give the necessary examples which are required to prove Lemma 4.4(b and c). Finally, the maximal torus of $U(\frac{1}{2}(m-n))$ is the same as the maximal torus of $SO(m-n)$. The manifolds and isometries which were constructed in the proof of Lemma 3.11(c) were all algebraic varieties and algebraic maps. Therefore, these examples give the necessary examples which are required to prove Lemma 4.4(d).

This completes the proof of Lemma 4.2 and completes the proof of Theorem 0.4(2a and 2b) for the spin complex. It is worth noting that $\mathcal{T}^r_{n,m,n}$ is a subspace of $\mathcal{U}_{n,m,n}$, so Theorem 4.3 also yields a uniqueness result for the formulas of $\mathcal{T}^r_{n,m,n}$.

In the remainder of this section, we will discuss the Dolbeault complex by using the spin^c complex. Let $d: C^\infty(\Lambda M) \to C^\infty(\Lambda M)$ be exterior differentiation. Extend d to be complex linear on $C^\infty \Lambda_c M$. Let $\bar\partial$ be the part of d mapping $C^\infty(\Lambda^{p,q}M) \to C^\infty(\Lambda^{p,q+1}M)$. The Nirenberg-Neulander integrability condition states that M is a holomorphic manifold if and only if $\bar\partial^2 = 0$. Therefore, the Dolbeault complex $\bar\partial: C^\infty(\Lambda^{0,q}M) \to C^\infty(\Lambda^{0,q+1}M)$ is elliptic if and only if M is holomorphic.

If M is only almost complex, we can still define an elliptic complex by rolling up the Dolbeault complex. Let δ'' be the formal adjoint of the operator $\bar\partial$ and let

$$\bar\partial + \delta'': C^\infty(\Lambda^{0,+}M) \to C^\infty(\Lambda^{0,-}M)$$

be a two term elliptic complex. If E is a coefficient bundle, we can define the corresponding complex with coefficients in E. T acts on this elliptic complex. Let $L(T,E)_{DOL}$ be the Lefschetz number of this complex and let $a^{DOL}_{p,n}(T,E)$ be the invariant of the heat equation. It is known (see Gilkey, Ph.D. dissertation, 1972) that even if $T = I$, $a^{DOL}_{p,n}$ is not in general the integrand of the Riemann-Roch formula.

THEOREM 4.5. There exists a unique local formula $a_{DOL} = \{a^r_{n,m}\} \in \mathcal{T}^r_{n,m,n}$ such that $a_{DOL}(T,E)[M] = L(T,E)_{DOL}$.

In evaluating the formula of Theorem 4.5, we choose a Riemannian connection ∇ for T_cM which commutes with the natural action of T. By Lemma 3.2, the integral of the generalized characteristic forms is independent of the particular ∇ chosen. Let J^i_j be the components of the almost complex structure. We say that ∇ is natural if the connection form of ∇ can be expressed functorially in any coordinate system as a linear combination of the $\{g_{ij/k}, J^i_{j/k}\}$ variables with coefficient functions which are smooth in the $\{g_{ij}, J^i_j\}$ variables. The holomorphic connections and the projection of the Levi-Civita connection to T_cM are both examples of natural connections. Since T preserves both the metric and J, T commutes with any natural connection. If the almost complex structure is integrable (so M is holomorphic) and if the metric is Kähler, there is only one natural connection, but in general there will be many.

We prove Theorem 4.5 using the spin^c complex. Let a_{DOL} be the sequence $a^{spc}_{p,n}$ relative to the natural spin^c structure. We showed earlier that $\Delta^{\pm}_c M = \Lambda^{0,\pm}M$ and that the two operators $d^{\pm}_{\text{spin}^c}$ and $(\bar{\partial} + \delta'')$ have the same leading order symbol. Both commute with T. In general, these two operators will not agree, but they are equal for a Kähler metric. We can construct a homotopy $d_s = s(d^{\pm}_{\text{spin}^c}) + (1 - s)(\bar{\partial} + \delta'')$ to connect these two operators. d_s yields an elliptic complex which commutes with T for $s \in [0,1]$. We will show that this implies $L(T,E,d_s)$ is constant and therefore $L(T,E)_{DOL} = L(T,E)_{spc}$. This proves the first part of Theorem 4.5. The uniqueness assertion follows from Theorem 4.4. This completes the proof of Theorem 0.4.

We have now reduced the proof of Theorem 4.5 to the following assertion which is the analogue of Theorem 3.2.

THEOREM 4.6. Let (V,d_s) be a family of elliptic complexes over M for $s \in [0,1]$. Let T be an isometry of M and let $A(T)$ be a fixed action of T on V which preserves the fibre metric of V. Let T commute with d_s for $s \in [0,1]$. Then $L(T,V,d_s)$ is constant.

Proof. Let d_s^* be the adjoint and let $D_{s,i} = (d_s^* d_s + d_s d_s^*)_i$ be the associated Laplacians. T is an isometry and A(T) acts on V by isometries. This implies A(T) commutes with d_s^* and $D_{s,i}$ as well. Let $E(s,i,\lambda)$ be the finite dimensional eigenspaces of $D_{s,i}$. $A(T)_i$ preserves $E(s,i,\lambda)$. By decomposing each $E(s,i,\lambda)$ into a direct sum of eigenspaces of $A(T)_i$, we obtain a direct sum decomposition of $L^2(V_i) = \oplus_\theta A(i,\theta)$, where the $A(i,\theta)$ are the eigenspaces of A(T) which correspond to the eigenvalue $\exp(i\theta)$. $D_{s,i}$ preserves $A(i,\theta)$; let $D_{s,i,\theta}$ be the restriction of $D_{s,i}$ to $A(i,\theta)$.

Since d_s commutes with A(T), d_s induces map $d_{s,i,\theta}: A(i,\theta) \to A(i+1,\theta)$. We compute

$$L(T,V,d_s) = \sum_i (-1)^i \mathrm{Tr}(A(T)_i \exp(-tD_{s,i}))$$

$$= \sum_{i,\theta} \exp(i\theta)(-1)^i \mathrm{Tr}\ \exp(-tD_{s,i,\theta}) = \sum_\theta \exp(i\theta)\,\mathrm{index}(d_{s,\theta})$$

$\mathrm{Index}(d_{s,\theta})$ is a continuous integer valued function of s which does not vanish for only finitely many θ. This implies $\mathrm{index}(d_{s,\theta})$ is constant, which proves $L(T,V,d_s)$ is independent of the parameter s.

REFERENCES

Atiyah, M. F. and Bott, R. A Lefschetz fixed point formula for elliptic complexes, I. Ann. of Math., 86 (1967), pp. 374-407.

Atiyah, M. F., Bott, R. and Patodi, V. K. On the heat equation and the index theorem. Inv. Math., 19, Fasc. 4 (1973), pp. 279-330.

Atiyah, M. F., Bott, R. and Shapiro, A. Clifford Modules. Topology, 3, Supp. 1 (1964), pp. 3-38.

Atiyah, M. F., Patodi, V. K., and Singer, I. M. Spectral Asymmetry and Riemannian Geometry. Bull. London Math. Soc., 5 (1973) pp. 220-234.

Atiyah, M. F., and Singer, I. M. The index of elliptic operators, III. Ann. of Math., 87 (1968) pp. 546-604.

Berger, M. Sur les spectres d'une variété Riemanniene. C. R. Acad. Sci, Paris, 263 (1963), pp. 13-16.

Chern, S. A simple intrinsic proof of the Gauss-Bonnet formula for closed Riemannian manifolds. Ann. of Math., 45 (1944), pp. 747-752.

Donnelly, H. Spectrum and the fixed point set of isometries, I. Ann. of Math., 244 (1976), pp. 161-170.

Donnelly, H. Spectrum and the fixed point set of isometries, III. (Preprint.

Donnelly, H. (to appear)

Donnelly, H. and Patodi, V. K. Spectrum and the fixed point set of isometries, II. Topology, 161 (1977), pp. 1-11.

Gilkey, P. Curvature and the eigenvalues of the Laplacian for geometrical elliptic complexes. Ph.D. dissertation, Harvard (1972).

Gilkey, P. Curvature and the eigenvalues of the Laplacian for elliptic complexes. Adv. in Math., 10 (1973), pp. 344-382.

Gilkey, P. Curvature and the eigenvalues of the Dolbeault complex for Kähler manifolds. Adv. in Math., 11 (1973), pp. 311-325.

Gilkey, P. The boundary integrand in the formula for the signature and Euler characteristic of a Riemannian manifold with boundary. Adv. in Math., 15 (1975), pp. 344-360.

Gilkey, P. Curvature and the eigenvalues of the Dolbeault complex for Hermitian manifolds. Adv. in Math., 21 (1976)

Gilkey, P. Spectral geometry of a Riemannian manifold. J. Diff. Geo., 10 (1975), pp. 601-618.

Gilkey, P. Recursion relations and the asymptotic behavior of the eigenvalues of the Laplacian (to appear).

Greiner, P. An asymptotic expansion for the heat equation. Arch. Rat. Mech. Anal., 41 (1971), pp. 163-218.

Hitchin, N. Harmonic Spinors. Adv. in Math., 14 (1974), pp. 1-55.

Kac, N. Can one hear the shape of a drum? Am. Math. Mo., 73 (1965), pp. 1-33.

Kawasaki, T. An analytic Lefschetz fixed point formula and its application to V manifolds, general defect formula, and Hirzebruch signature theorem. Ph.D. dissertation, Johns Hopkins University (1976).

Kotake, T. The fixed point theorem of Atiyah-Bott via parabolic operators. Comm. Pure and Appl. Math., 22 (1969), pp. 789-806.

Lee, S. C. A Lefschetz formula for higher dimensional fixed point sets. Ph.D. dissertation, Brandeis University (1975).

McKean, H. P. and Singer, I. M. Curvature and the eigenvalues of the Laplacian. J. Diff. Geo., 1 (1967), pp. 43-69.

Patodi, V. K. Curvature and the eigenforms of the Laplace operator. J. Diff. Geo., 5 (1971), pp. 233-249.

Patodi, V. K. An analytic proof of the Riemann-Roch-Hirzebruch theorem for Kähler manifolds. J. Diff. Geo., 5 (1971), pp. 251-283.

Patodi, V. K. Holomorphic Lefschetz fixed point formula. Bull. A. M. S. vol. 81, 6 (1971), pp. 1133-1134.

Patodi, V. K. Curvature and the findamental solution of the heat equation. J. Indian Math. Soc., 34 (1970), pp. 269-285.

Sakai, T. On the eigenvalues of the Laplacian and curvature of Riemannian manifolds. Tohuku Math. J., 23 (1971), pp. 585-603.

Seeley, R. T. Complex power of an elliptic operator. Am. Math. Soc. Proc. Symp. Pure Math., 10 (1967), pp. 288-307.

Seeley, R. T. Analytic extension of trace assocaited to elliptic boundary problems. Am. J. Math., 91 (1969), pp. 963-983.

Toledo, D. and Tong, Y. L. The holomorphic Lefschetz formula. Bull. A. M. S., 81 (1975), pp. 1133-1136.

Weyl, H. The Classical Groups. Princeton University Press, Princeton, New Jersey (1946).

Princeton University
Princeton, New Jersey

ON THE SPECTRA OF COMMUTING PSEUDODIFFERENTIAL OPERATORS: RECENT WORK OF KAC-SPENCER, WEINSTEIN AND OTHERS

V. Guillemin and S. Sternberg

Introduction

This report is expository in intent. Its purpose is to describe recent work by various authors on the spectra of commuting pseudodifferential operators. We begin with some remarkable results of Kac-Spencer (to appear) and Weinstein (to appear) on the spectrum of $\Delta + q$ where Δ is the Laplace operator on S^n and q a smooth potential. To describe these results we need a few preliminary definitions. Let Z be the space of all oriented closed geodesics on S^n. This is canonically a $(2n - 2)$-dimensional symplectic manifold and the potential $q \in C^\infty(S^n)$ determines a smooth function $\hat{q}$ on Z by the rule which to each oriented closed geodesic $\gamma \in Z$ assigns the integral

$$\hat{q}(\gamma) = \frac{1}{2\pi} \int_0^{2} q(\gamma(s))ds$$

(s being arc-length). We will assume that Δ is the normalized Laplacian on S^n; so that its eigenvalues are the points

$$\left(k + \frac{n-1}{2}\right)^2, \qquad k = 0,1,2,... \tag{1.1}$$

each eigenvalue occurring with multiplicity

$$N_k = (2/(n-1)!)k^{n-1} + O(k^{n-2}) \tag{1.2}$$

The eigenvalues of the operator $\Delta + q$ form clusters:

$$\lambda_i^{(k)}, \qquad i = 1,...,N_k \tag{1.3}$$

where

$$\left|\lambda_i^{(k)} - \quad k + \frac{n-1}{2} \quad ^2\right|$$

is bounded by a constant independent of k. Let $u^{(k)}$ be the measure on the real line

$$u^{(k)} = 1/N_k \sum \delta\left(\lambda - \lambda_i^{(k)} - \left(k + \frac{n-1}{2}\right)^2\right)$$

which describes how the perturbed eigenvalues are distributed in the kth cluster. (We have normalized this measure so that the total measure is one.) The main result of Kac-Spencer and Weinstein is

THEOREM. Let vol be the measure on Z associated with the intrinsic symplectic structure, normalized so that the total measure of Z is one. Let $u = (\hat{q})_* d\mathrm{vol}$. Then

$$u^{(k)} = u + O(1/k) \tag{1.4}$$

This result is related to another result, of a general nature, concerning the spectrums of two commuting pseudodifferential operators: Let X be an arbitrary compact manifold, $P: C^\infty(X) \to C^\infty(X)$ a positive selfadjoint elliptic pseudodifferential operator (which for simplicity we will take to be of order one) and Q a bounded selfadjoint pseudodifferential operator which commutes with P. Let V_λ be the subspace of $C^\infty(X)$ spanned by the eigenfunctions of P with eigenvalue $< \lambda$. Let $Q_\lambda = Q$ restricted to V_λ and let u_λ be the spectral measure of Q_λ; i.e., for every $f \in C(\mathbb{R})$ let

$$u_\lambda(f) = \mathrm{tr}\, f(Q_\lambda)$$

Let u be the measure on the real line defined by

$$u(f) = (2\pi)^{-n} \int_{\sigma(P) \leq 1} f(\sigma(Q)(x,\xi))\, dx\, d\xi$$

THEOREM 1. For all functions $f \in C^\infty(R)$

$$u_\lambda(f) = \lambda^n u(f) + O(\lambda^{n-1}) \tag{1.5}$$

the error term being uniform on C^2-bounded subsets of C^∞.

The proof of this theorem will be given below. However, let us first see how the result of Kac-Spencer-Weinstein is related to this result. With the same notation as above, let $X = S^n$, $P = \sqrt{\Delta + q}$, and $Q = (-1)^{n-1/2} P \cdot \sin 2\pi P$. In Guillemin (to appear) we made a systematic study of the operator

Q which we called the "return operator." In particular, we showed that it is a zeroth order pseudodifferential operator whose leading symbol is real and equal to the function $\hat{q}$ defined at the beginning of this section. P and Q commute and are selfadjoint; therefore, we are in a position to apply (1.5) and read off information about the asymptotic behavior of the eigenvalues of Q. First, let us note that if

$$u + \left(k + \frac{n-1}{2}\right)^2$$

is an eigenvalue of $\Delta + q$ lying in the cluster centered at $\left(k - \frac{n-1}{2}\right)^2$ and ϕ is its corresponding eigenfunction, then ϕ is also an eigenfunction of Q and the eigenvalue associated with it is of the form

$$\pi u + O(1/k^3) \tag{1.6}$$

Let V_k be the subspace of $C^\infty(S^n)$ spanned by the eigenfunctions of P whose eigenvalues lie in the kth cluster, and let

$$u_k(\lambda) = \sum_u \delta(\lambda - u)$$

the sum running over the eigenvalues, u, of $\Delta + q - \left(k + \frac{n-1}{2}\right)^2$ restricted to V_k. Similarly, let

$$u_k^Q = \sum_u \delta(\lambda - u)$$

the sum running over the eigenvalues of $(\pi)^{-1}Q$ restricted to V_k. Then for $f \in C^1(\mathbb{R})$

$$u_k(f) = u_k^Q(f) + O(k^{n-4}) \tag{1.7}$$

by (1.6) and (1.2). On the other hand, by (1.5)

$$\sum_{r \leqslant k} u_r^Q(f) = k^n u(f) + O(k^{n-1}) \tag{1.8}$$

Together (1.7) and (1.8) give

$$\sum_{r \leqslant k} u_r(f) = k^n u(f) + O(k^{n-1}) \tag{1.9}$$

This is the "integrated" form of the Kac-Spencer-Weinstein result. We do not, at the moment, see how to obtain the strong (nonintegrated) form of their result from general results of the nature of (1.5). Indeed, for

$Q = I$, (1.5) is just the spectral theorem of Hormander (1968) and, as is well known, this theorem is not true in its nonintegrated form.

From (1.9) we can already obtain one of the most striking corollaries of the Kac-Spencer-Weinstein theory.

THEOREM. The measure u is a spectral invariant of $\Delta + q$.

As we will see below, it is the first of a countable sequence of such "nonlocal" spectral invariants.

There are a number of interesting generalizations and variants of formula (1.5). We will content ourselves with discussing just one. Let $P_1, \ldots, P_k$ be k commuting selfadjoint pseudodifferential operators of positive order. We will say that $(P_1, \ldots, P_k)$ is <u>elliptic</u> if the set of $(x, \xi) \in T^*X \backslash 0$ satisfying

$$p_1(x, \xi) = \ldots = p_k(x, \xi) = 0 \tag{1.10}$$

is empty. (Here p_i is the symbol of P_i.) We will call $\phi \in C^\infty(X)$ an eigenfunction of $(P_1, \ldots, P_k)$ if

$$P_i\phi = \lambda_i\phi, \qquad i = 1, \ldots, k$$

and we will call $\lambda = (\lambda_1, \ldots, \lambda_k)$ the associated eigenvalue. Let m_i be the order of P_i, let s be the least common multiple of the m_i's and let $a_i = s/m_i$. Then, for every smooth, bounded, strictly convex region, V, in $\mathbb{R}^k$ which is invariant under the reflections in the coordinate planes, one has

THEOREM 2. The number of eigenvalues, $N(\lambda)$, lying in the set

$$\left\{\lambda, \left(\lambda_1^{a_1}, \ldots, \lambda_k^{a_k}\right) \in \lambda^s V\right\}$$

and the symplectic volume, $V(\lambda)$, of the set

$$\{(x, \xi), \left(p_1^{a_1}, \ldots, p_k^{a_k}\right)(x, \xi) \in \lambda^s V\} \tag{1.11}$$

are related by

$$N(\lambda) = (2\pi)^{-n} N(\lambda) + 0(\lambda^{n-1}) \tag{1.12}$$

In the language of physics: The probability that a classical particle finds itself in the region defined by the equations (1.11) is, for large

λ, approximately equal to the probability that the corresponding quantum particle finds itself in this region.

The situation we have just described, of k commuting pseudodifferential operators, has been encountered by Kolk and Varadarajan (1977) in their study of the Selberg trace formula for symmetric spaces of rank greater than one. Let $G = KAN$ be a real semisimple Lie group of rank k and Γ a discrete cocompact subgroup of G acting in a fixed point free way on the generalized Siegel halfplane $K\backslash G$. The K-invariant differential operators on $K\backslash G$ form a commuting algebra with k generators $P_1,\ldots,P_k$ the lowest generator being the Laplacian associated with the K-invariant metric on $K\backslash G$. These commuting operators also live on the quotient space $K\backslash G/\Gamma$, and satisfy the ellipticity condition (1.10), since Δ by itself is elliptic. Hence, (1.12) is applicable. We will not bother to describe the explicit form which (1.12) assumes in this case. In their forthcoming paper, Kolk and Varadarajan derive an analogue of (1.12) from their generalized Selberg trace formula. Their result is slightly more agreeable than ours, both as regards the hypotheses on V and the conclusion about the distribution of eigenvalues.

THE PROOF OF THEOREM 1

With small modifications, this proof is identical with the proof of the spectral theorem for selfadjoint elliptic operators obtained by Hormander in 1968. We will need the following lemma, whose proof will be given in an appendix.

Lemma 1. Let Q be a selfadjoint zeroth order pseudodifferential operator. Then, for all $f \in C^\infty(R)$, the operator $f(Q)$ is also a pseudodifferential operator of order zero and its symbol is $f(\sigma(Q))$.

Consider the operator $\exp\sqrt{-1}\, tP$. By Duistermaat-Hormander (1972), this is an elliptic Fourier integral operator of order zero associated with the canonical transformation $\exp tH_p$, p being the symbol of P and H_p the Hamiltonian system having p as its generating function. It follows (see Hormander, 1971) that the Schwartz kernel of $\exp\sqrt{-1}\, tP$ can, for t small, be represented as an oscillatory integral

$$\int a(x,t,\xi)e^{i(p(x,\xi)t + \phi(x,y,\xi))}d\xi$$

where a is a classical symbol of order zero and ϕ a nondegenerate phase function, homogeneous of degree one in ξ, and of the form

$$\phi(x,y,\xi) = (x - y)\xi + O(|\xi||x - y|^2) \tag{2.1}$$

It is also not hard to see that the leading term in the asymptotic expansion of a is just the constant function 1. We refer to Duistermaat and Hormander (1972), Guillemin (to appear), and Hormander (1968) for details.

Next let us consider the operator

$$f(Q)\exp\sqrt{-1}\ tP, \quad f \in C^\infty(R) \tag{2.2}$$

By Lemma 1 and the usual composition formulas for pseudodifferential and Fourier integral operators, (2.2) is a Fourier integral operator, and its leading symbol is $f(q)$ times the leading symbol of $\exp\sqrt{-1}\ tP$. In particular, for t small, the Schwartz kernel of (2.2) can be expressed as an oscillatory integral

$$\int b(x,\xi,t)e^{i(p(x,\xi)t + \phi(x,y,\xi))}d\xi \tag{2.3}$$

$b(x,\xi,t)$ being a classical symbol of order zero whose leading term is $f(q(x,\xi))$. Following Hormander (1968), we compute the trace of (2.2) by setting $x = y$ in (2.3) and integrating over X. Taking note of (2.1), we get for the trace of (2.2)

$$\int b(x,\xi,t)e^{ip(x,\xi)t}dxd\xi \tag{2.4}$$

when t is small. We will now simplify (2.4) following the line of argument of Section 4 of Hormander (1968). First, replacing $b(x,\xi,t)$ by its Taylor series with remainder

$$\sum_{i=0}^{N} (\partial^i b/\partial t^i)(x,\xi,0)(t^i/i!) + b_N(x,\xi,t)t^N, \quad N >> 0$$

in (2.4) and integrating by parts with respect to the ξ variable, we can arrange that b is independent of t. We next replace b by its asymptotic series

$$\sum_{m=0}^{\infty} b_m(x,\xi) \tag{2.5}$$

with $b_o = f(q(x,\xi))$ and b_m homogeneous of degree $-m$ in ξ. This gives us an asymptotic expansion for (2.4)

$$\sum\int b_m(x,\xi)e^{itp(x,\xi)}dxd\xi \tag{2.6}$$

We now introduce polar coordinates (s,ω) in ξ space with $s = p$ and rewrite (2.6) as

$$\sum c_m \int_0^\infty e^{its} s^{n-m-1} ds \tag{2.7}$$

where $c_m = \int b_m(x,\omega)dxd\omega$. Notice that the leading term in (2.17) is

$$nu(f) \int_0^\infty e^{ist} s^{n-1} ds \tag{2.8}$$

where $u(f)$ is as in Section 1, namely

$$u(f) = (1/n) \int f(q(x,\omega))d\omega dx = \int_{p\leq 1} f(q(x,\xi))d\xi dx$$

Now let V_λ be the λth eigenspace of P and let $\nu_\lambda(f)$ be the trace of $f(Q)|V_\lambda$. Then, formally

$$\mathrm{tr}\, f(Q) \exp \sqrt{-1}\, tP = \sum_{\lambda\in \mathrm{spec}\, P} \nu_\lambda(f)e^{\sqrt{-1}\, t\lambda} \tag{2.9}$$

Let us now multiply (2.9) by a smooth function, $\rho(t)$, whose support is contained in a neighborhood of the origin for which (2.3) is valid, and take the Fourier transform of the resulting expression. From (2.9) and (2.7) we get

$$\sum \nu_\lambda(f)\hat{\rho}(s - \lambda) \sim \sum_m c_m\, \hat{\rho}* s_+^{n-m-1} \tag{2.10}$$

To simplify this expression we note

Lemma 2. If $\rho(0) = 1$, then $\hat{\rho}* s_+^k = s^k + 0(s^{k-1})$ for $s > 0$.

Proof. $\hat{\rho}* s_+^{n-1} = \int_{-\infty}^{s} (s - z)^{n-1}\hat{\rho}(z)ds$

$$= s^{n-1} \int_{-\infty}^{\infty} \hat{\rho}(z)dz + (1 - n)s^{n-2} \int_{-\infty}^{s} z\ (z)dx + \ldots \tag{2.11}$$

Since $\hat{\rho}(z)$ is rapidly decreasing, each of the integrals

$$\int_{-\infty}^{s} z^k \hat{\rho}(z)dz$$

can be written as

$$\int_{-\infty}^{\infty} a^k \hat{\rho}(z)dz$$

with an error term of order $O(s^{-N})$, $N >> 0$. Since

$$\int_{-\infty}^{\infty} \hat{\rho}(z)dz$$

is $\rho(0)$, this proves the lemma. *QED*

Choosing ρ so that $\rho(0) = 1$, (2.11) becomes

$$\sum \nu_\lambda(f)\hat{\rho}(s - \lambda) = nu(f)s^{n-1} + O(s^{n-2}) \tag{2.12}$$

for s large. If we now set

$$g(s) = u_s(f) = \sum_{\lambda \leqslant s} \nu_\lambda(f)$$

the sum taken over $\lambda \in \text{spec } P$, then the left hand side of (2.12) becomes

$$\hat{\rho} * (dg/ds) \quad \text{or} \quad (d/ds)(\hat{\rho} * g)$$

Integrating (2.12) from 0 to s we get

$$(\hat{\rho} * g)(s) = u(f)s^n + O(s^{n-1}) \tag{2.13}$$

To conclude the proof we will prove

Lemma 3. The following estimate holds.

$$|g(s) = (\hat{\rho} * g)(s)| = O(s^{n-1})$$

Proof.
$$(g - \hat{\rho} * g)(s) = g(s) - \int g(s - z)\hat{\rho}(z)dz$$
$$= \int (g(s) - g(s - z))\hat{\rho}(z)dz$$

In Section 4, Hormander (1968) shows that the number of eigenvalues of P lying on the interval between s and $s - z$ is bounded above by a constant multiple of $(|s| + |z|)^{n-1}|z|$; so, from the definition of g, we get the crude estimate

$$|g(s) - g(s - z)| \leqslant C(|s| + |z|)^{n-1}|z|$$

Hence

$$|(g - \hat{\rho}*g)(s)| \leq C \int (|s| + |z|)^{n-1}|z|\hat{\rho}(z)dz$$

Since $\hat{\rho}(z)$ is rapidly decreasing, the right hand side can be estimated by $O(s^{n-1})$. *QED*

Remark. The c_m's occurring in (2.7) are all spectral invariants of (P,Q,f). It is clear from (2.9) that $c_m(P,Q,f)$ depends linearly on f; so we can think of $c_m(P,Q)$ as a <u>distribution on the real line</u>. We will prove

THEOREM 3. The support of $c_m(P,Q)$ lies on the interval $a \leq s \leq b$, where $a = \min \sigma(Q)(x,\xi)$ and $b = \max \sigma(Q)(x,\xi)$

Proof. The operator $Q - aI$ has a nonnegative symbol, so we can write

$$Q - aI = Q' + K$$

where Q' is a nonnegative selfadjoint zeroth order pseudodifferential operator and K a pseudodifferential operator of order -1. K being compact (as an operator from L^2 to L^2) it follows that, for all $\varepsilon > 0$, only a finite part of the spectrum of Q lies to the left of $a - \varepsilon$, for all $\varepsilon > 0$. Similarly, only a finite part of the spectrum lies to the right of $b + \varepsilon$. Hence, if $f_1 = f_2$ outside the interval $-\varepsilon + a < s < b + \varepsilon$, $f_2(Q) - f_1(Q)$ is a finite rank smoothing operator. In particular, $\mathrm{tr}(f_2(Q) - f_1(Q))\exp\sqrt{-1}\, tP$ is smooth as a function of t. It follows from (2.9) and (2.7) that $c_m(P,Q,f_1) = c_m(P,Q,f_2)$. *QED*

Using results about $f(Q)$ proved in the appendix, one can actually show that $c_m(P,Q)$ is of the form

$$(d/dt)^{2m}u_m(P,Q)$$

where $u_m(P,Q)$ is a measure. We will not, however, bother to prove this here.

The Proof of Theorem 2

Let V be a convex region in R^k satisfying the hypotheses of Theorem 2. We will need the following.

Lemma 1. There exists a smooth everywhere positive function, f, on

$\mathbb{R}^k - 0$ such that f is homogeneous of degree $1/2$ and such that $x \in V - 0 \Leftrightarrow f(x_1^2,\dots,x_k^2) \leqslant 1$.

Proof. For each $x \in \mathbb{R}^k - 0$, let $r(x)$ be the distance from the origin to the point on the ray through s where this ray cuts the boundary of V. Let $g(x) = |x|r(x)^{-1}$. Then V is defined by the inequality $g(x) < 1$. Let W be the dihedral group generated by the reflections in the coordinate planes of $\mathbb{R}^k$. Then g is W-invariant because of the hypotheses on V. The ring of W-invariant polynomials on $\mathbb{R}^k$ is generated by $x_1^2, x_2^2, \dots, x_k^2$; so, by a theorem of G. Schwartz (1975), $\exists f \in C\ (\mathbb{R}^k - 0)$ such that $g(x_1,\dots,x_k) = g(x_1^2,\dots,x_k^2)$. *QED*

We will also need

Lemma 2. With f as in Lemma 1, the operator

$$f\left(P_1^{2a_1},\dots,P_k^{2a_k}\right)$$

is a positive elliptic pseudodifferential operator of degree s with leading symbol

$$f\left(P_1^{2a_1},\dots,P_k^{2a_k}\right)$$

Proof. Let $P = P_1^{2a_1} + \dots + P_k^{2a_k} + 1$. P is a positive elliptic pseudodifferential operator of degree $2s$ and hence

$$P_i^{2a_i}/P$$

is a pseudodifferential operator of degree zero. The operator

$$f\left(P_1^{2a_1},\dots,P_k^{2a_k}\right)$$

has the same eigenvalues as the operator

$$\sqrt{P}\, f\left(P_1^{2a_1}/P,\dots,P_k^{2a_k}/P\right) \tag{3.1}$$

on the λth eigenspace of $\{P_1,\dots,P_k\}$, namely

$$f\left(\lambda_1^{2a_1},\dots,\lambda_k^{2a_k}\right)$$

so it is identical with (3.1). In the appendix, we show that the second factor of (3.1) is a pseudodifferential operator of order zero; so we conclude that (3.1) is of order s as claimed. *QED*

We will now prove the theorem. Let

$$A = f\left(P_1^{2a_1},\dots,P_k^{2a_k}\right)$$

The number of eigenvalues of

$$\left\{P_1^{a_1},\dots,P_k^{a_k}\right\}$$

occurring in the set $\lambda^s V$ is equal to the number of eigenvalues of A lying on the interval $(0,\lambda^s)$. By Hormander (1968), this is equal, modulo $0(\lambda^{n-1})$, to $(2\pi)^{-n}$ times the symplectic volume of the set $\sigma(A) \leqslant \lambda^2$. Since this is the same as the symplectic volume of (1.11), we are done.

Appendix

Our goal is to prove the following.

THEOREM I. Let Q be a selfadjoint zeroth order pseudodifferential operator and $f \in C^\infty(R)$. Then $f(Q)$ is again a zeroth order pseudodifferential operator and $\sigma(f(Q)) = f(\sigma(Q))$.

We will first of all prove the theorem for the special case when f is an exponential polynomial and then use the Fourier inversion formula to prove it in general. Let $U(t) = \exp\sqrt{-1}\, tQ$. $U(t)$ is a one-parameter group of unitary operators on L^2. We will need

Lemma 1. $U(t)$ is a bounded operator from H^k to H^k for all $k > 0$ and satisfies

$$\|U(t)w\|_k \leqslant C(k)(\|w\|_k + t^k\|w\|_0) \quad \text{for all} \quad w \in H^k$$

Proof. (By induction.) We will prove inductively the inequality

$$\|U(t)w\|_k \leqslant C(k)\sum_{i=0}^{k} t^{k-i}\|w\|_i \tag{A.1}$$

The case $k = 0$ is obvious, since $U(t):L^2 \to L^2$ is unitary. Assuming

(A.1) is true for $k - 1$, we will prove it for k. Let $P:H^k \to H^o$ be a kth order elliptic pseudodifferential operator such that the H^k norm of w is given by

$$\|w\|_k = \|Pw\|_o \tag{A.2}$$

Now, the derivative of $U(-t)PU(t)$ is formally just $U(-t)[P,Q]u(t)$. Integrating, we get

$$U(-t)PU(t)w - Pw = \int_0^t U(-s)[P,Q]U(s)w\,ds$$

Since $[P,Q]$ is of order one less than the order of P itself, we can estimate the L^2 norm on the right by

$$\int_0^t \|U(s)w\|_{k-1} ds$$

which, by induction, is majorized by a constant times

$$\sum_{i=0}^{k-1} t^{k-i}\|w\|_i$$

Taking L^2 norms on the left, we get

$$\|PU(t)w\|_0 \leqslant \|Pw\|_0 + \sum_{i=0}^{k-1} t^{k-i}\|w\|_i$$

In view of (A.2), this is the estimate (A.1) which we are trying to prove. Next, we use the Hadamard inequality (see Hormander, 1963, Chapter 1):

$$\|w\|_i \leqslant C\|w\|_k^{i/k}\|w\|_0^{k-i/k}$$

to estimate the right hand side of (A.1) by a constant times

$$\sum\|w\|_k^{i/k}(t\|w\|_o)^{k-i/k}$$

which, by the binomial theorem, can be estimated by

$$(\|w\|_k^{1/k} + (t\|w\|_o)^{1/k})^k$$

or, finally, by a constant times $\|w\|_k + t^k\|w\|_0$. *QED*

We will now show that $U(t)$ is a pseudodifferential operator, and, at the same time, give an inductive procedure for computing its total symbol. We will assume that a global phase function, in the sense of Hormander (1971) has been chosen on (x,y,ξ) space so that to each pseudodifferential operator corresponds a global symbol which determines it uniquely up to smooth operators. We will assume by induction that we can construct a sequence of zeroth order pseudodifferential operators $U^{(k)}(t)$ such that

$$\dot{U}^{(k)} - \sqrt{-1}\, QU^{(k)} \tag{A.3}$$

is a pseudodifferential operator of order $-(k + 1)$ and such that $U^{(k)}{}_0 = I$. We will also assume that the symbol of $U^{(k)}$ has the form

$$e^{itq_0} \sum_{r=0}^{-k} a_r(x,\xi,t) \qquad q_0 = \sigma(Q) \tag{A.4}$$

where a_r is homogeneous in ξ of degree r and is a polynomial in t of degree $\leqslant 2r$. The first step in the induction being clear, let us assume the theorem true for $k - 1$ and prove it for k. Setting

$$U^{(k)} = U^{(k-1)} + V^{(k)}$$

where the total symbol of $V^{(k)}$ is

$$e^{\sqrt{-1}\, tq_0} a_k(x,\xi,t)$$

and plugging $U^{(k)}$ into (A.3), we get

$$\begin{aligned} \dot{a}_k &= -[\dot{U}^{k-1)} - \sqrt{-1}\, QU^{(k-1)}]_k\, e^{-\sqrt{-1}\, tq_0} \\ &= -\sqrt{-1}\, [Q \cdot U^{(k-1)}]_k\, e^{-\sqrt{-1}\, tq_o} \end{aligned} \tag{A.5}$$

(Here we have introduced the provisional notation $[A]_k$ for the homogeneous term of order $-k$ in the total symbol of a pseudodifferential operator A.) It is easy to check that the right hand side of (A.5) is a polynomial in t of degree at most $2k - 1$. For instance, the contribution of

$$a_r(x,\xi,t)e^{itq_0}, \qquad r \leqslant k - 1$$

to the right hand side of (A.5) involves derivatives of

$$a_r(x,\xi,t)e^{itq_o}$$

in x and ξ of order at most $k - r$, and these derivatives introduce at most $k - r$ new powers of t into (A.5). Integrating (A.5), we get for $a_k(x,\xi,t)$ a polynomial of degree $\leqslant 2k$ as claimed.

Now let $U^{\infty}(t)$ be a pseudodifferential operator with total symbol

$$e^{itq_o}\sum_{r=0}^{\infty} a_r(x,\xi,t)$$

Then, $U^{\infty}(0) = I$ and

$$\dot{U}^{\infty}(t) = \sqrt{-1}\, QU^{\infty}(t) = R(t)$$

is smoothing. Let $V(t) = U(t) - U^{\infty}(t)$. Then $V(0) = 0$ and, by variation of constants,

$$V(t) = -U(t)\int_0^t U(-s)R(s)ds$$

Now R maps H^s boundedly into H^{s+N} for all s and N, and $U(t)$ and $U(-t)$ both act as bounded operators on H^s and H^{s+N}. Hence, $V(t)$ maps H^s boundedly into H^{s+N} for arbitrary s and N is therefore smoothing. This proves that $U(t)$ is a zeroth order pseudodifferential operator as claimed. Summarizing we have proved

Proposition 2. Let Q be a selfadjoint zeroth order pseudodifferential operator with leading symbol q_o. Then $U(t) = \exp\sqrt{-1}\, tQ$ is a zeroth order pseudodifferential operator with leading symbol

$$e^{itq_o}$$

Moreover, there exists a pseudodifferential operator $U^{(k)}(t)$ with leading symbol of the form (A.4) such that $U(t) = U^{(k)}(t)$ is a pseudodifferential operator of order $-(k + 1)$. In addition, for each $s > 0$, one has an estimate of the form

$$\|(U(t) - U^{(k)}(t))w\|_{s+k+1} \leqslant C(k,s)(|t| + 1^N\|w\|_s \tag{A.6}$$

with $N = 4k + 2n + s$.

Proof. We have proved everything except (A.6). To prove (A.6), set $R = U^{(k)}(t) = \sqrt{-1}\, QU^{(k)}(t)$. Then, by variation of constants

$$U^{(k)}(t) - U(t) = U(t) \int_0^t U(-s)R(s)ds \tag{A.7}$$

By (5.4), R maps H^s boundedly into H^{s+k+1} for all $s > 0$ and satisfies an estimate of the form

$$\|Rw\|_{s+k+1} \leqslant C(|t| + 1)^{2k+s}\|w\|_s$$

for $w \in H_s$. So, by (A.7) and Lemma 1,

$$\|(U(t) - U^{(k)}(t))w\|_{s+k+1} \leqslant C(|t| + 1)^N\|w\|_s$$

provided $N \geqslant 4k + 2n + s$. *QED*

We now define

$$f(Q) = \int \hat{f}(t)U(t)dt \tag{A.8}$$

for $f \in C_0^\infty(R)$. Since $U(t)$ is unitary as an operator from L^2 to L^2, and $\hat{f}$ is rapidly decreasing, the right hand side of (A.8) converges in the L^2 operator norm, so (A.8) is well defined. Let us show that it is a pseudodifferential operator of order zero with leading symbol $f(\sigma(Q))$. Let

$$a_r(x,\xi,f) = \int e^{itq_0}\hat{f}(t)a_r(x,\xi,f)dt \tag{A.9}$$

where $a_r(x,\xi,t)$ is given by (A.4). Since $a_r(x,\xi,t)$ is a polynomial of degree $\leqslant 2r$ in t, the integral (A.9) is well defined. Let $f^\infty(Q)$ be a pseudodifferential operator with the total symbol

$$\sum a_r(x,\xi,f)$$

We will show that $f(Q) - f^\infty(Q)$ is smoothing. By (A.6), the difference

$$\int \hat{f}(t)(U(t) - U^{(k)}(t))dt$$

is bounded as an operator from H^s to H^{s+k} for all s; and, by construction, the operator

$$f^\infty(Q) - \int \hat{f}(t)U^{(k)}(t)dt$$

is a pseudodifferential operator of order $-k$; so it also has this property. Therefore, the difference $f(Q) - f^{\infty}(Q)$ has this property as well, and since k is arbitrary, this difference is smoothing. The leading symbol of $f(Q)$ is the same as the leading symbol of $f^{\infty}(Q)$, and this is just

$$\int \hat{f}(t) e^{itq_0} dt$$

or $f(q_0)$; so the proof of the theorem is complete.

Theorem I has the following simple generalization.

THEOREM II. Let $Q_1,\dots,Q_k$ be commuting selfadjoint zeroth order pseudodifferential operators and let $f \in C^{\infty}(\mathbb{R}^k)$. Then $f(Q_1,\dots,Q_k)$ is again a zeroth order pseudodifferential operator and

$$\sigma(f(Q_1,\dots,Q_k)) = f(\sigma(Q_1),\dots,\sigma(Q_k))$$

Proof. We first observe that $\exp \sqrt{-1}\,(t_1Q_1 + \dots + t_kQ_k)$ is a zeroth order pseudodifferential operator and that its total symbol has the property that each homogeneous component is of the form

$$e^{i(t_1q_1+\dots+t_kq_k)}$$

times a polynomial in $t_1,\dots,t_k$ where q_i is the leading symbol of Q_i. This follows immediately from what we have proved above, in view of the identity

$$\exp \sqrt{-1}\,(t_1Q_1 + \dots + t_kQ_k) = (\exp \sqrt{-1}\,t_1Q_1)\dots(\exp \sqrt{-1}\,t_kQ_k)$$

We now conclude, as above, that

$$f(Q) = \int f(t_1,\dots,t_k) \exp \sqrt{-1}\,(t_1Q_1 + \dots + t_kQ_k)\,dt_1\dots dt_k$$

is a pseudodifferential operator with leading symbol equal to $f(q_1,\dots,q_k)$.

Acknowledgement: The material in this appendix is the joint work of Richard Melrose, Bob Seeley, and Gunther Uhlman.

REFERENCES

Duistermaat, J. J. and Hormander, L. Fourier Integral Operators, II. Acta Math., 128 (1972), pp. 183-269.

Guillemin, V. Some spectral results on rank one symmetric spaces. To

Guillemin, V. and Sternberg, S. Geometric Asymptotics. A.M.S. Survey Publications, A.M.S., Providence, Rhode Island (1977).

Hormander, L. The spectral function of an elliptic operator. Acta Math., 121 (1968), pp. 193-218.

Hormander, L. Fourier Integral Operators, I. Acta Math., 127(1971), pp. 79-183.

Hormander, L. Linear Partial Differential Operators. Grundlehren d. Math. Wiss., 116, Springer Verlag, Berlin (1963).

Kac, M. and Spencer T. To appear.

Kolk, J. Formule de Poisson et distribution asymptotique du spectre commun des opérateurs différentiels. Univ. of Utrecht., Dept. of Math., preprint no. 46 (1977).

Schwartz, G. Smooth functions invariant under the action of a compact Lie group. Topology, 14 (1975), pp. 63-68.

Weinstein, Alan. To appear.

Massachusetts Institute of Technology
Cambridge, Massachusetts
Harvard University
Cambridge, Massachusetts

A CONSTELLATION OF MINIMAL VARIETIES DEFINED OVER THE GROUP G_2

Reese Harvey* and H. Blaine Lawson, Jr.†

Introduction

This paper constitutes an interesting special study in a general program undertaken by the authors two years ago. The initial purpose of the program was to develop and analyse new classes of subvarieties of Euclidean space which were absolutely area minimizing. Apart from a handful of special examples (see Bindschadler, to appear, Bombieri, diGiorgi and Giusti, 1969 and Lawson, 1972), the only such class of varieties known was the set of complex analytic subvarieties of complex Euclidean space.

In this paper, we shall be concerned with a special geometry of three-folds, and its dual geometry of four-folds, in Euclidean 7-space. This geometry is intimately connected with the exceptional Lie group, G_2. Our discussion will be principally concerned with defining the subvarieties, proving that they are area minimizing, and then establishing the system of nonlinear partial differential equations (analogues of the Cauchy-Riemann equations) whose solutions precisely represent these varieties. We will then examine some special solutions to these equations. In particular, it will be proved that the nonparametric minimal cone of Lawson and Osserman (1977), which gives a Lipschitz, non C^1 solution to the minimal surface system, is, in fact, absolutely area minimizing in R^7.

Some Generalities

Let M be a Riemannian n-manifold and denote by $G_p(M)$ the bundle of oriented tangent p-planes to M. This bundle will be considered a subset of the pth exterior power of the tangent bundle as follows:

$$G_p(M) = \{\xi \in \Lambda_p(M) : \xi \text{ is a unit simple vector}\}$$

*Research supported by N. S. F. Grant MPS75-05270.
†Research supported by N. S. F. Grant GP-23785X2.

Any exterior p-form ϕ on M gives, by restriction, a function $\phi: G_p(M) \to \mathbb{R}$ and we define its comass to be

$$\|\phi\|^* = \sup_{\xi \in G_p(M)} \phi(\xi)$$

following Whitney and Federer. Assuming $\|\phi\|^* = 1$, we then consider the set

$$G(\phi) = \{\xi \in G_p(M) : \phi(\xi) = 1\}$$

Using this set, we can now define a geometry of p-dimensional subvarieties of M as follows. A p-dimensional, oriented, C^1 submanifold S of M is called a ϕ-manifold if the oriented tangent plane $\vec{S}_x \in G(\phi)$ for all $x \in S$. More generally, an integral p-current S is called a ϕ-variety if $\vec{S}_x \in G(\phi)$ for $\|S\|$ - almost all $x \in M$.

The main observation concerning ϕ-varieties is the following. The essential idea goes back to Federer (1965). Assume ϕ is a nonzero p-form on M, normalized so that $\|\phi\|^* = 1$.

THEOREM 1.1. If $d\phi = 0$, then any ϕ-variety with compact support is homologically mass minimizing in M.

Note: Here the mass is the "weighted area" in the sense of Federer and Fleming (1960). For those unfamiliar with geometric measure theory, "ϕ-variety" can be replaced by "ϕ-manifold" and "mass" by "area". The proof is exactly the same.

Proof. Suppose that S' is any other integral p-current with compact support such that $dS' = dS$ and $S' - S$ is homologous to zero in M. Then, since $d\phi = 0$, we have that $S'(\phi) = S(\phi)$. However, letting $\|S\|$ denote the mass or "weighted area" measure on S, we have that

$$S(\phi) = \int \phi(\vec{S}_x)\, d\|S\|(x) = \int d\|S\|(x) = \underline{\mathbb{M}}(S')$$

and

$$S'(\phi) = \int \phi(\vec{S}'_x)\, d\|S'\|(x) \leq \int d\|S'\|(x) = \underline{\mathbb{M}}(S')$$

Hence, $\underline{\mathbb{M}}(S) \leq \underline{\mathbb{M}}(S')$ as asserted.

EXAMPLE. Let M be a Kähler manifold of complex dimension n with Kähler form ω. Fix an integer p, $0 < p < n$ and set $\phi = (1/p!)\omega p$. The (algebraic) Wirtinger inequality (see Federer and Fleming, 1960 or Lawson, 1973) states that, for any unit simple $2p$-vector ξ, $\phi(\xi) \leqq 1$ and equality holds if and only if ξ corresponds to a canonically oriented complex p-plane. Thus, $\|\phi\|^* = 1$, and $G(\phi)$ is the bundle of canonically oriented complex p-planes on M. It then follows from the structure theory (see King, 1971, and Harvey and Shiffman, 1974) that the ϕ-varieties in M are the (canonically oriented) complex analytic subvarieties of dimension p in M.

One might guess that the complex geometries are the only interesting ones to be obtained in this manner. One of the main points of this article is to demonstrate that other interesting geometries of this type exist.

From this point on, we shall be concerned with the special case $M = R^7$ and with forms ϕ which are parallel in R^7. The algebra involved in discussing these forms is complicated. The most efficient method we have found for treating them uses the octonions (or Cayley numbers), which we shall now examine in some detail.

Remarks Concerning the Octonions

Recall that the quaternions are a four dimensional noncommutative algebra $\mathbb{H} = C \oplus C$ with multiplication defined by the rule

$$q\cdot q' = (z_1,z_2)\ (z_1',z_2') \equiv (z_1z_1' - \bar z_2' z_2, z_2' z_1 + z_2\bar z_1')$$

There is a conjugation operation $\bar q = \overline{(z_1,z_2)}\quad (\bar z_1,-z_2)$ such that (qq') $\overline{(qq')} = \bar q'\cdot\bar q$ and $q\bar q = \|q\|^2$. Here $\|\cdot\|$ is the standard Euclidean norm coming from the inner product $(q,q') = \mathrm{Re}(q\bar q')$. In particular, $\bar q/\|q\|^2$ is both a left and right inverse for q. Multiplication has the property that $\|qq'\| = \|q\|\,\|q'\|$.

In analogy, the octonions are defined as the 8-dimensional algebra $\mathbb{O} = \mathbb{H} \oplus \mathbb{H}$ with multiplication defined by the rule

$$x\cdot x' = (q_1,q_2)(q_1',q_2') \equiv (q_1q_1' - \bar q_2' q_2, q_2' q_1 + q_2\bar q_1')$$

Again there is a conjugation $\bar x = \overline{(q_1,q_2)}\quad (\bar q_1,-q_2)$ with the properties that $\overline{(x\cdot x')} = \bar x'\cdot\bar x$ and $x\bar x = \|x\|^2$, so that $\bar x/\|x\|^2$ is both a left and a right inverse for x. Here $\|\cdot\|$ is the standard Euclidean norm coming from the inner product $\langle x,x'\rangle = \mathrm{Re}(x\bar{x}')$. It has the property that $\|xx'\| = \|x\|\|x'\|$. This algebra is both noncommutative and nonassociative. However, a weak form of associativity is valid. Namely, $\bar{x}(xy) - (\bar{x}x)y$. In particular, the equation $xy = z$ with x,z given and $x \neq 0$, may be solved for y by left multiplication by x^{-1}.

A major fact concerning the octonions is the following (see Schafer, 1966, pg. 29).

THEOREM 2.1. The subalgebra with unit generated by any two elements in $\mathbb{O}$ is associative.

We now consider a fundamental trilinear map $A:\mathbb{O} \times \mathbb{O} \times \mathbb{O} \to \mathbb{O}$ defined by setting

$$A(x,y,z) = \frac{1}{2}\{x(\bar{y}z) - z(\bar{y}x)\}$$

PROPOSITION 2.2. A is an alternating 3-form and has the property that

$$\|A(x,y,z)\| = \|x\wedge y\wedge z\|$$

for all $x,y,z \in \mathbb{O}$.

Proof. To see that A is alternating, it suffices to show that $A(x,y,z) = 0$ whenever two of the arguments are equal. This follows immediately from Theorem 2.1.

To prove the second property, one proceeds as follows. Assume that $\|x\wedge y\wedge z\| = 1$ and choose orthonormal vectors e_1,e_2,e_3 such that $x\wedge y\wedge z = e_1\wedge e_2\wedge e_3$. Since A is alternating, $A(x,y,z) = A(e_1,e_2,e_3)$.

We now observe that, for all $u,v \in \mathbb{O}$,

$$u(\bar{v}x) + v(\bar{u}x) = 2(u,v)x \tag{2.1}$$

To see this, we consider the equation

$$(u + v)[(\bar{u} + \bar{v})x] = [(u + v)(\bar{u} + \bar{v})]x \tag{2.2}$$

which follows from Theorem 2.1. The first term of (2.2) equals $(\|u\|^2 + \|v\|^2)x + u(\bar{v}x) + v(\bar{u}x)$, and the second equals $\|u + v\|^2 x$. This proves (2.1).

It now follows that $e_3(\bar{e}_2 e_1) = -e_2(\bar{e}_3 e_1) = e_2(\bar{e}_1 e_3) = -e_1(\bar{e}_2 e_3)$.

Hence, $A(e_1,e_2,e_3) = \frac{1}{2}[e_1(\bar{e}_2e_3) - e_3(\bar{e}_2e_1)] = e_1(\bar{e}_2e_3)$, and so $\|A(e_1,e_2,e_3)\| = 1$. This proves Proposition 2.2.

Recall that the algebra $\mathbb{O}$ is not associative. Therefore, it is of interest to consider the trilinear map $[\cdot,\cdot,\cdot]:\mathbb{O}\times\mathbb{O}\times\mathbb{O} \to \mathbb{O}$, called the associator, which is given by the expression

$$[x,y,z] = \frac{1}{2}[(xy)z = x(yz)]$$

PROPOSITION 2.3. The associator is an alternating 3-form. Moreover, if x,y,z are purely imaginary (i.e., equal to their conjugates), then, so is $[x,y,z]$.

Proof. By Theorem 2.1, $[x,y,x] = 0$ whenever any two of the arguments x, y and z are equal. Hence, $[x,y,z]$ is alternating.

Suppose now that x, y and z are purely imaginary. Then $\overline{[x,y,z]} = \frac{1}{2}[\bar{z}(\overline{yx}) - (\overline{zy})\bar{x}] = \frac{1}{2}[(zy)x = z(yx)] = [z,y,x] = -[x,y,z]$. Hence, $[x,y,z]$ is also purely imaginary. This completes the proof.

A fundamental relationship between these forms is the following.

PROPOSITION 2.4. If x, y and z are purely imaginary, then

$$A(x,y,z) = \langle x,y,z\rangle + [x,y,z].$$

In particular,

$$\operatorname{Re} A(x,y,x) = \langle x,y,z\rangle$$

$$\operatorname{Im} A(x,y,z) = [x,y,z]$$

for all $x,y,z \in \operatorname{Im}(\mathbb{O})$.

Proof. Observe that

$$2A(x,y,z) = x(\overline{y}z) = z(\overline{y}x) = -x(yz) + z(yx) = -x(yz) + (zy)x = 2[z,y,x]$$

$$= \bar{x}(yz) + (\overline{yz})x + 2[x,y,z] = 2(\langle x,y,z\rangle + [x,y,z])$$

The second statement follows from Proposition 2.3. This completes the proof.

The fundamental 3-form and its Dual

We now consider the real valued 3-form ϕ defined on $R^7 = \mathrm{Im}(\mathbb{O})$ by setting

$$\phi(x,y,z) = \langle x, y\cdot z\rangle \tag{3.1}$$

for $x,y,z \in \mathrm{Im}(\mathbb{O})$. Propositions 2 and 4 immediately give the following result.

THEOREM 3.1. The form ϕ has comass one. In fact, for all unit simple 3-vectors ξ,

$$|\phi(\xi)| \leqq 1$$

with equality if and only if the associator vanishes on ξ.

We are now interested in studying the set $G(\phi) \subset \Lambda_3 R^7$ of oriented 3-planes on which $\phi = 1$. Such oriented 3-planes will be called associative 3-planes. It is clear from the definition (3.1) that $\phi(\xi) = 1$ if and only if there is an oriented orthonormal basis (e_1, e_2, e_3) for ξ such that $e_i e_j = e_k$ for any cyclic permutation (i,j,k) of $(1,2,3)$. It follows immediately that the $\mathrm{span}_{\mathbb{R}}(1,\xi) \subset \mathbb{O}$ is a subalgebra isomorphic to the quaternions. Conversely, if $H \subset \mathbb{O}$ is any subalgebra isomorphic to the quaternions, then $\xi = \mathrm{Im}(H) \subset \mathrm{Im}(\mathbb{O})$, with the orientation determined by the condition that $e_1 e_2 = e_3$, is a plane on which $\phi = 1$. We have proved the following.

PROPOSITION 3.2. $\xi \in G(\phi)$ if and only if ξ is the canonically oriented imaginary part of a quaternion subalgebra of $\mathbb{O}$.

Recall now that the group of automorphisms of $\mathbb{O}$ is a compact simple Lie group conventionally denote by G_2. This group commutes with conjugation (since it commutes with inverses), and so it preserves the subspace $\mathrm{Im}(\mathbb{O})$. It is obvious from the definitions, that G_2 fixes the form ϕ and acts equivariantly on the associator. In particular, G_2 acts on the space $G(\phi)$.

THEOREM 3.3. The action of G_2 on $G(\phi)$ is transitive. There is an equivariant diffeomorphism

$$G(\phi) \cong G_2/SO_4$$

Proof. Suppose $\xi = e_1 \wedge e_2 \wedge e_3 \in G(\phi)$ where e_1, e_2, e_3 are orthonormal, and set $H = \text{span}(1,\xi) \subset \mathbb{O}$. As we saw above, H is a subalgebra of $\mathbb{O}$; in fact, the linear map

$$g_o : H \to \mathbb{H}$$

sending $1 \mapsto 1$, $e_1 \mapsto i$, $e_2 \mapsto j$, $e_3 \mapsto k$ is an algebra isomorphism of H with the canonical quaternion subalgebra $\mathbb{H} = \mathbb{H} \times \{0\} \subset \mathbb{H} \oplus \mathbb{H} = \mathbb{O}$.

Both left and right multiplication by unit vectors in $\mathbb{O}$ are orthogonal transformations. Since H is invariant under left and right multiplication by unit elements in H, so is $H^{\perp}$. That is, $H^{\perp}$ is both a right and left H-module.

Let ε be any unit vector in $H^{\perp}$. Then, since H is a simple algebra of dimension 4, $H^{\perp} = H \cdot \varepsilon$. Thus we have an orthogonal decomposition

$$\mathbb{O} = H \oplus H \cdot \varepsilon \tag{3.1}$$

LEMMA 3.4. With respect to the decomposition (3.1), multiplication in $\mathbb{O}$ is given by the formula

$$(q_1 + q_2\varepsilon)\,(q_1' + q_2'\varepsilon) = (q_1q_1' - \bar{q}_2'q_2) + (q_2'q_1 + q_2\bar{q}_1')\varepsilon$$

Proof. Recalling (2.1) and its conjugate, we have

$$u(\bar{v}x) + \bar{v}(ux) = 0 \quad \text{if} \quad u \perp v, \tag{3.2}$$

and

$$(x\bar{u})v + (x\bar{v})u = 0 \quad \text{if} \quad u \perp v \tag{3.3}$$

Taking $x = 1$ in (3.2), we have

$$u\bar{v} + \bar{v}u = 0 \quad \text{if} \quad u \perp v \tag{3.4}$$

Now $(q_1 + q_2\varepsilon)(q_1' + q_2'\varepsilon) = q_1q_2' + (q_2\varepsilon)(q_2'\varepsilon) + q_1(q_2'\varepsilon) + (q_2\varepsilon)q_1'$, and (3.2), (3.3), (3.4) together with the fact that $\varepsilon^2 = -1$ can be used to rearrange the last three forms yielding $(q_1q_1' - \bar{q}_2'q_2) + (q_2'q_1 + q_2\bar{q}_1')\varepsilon$. This completes the proof of the lemma.

Consider now the canonical definition of $\mathbb{O} = \mathbb{H} \oplus \mathbb{H}$ given at the beginning of section 2. Set $\varepsilon_o = (0,1)$ and let $g : \mathbb{O} \to \mathbb{O}$ be the $\mathbb{H}$-module map extending g_o and sending ε to ε_o. That is, $g(e_1) = i$,

$g(e_1 \cdot \varepsilon) = i\varepsilon_0$, $g(e_2) = j$, $g(e_2 \cdot \varepsilon) = j\varepsilon_0$, etc. It follows directly from Lemma 3.4 that g is an algebra automorphism, i.e., $g \in G_2$. Clearly, $g(\xi) = i \wedge j \wedge k$, so we have proved that G_2 is transitive on $G_0(\phi)$.

It remains to compute the subgroup K of G_2 which fixes the plane $\xi_0 \equiv i \wedge j \wedge k$. Clearly, every element $g \in K$ can be expressed, with respect to the canonical splitting $\mathbb{O} = \mathbb{H} \oplus \mathbb{H}$, as $g = (g_1, g_2)$ where $g_1 \in \mathrm{Aut}(\mathbb{H}) \cong SO_3$ and $g_2 \in SO_4$. Consider now the subgroup K_0 $K_0 \cong Sp_1 \times Sp_1/\mathbb{Z}_2 \cong SO_4$ of K given by assigning to each pair (q_1, q_2) of unit quaternions the map

$$g(x,y) = (q_1 x \bar{q}_1, q_2 y \bar{q}_1) \tag{3.5}$$

One can easily check that this map is in G_2 and preserves ξ_0. These maps are transitive on pairs (F,ε) where F is an oriented orthonormal frame in ξ_0 and ε is a unit vector in $\{0\} \times \mathbb{H}$.

Now, let $g = (g_1, g_2)$ be any element of K. After conjugation by an element in K_0 we can assume g_1 = identity and $g_2(\varepsilon_0) = \varepsilon_0$. Since g is an automorphism, we have that $g(0, q\varepsilon_0) = g((q,0)(0,\varepsilon_0)) = g(q,0)g(0,\varepsilon_0)$. Hence, $g_2\,(q\varepsilon_0) = g_1(q)g_2(\varepsilon_0) = q\varepsilon_0$, and so g = identity. It follows that $K = K_0$ and the proof is complete.

REMARK. The above proof actually shows that if (e_1, e_2, e_3) and (e_1', e_2', e_3') are orthonormal triples in $\mathrm{Im}(\mathbb{O})$ such that $e_1 \wedge e_2 \wedge e_3$, $e_1' \wedge e_2' \wedge e_3' \in G_0(\phi)$, then there exists $g \in G_2$ such that $g(e_j) = e_j'$ for $j = 1,2,3$. Now, for any orthonormal pair (e_1, e_2) in $\mathrm{Im}(\mathbb{O})$, we have $e_1 \wedge e_2 \wedge (e_1 e_2) \in G_0(\phi)$. This gives the following

COROLLARY 3.5. The group G_2 acts transitively on oriented pairs of orthonormal vectors in R^7, i.e., G_2 acts transitively on the Stiefel manifold $V_{7,2}$. In particular, G_2 acts transitively on S^6. Moreover, there are equivariant diffeomorphisms

$$V_{7,2} \cong G_2/SU_2$$

$$S^6 \cong G_2/SU_3$$

Proof. Given the remark above, it remains only to prove the explicit diffeomorphisms. The first follows directly from the computation of the isotropy subgroup of G_2 acting on $G(\phi)$ given in the proof of Theorem

3.3. For the second, we consider the subgroup K of G_2 which fixes $i = (i,0) \in \mathbb{H} \times \{0\}$. Let V be the orthogonal compliment of i in $\mathrm{Im}(\mathbb{O})$. Clearly, $K \subseteq O(V)$. Now, left multiplication by i gives a complex structure on V. For any $\tau = t_1 + it_2$, where $t_1, t_2 \in R$, we have $g(\tau x) = g(\tau)g(x) = \tau g(x)$ for all $x \in V$. Hence, $K \subseteq U(V) \cong U_3$. Note now that the fiber bundle $V_{7,2} \to S^6$ has fiber $S^5 = K/SU_2$. It follows that the dimension of K is 8, and therefore, $K \cong SU_3$ (the only compact, connected, simply connected Lie group of dimension 8). This completes the proof.

Observe that, in the above proof, we have constructed a G_2-invariant almost complex structure on S^6.

It is clear that any G_2-invariant 3-form on R^7 will give rise to an interesting geometry. However, we have the following.

THEOREM 3.6. The form ϕ is the unique 3-form (up to scalar multiples) on R^7 which is G_2-invariant.

Proof. It follows from the standard formulas of Weyl that the irreducible representation of highest weight in $\Lambda^2\rho$, where ρ is the 7-dimensional representation of G_2, has dimension 27. The associator, $[x,y,z]$, gives a copy of the 7-dimensional representation in $\Lambda^3\rho$. Since $\Lambda^3 R^7$ has dimension 35, this completes the proof.

Recall now that the $*$-operator gives an isometry of the Euclidean spaces $*: \Lambda^3 R^7 \to \Lambda^4 R^7$ which maps simple vectors to simple vectors. It therefore preserves the comass norm, and clearly it is equivariant with respect to the action of G_2 induced on these spaces. Consequently, we have, from Theorem 3.6, that $\psi \equiv *\phi$ is, up to sign, the unique G_2-invariant 4-form of comass 1 on R^7. This form can be succinctly described.

THEOREM 3.7. Let $\psi = *\phi$ be the 4-form dual to ϕ. Then

$$\psi(x,y,z,w) = \langle x, [y,z,w]\rangle \tag{3.6}$$

for all $x,y,z,w \in \mathrm{Im}(\mathbb{O})$.

Proof. The multilinear form $\psi_0(x,y,z,w) \equiv \langle x, [y,z,w]\rangle$ is evidently G_2-invariant, and by Proposition 2.3, it is alternating in the variables y, z, and w. Moreover, $\psi_0(x,x,z,w) = \frac{1}{2}\langle x, (xz)w - x(zw)\rangle = \frac{1}{2}\{\langle x\bar{w}, xz\rangle - \langle \bar{x}x, zw\rangle\} = \frac{1}{2}\|x\|^2\{\langle \bar{w}, z\rangle - \langle \bar{w}, z\rangle\} = 0$, and so ψ_0 is alternating in all its variables. From the uniqueness quoted in the paragraph above, we conclude that $\psi = r\psi_o$ for some $r \in R$. However, letting $\eta_o = *(i \wedge j \wedge k)$, we have that $\psi(\eta_o) = 1$, and using Lemma 3.4, one can straightforwardly compute that $\psi_o(\eta_o) = \langle \varepsilon_o, [i\varepsilon_o, j\varepsilon_o, k\varepsilon_o] \rangle = 1$. Hence, $r = 1$ and the proof is complete.

COROLLARY 3.8. For all unit simple 4-vectors η in $R^7 = \mathrm{Im}(\mathbb{O})$,

$$|\psi(\eta)| \leqq 1$$

with equality if and only if η satisfies the condition that $[x,y,z] \in \eta$ for all $x,y,z \in \eta$.

Proof. This is an immediate consequence of Theorem 3.7.

Note that the *-operator gives a natural diffeomorphism

$$G(\psi) \cong G(\phi) \cong G_2/SO_4$$

Since the 3-planes in $G(\phi)$ are called associative, the dual 4-planes in $G(\psi)$ will be called coassociative.

Note that, for any orthogonal decomposition $\mathrm{Im}(\mathbb{O}) = \xi \oplus \eta$ where ξ is an associative 3-plane, we have the relations

$$\begin{aligned} \xi \cdot \xi &= \xi \oplus R \cdot 1 \\ \xi \cdot \eta &= \eta \cdot \xi = \eta \qquad (3.7) \\ \eta \cdot \eta &= \xi \oplus R \cdot 1 \end{aligned}$$

Each of the conditions in (3.7) is equivalent to the remaining two. Therefore, we may conclude the following.

COROLLARY 3.9. For all unit simple vectors η in $R^7 = \mathrm{Im}(\mathbb{O})$

$$|\psi(\eta)| \leqq 1$$

with equality if and only if η satisfies the condition that $x \cdot y \in \eta$ for all orthogonal pairs $x,y \in \eta$.

The Associated Systems of Partial Differential Equations

As explained in section 1, the form ϕ given by (3.1) defines a geometry of 3-dimensional varieties in R^7. These varieties will be called associative threefolds. The 4-dimensional varieties similarly determined by the dual 4-form ψ (see (3.6)) will be called coassociative fourfolds. By Theorem 1.1, each of these varieties is absolutely mass minimizing in R^7.

It is clear, in principle, that all such varieties represent solutions to a first order, nonlinear partial differential equation. Suppose, for example, that $F:M^3 \to R^7$ is an oriented, 3-dimensional submanifold of R^7. Then, this submanifold is associative if and only if

$$F^* \phi = dv_F \tag{4.1}$$

where dv_F is the oriented volume form for the metric induced by F.

The equation (4.1) can be made more intrinsic to Euclidean space by considering submanifolds locally as graphs of functions over their tangent planes. We do this as follows. Recall that any two planes $\xi_1, \xi_2 \in G(\phi)$ are equivalent under G_2 (Theorem 3.3). Therefore, it suffices to consider the case $\xi = \mathrm{Im}(\mathbb{H}) \subset \mathrm{Im}(\mathbb{O})$. Let $\Omega \subseteq \mathrm{Im}(\mathbb{H})$ be an open set and consider a C^1 map

$$f:\Omega \to \mathbb{H} \tag{4.2}$$

Then if we choose coordinates (x_1, x_2, x_3) with respect to the basis (i,j,k) of $\mathrm{Im}(\mathbb{H})$, we have that the graph of f in R^7 is associative (i.e., the immersion $F(x) = (x, f(x))$ satisfies (4.1)) if and only if

$$\begin{aligned}\phi&\left(i + \frac{\partial f}{\partial x_1},\ j + \frac{\partial f}{\partial x_2},\ k + \frac{\partial f}{\partial x_3}\right) \\ &= \left\| \left(i + \frac{\partial f}{\partial x_1}\right) \wedge \left(j + \frac{\partial f}{\partial x_2}\right) \wedge \left(k + \frac{\partial f}{\partial x_3}\right) \right\|\end{aligned} \tag{4.3}$$

It is clear that equation (4.3) is, in general, a highly unsatisfactory formulation of the problem. To see this, we consider the analogous case of the Kähler form ω. Here, one has a map $g:C^p \to C^{n-p}$ and the condition that the graph of g be a complex submanifold, i.e., that g be holomorphic, is expressed by a single equation analogous to (4.3) with ϕ replaced by $\omega^p/p!$. A direct examination of this single equations shows it to be a mess. However, in this case, the single equation can be shown to be equivalent to a system of linear equations, namely the Cauchy-Riemann equations.

The purpose of this section is to extract a simple and rather beautiful system of equations for f which is equivalent to condition (4.3), and should be considered as the analogue of the Cauchy-Riemann equations.

To express this system concisely, we shall define two distinct first order operators on C^1 functions $f:\mathrm{Im}(\mathbb{H}) \to \mathbb{H}$.

The first is a linear operator given by the formula

$$Df = -\left(\frac{\partial f}{\partial x_1}\cdot i + \frac{\partial f}{\partial x_2}\cdot j + \frac{\partial f}{\partial x_3}\cdot k\right) \tag{4.5}$$

where (x_1,x_2,x_3) are coordinates with respect to the basis (i,j,k). It is easy to check that

$$Df = \sum_{j=1}^{3} (\nabla_{e_j} f)\cdot\bar{e}_j \tag{4.6}$$

where (e_1,e_2,e_3) is any oriented orthonormal basis of $\mathrm{Im}(\mathbb{H})$. In fact, D is just the <u>Dirac Operator</u> defined by considering $\mathbb{H}$ appropriately as a left module over the Clifford algebra on R^3. It is a formally selfadjoint, elliptic operator with the property that

$$D^2 = \Delta\cdot 1$$

where

$$\Delta = -\sum_j \frac{\partial^2}{\partial x_j^2}$$

is the positive scalar Laplacian and 1 is the identity matrix.

To define the second operator, we consider the alternating trilinear map $A:\mathbb{H} \times \mathbb{H} \times \mathbb{H} \to \mathbb{H}$ given (in analogy with the one in section 2) by setting

$$A(x,y,z) = \frac{1}{2}(x\bar{y}z - z\bar{y}x) \tag{4.7}$$

for $x,y,z \in \mathbb{H}$. We then define a <u>generalized Monge-Ampère operator</u> σ by the formula

$$\sigma(f) = A\left(\frac{\partial f}{\partial x_1},\frac{\partial f}{\partial x_2},\frac{\partial f}{\partial x_3}\right) \tag{4.8}$$

Again, if (e_1,e_2,e_3) is any oriented orthonormal basis of $\mathrm{Im}(\mathbb{H})$, we have

$$\sigma(f) = A(\nabla_{e_1} f,\ \nabla_{e_2} f, \nabla_{e_3} f) \tag{4.9}$$

Note that σ is homogeneous of degree three in the first derivatives of f. It depends linearly on the determinants of the 3×3 minors of the Jacobian matrix.

THEOREM 4.1. Let $f:\Omega \to \mathbb{H}$ be a C^1 map where Ω is a domain in $\mathrm{Im}(\mathbb{H})$. Then, the graph of f in $R^7 = \mathrm{Im}(\mathbb{H}) \oplus \mathbb{H} = \mathrm{Im}(\mathbb{O})$ is an associative manifold if and only if f satisfies the equation

$$Df = \sigma(f) \tag{4.10}$$

COROLLARY 4.2. The graph of any solution of equation (4.10) is absolutely area minimizing in R^7. In particular, from the regularity theory of C. B. Morrey (1954), any C^1 solution to (4.10) is real analytic.

We conjecture that, in analogy with the complex case, a stronger regularity theory exists for associative and coassociative varieties. However, one must be careful. While the statement analogous to Corollary 4.2 holds for the dual equations (see below), the slightly more general statement that Lipschitz solutions are real analytic is false. This will be proved in section 5.

REMARK. Before proving Theorem 4.1, we observe that A has the obvious properties that

$$A(ex,ey,ez) = eA(x,y,z)$$

$$A(xe,ye,ze) = A(x,y,z)e$$

for any unit quaternion e and for all $x,y,z \in \mathbb{H}$. From this it follows easily that the equations (4.10) are essentially preserved by the isotropy representation of SO_4 given in (3.5).

The real part of equation (4.10) can be expressed in fairly transparent terms. Let $\hat{f}:\mathrm{Im}(\mathbb{H}) \to \mathrm{Im}(\mathbb{H})$ be obtained from f by orthogonal projection. Then, the real part of (4.10) is equivalent to the equation

$$\mathrm{trace}(\hat{f}_*) = \det(\hat{f}_*)$$

there $\hat{f}_*$ denotes the Jacobian matrix of $\hat{f}$. Using the invariance of the equations, we then obtain the result that (4.10) is equivalent to the equation

$$\mathrm{trace}[(\widehat{qf})_*] = \det[(\widehat{qf})_*] \tag{4.10'}$$

for all $q \in \mathbb{H}$.

Proof of Theorem 4.1. For notational convenience, we set $f_i = \partial f/\partial x_1$, $f_j = \partial f/\partial x_2$ and $f_k = \partial f/\partial x_3$. From Theorem 3.1, it follows immediately that the graph of f (properly oriented) is associative if and only if

$$[i + f_i, \ j + f_j, \ k + f_k] = 0 \tag{4.11}$$

Expanding and taking real and imaginary parts, we find that

$$[i,\overline{f}_j f_k] + [j,\overline{f}_k f_i] + (k,\overline{f}_i f_j] = 0 \tag{4.12}$$

$$f_i i + f_j j + f_k k + A(f_i,f_j,f_k) = 0 \tag{4.13}$$

where $[\cdot,\cdot]$ denotes the commutator $[x,y] = xy - yx$. Taking the inner product of the first equation with 1, i, j, and k successively, and using the invariance property $\langle [x,y],z\rangle = -\langle y,[x,z]\rangle$ we find that equation (4.12) is equivalent to the three equations

$$\langle f_i,\tau\rangle = \langle f_j,\tau\rangle = \langle f_k,\tau\rangle = 0$$

where $\tau = f_i i + f_j j + f_k k$. These equations are equivalent to the statement

$$\tau \perp \mathrm{image}(f) \tag{4.12'}$$

Observe now that

$$2\langle f_i, A(f_i,f_j,f_k)\rangle = 2\langle f_i, f_i\overline{f}_j f_k - f_k\overline{f}_j f_i\rangle$$

$$= 2\|f_i\|^2\langle 1,\overline{f}_f f_k - f_k\overline{f}_j\rangle = 0$$

By skew symmetry, it follows that

$$A(f_i,f_j,f_k) \perp \mathrm{image}(f)$$

We conclude that equation (4.12) is a consequence of equation (4.13). This completes the proof.

Note that, while the equation (4.10) is not linear, it is determined, that is, it is a system of four equations for four dependent variables.

The linearization of this system at the zero function is just the elliptic operator D.

We now consider the question of the equations defining coassociative fourfolds, that is, 4-manifolds determined by the dual form $\psi = *\phi$. In the spirit of our discussion above, we consider C^1 maps

$$f:\mathbb{H} \to \mathrm{Im}(\mathbb{H})$$

and construct a pair of first order operators. Let $(e_0,\ldots,e_3)$ be any orthonormal basis of $\mathbb{H}$, we define the dual Dirac operator on f by setting

$$\tilde{D}f = -\sum_{j=0}^{3} e_j(\nabla_{e_j} f) \tag{4.14}$$

This definition is evidently independent of the choice of orthonormal basis. Note that, while f has vlaues in $\mathrm{Im}(\mathbb{H})$, $\tilde{D}f$ has values in $\mathbb{H}$.

If we consider the conjugate operator on $\mathbb{H}$-valued functions g given by

$$\tilde{D}^*g = -\sum_{j=0}^{3} \bar{e}_j(\nabla_{e_j} g)$$

then, one easily sees from (2.1) that

$$\tilde{D}^*\tilde{D}f = \Delta\cdot 1$$

where Δ is the positive scalar Laplacian, as before.

To express the second operator, we choose an oriented orthonormal basis (e_1',e_2',e_3') for $\mathrm{Im}(\mathbb{H})$ and write $f = \sum f^j e_j'$ where the functions f^j are real valued. We then define a dual Monge-Ampère operator $\tilde{\sigma}$ by setting

$$\tilde{\sigma}(f) = A(\nabla f^1,\nabla f^2,\nabla f^3) \tag{4.15}$$

where ∇f^j is interpreted as a quaternion in the obvious way.

THEOREM 4.3. Let $f:\Omega \to \mathrm{Im}(\mathbb{H})$ be a C^1 map where Ω is a domain in $\mathbb{H}$. Then, the graph of f in $R^7 \cong \mathrm{Im}(\mathbb{H})\oplus\mathbb{H} = \mathrm{Im}(\mathbb{O})$ is coassociative if and only if f satisfies the equation

$$\tilde{D}f = \tilde{\sigma}(f) \tag{4.16}$$

COROLLARY 4.4. The graph of any solution of equation (4.16) is absolutely area minimizing in R^7. In particular, any C^1 solution to (4.16) is real analytic.

Proof. We begin by recalling the following elementary fact. Let $L:R^4 \to R^3$ be a linear map and consider the 4-plane

$$\xi = \{(L(x),x) \in R^7 Lx \in R^4\}$$

Then

$$\xi^{\perp} = \{(x,-L^t(x)) \in R^7 : x \in R^3\}$$

where L^t denotes the adjoint map. It follows from this principle and from duality, that the graph of f is coassociative if and only if the negative transpose of its Jacobian transformation satisfies the condition for associativity at each point. It remains only to reexpress this condition.

Choose an oriented orthonormal basis $(e_0,\ldots,e_3)$ of $\mathbb{H}$ so that $e_0 = 1$. Then (e_1,e_2,e_3) is a basis of $\mathrm{Im}(\mathbb{H})$. Let

$$f_{\alpha j} = \langle \nabla_{e_\alpha} f, e_j \rangle$$

for $\alpha = 0,\ldots,3$ and $j = 1,2,3$ be the Jacobian matrix of f, and set $g_{j\alpha} = -f_{\alpha j}$. The condition for associativity is that $\tau(f) = \alpha(f)$ where

$$\tau(g) = -\sum_{j=1}^{3} g_j e_j = -\sum_{\alpha=0}^{3}\sum_{j=1}^{3} g_{j\alpha} e_\alpha \cdot e_j = \sum_{\alpha=0}^{3}\sum_{j=1}^{3} \langle \nabla_{e_\alpha} f, e_j\rangle e_\alpha e_j$$

$$= \sum_{\alpha,j} e_\alpha \langle \nabla_{e_\alpha} f, e_j \rangle e_j = \sum_{\alpha} e_\alpha (\nabla_{e_\alpha} f) = -\tilde{D}f$$

and where

$$\alpha(g) = A(g_1,g_2,g_3) = -A(\nabla f^1, \nabla f^2, \nabla f^3) = -\sigma(f)$$

since

$$g_j = \sum g_{j\alpha} e_\alpha = -\sum \langle \nabla_{e_\alpha} f, e_j \rangle e_\alpha = -\nabla\langle f, e_j \rangle = -\nabla f^j$$

This completes the proof.

We conclude this section with a remark concerning the Dirac operators. Let $(e_0,e_1,e_2,e_3) = (1,i,j,k)$ be the standard basis for $\mathbb{H}$. With respect to this basis, we choose coordinates (x_0,x_1,x_2,x_3) and set $\partial_j \equiv \partial/\partial x_j$.

Then the Dirac operators can be expressed by the matrix operators

$$D = \begin{pmatrix} 0 & \partial_1 & \partial_2 & \partial_3 \\ -\partial_1 & 0 & -\partial_3 & \partial_2 \\ -\partial_2 & \partial_3 & 0 & -\partial_1 \\ -\partial_3 & -\partial_2 & \partial_1 & 0 \end{pmatrix} \tag{4.17}$$

$$\tilde{D} = \begin{pmatrix} \partial_1 & \partial_2 & \partial_3 \\ -\partial_0 & \partial_3 & -\partial_2 \\ -\partial_3 & -\partial_0 & \partial_1 \\ \partial_2 & -\partial_1 & -\partial_0 \end{pmatrix} \tag{4.18}$$

Note that functions in the kernel of D or $\tilde{D}$ are harmonic, since $D^2 = \Delta 1$ and $\tilde{D}^*\tilde{D} = \Delta \cdot 1$.

Some Examples and Applications

We now consider the problem of solving the equations (4.10) and (4.16). We begin with some special cases of (4.10).

Suppose we consider functions f which depend only on x_2 and x_3. Then $\sigma(f) \equiv 0$ and (4.10) becomes the equation

$$\frac{\partial f}{\partial x_2} j + \frac{\partial f}{\partial x_3} k = \left(\frac{\partial}{\partial x_2} + i \frac{\partial}{\partial x_3}\right) f \cdot j = 0 \tag{5.1}$$

Hence, setting $z = x_2 + ix_3$ and writing $f = U + Vj$ where $U = u_1 + iu_2$ and $V = v_1 + iv_2$, we see that (5.1) is equivalent to the fact that

$$\frac{\partial U}{\partial \bar{z}} = \frac{\partial V}{\partial \bar{z}} = 0$$

i.e., $U(z)$ and $V(z)$ are holomorphic. In this case, the graph is of the form $R \times \Sigma \subset R \times C^3 = R^7$ where Σ is a holomorphic curve in C^3.

A second special case is given by considering functions f such that $\mathrm{Re}(f) \equiv 0$. In this case, $f: \mathrm{Im}(\mathbb{H}) \to \mathrm{Im}(\mathbb{H})$. Since A is alternating and $A(i,j,k) = 1$, we have that $\sigma(f)$ is real. Hence, equation (4.10) reduces to the system

$$\begin{aligned} \mathrm{Im}(Df) &= 0 \\ \mathrm{tr}(f_*) &= \det(f_*) \end{aligned} \tag{5.2}$$

where f_* is the Jacobian determinant of f as a map from R^3 to R^3. Writing $f = f^1 i + f^2 j + f^3 k$, one sees directly from (4.17) that $\mathrm{Im}(Df) = 0$ is equivalent to

$$\frac{\partial f^\alpha}{\partial x_\beta} = \frac{\partial f^\beta}{\partial x_\alpha} \qquad 1 \leqq \alpha, \beta \leqq 3 \tag{5.3}$$

Assuming that f is defined on a simply connected domain, this is equivalent to the existence of a real valued function F such that

$$f^\alpha = \frac{\partial F}{\partial x_\alpha} \qquad \alpha = 1,2,3 \tag{5.4}$$

The second equation in (5.2) then becomes

$$\nabla^2 F = \mathbb{H}(F) \tag{5.5}$$

where $\nabla^2 = \sum \partial^2/\partial x_\alpha^2$ and $\mathbb{H}(F)$ is the determinant of the Hessian matrix of F. By the Implicit Function Theorem, one can solve the Dirichlet Problem for (5.5) over a compact domain in R^3 for all boundary data sufficiently close to zero.

Let us now consider the dual equations for a function $f: \mathbb{H} \to \mathrm{Im}(\mathbb{H})$. If, as above, we consider functions which are independent of the first variable x_0, i.e., functions $f: \mathrm{Im}(\mathbb{H}) \to \mathrm{Im}(\mathbb{H})$, the equations (4.16) reduce again to (5.4) and (5.5).

One of the most interesting examples of a coassociative manifold is found by looking for symmetric solutions. Consider the action of $Sp_1 \cong \{q \in \mathbb{H} : \|q\| = 1\}$ on $\mathrm{Im}(\mathbb{H}) \times \mathbb{H}$, given by defining

$$\Phi_q(x,y) = (qx\bar{q}, yq) \tag{5.6}$$

Recall from (3.5) that this embeds Sp_1 into G_2, and, therefore, this action preserves the form ψ. Note that the orbit of any point (x,y), where $y \neq 0$, is diffeomorphic to S^3.

We now consider the fourfolds obtained by taking the cone over these 3-dimensional orbits of Sp_1. Specifically, for $0 \leqq a < 1$ and $b = \sqrt{1-a^2}$, we consider the set

$$C_a = \{t(aqi\bar{q}, bi\bar{q}) : t \geqq 0 \quad \text{and} \quad q \in Sp_1\} \tag{5.7}$$

which is a manifold outside the origin.

THEOREM 5.1. The cone

$$C_{\sqrt{5}/3}$$

is a coassociative fourfold. In particular, any relatively compact domain on

$$C_{\sqrt{5}/3}$$

represents an absolutely area minimizing 4-current in R^7.

Proof. To show that C_a is coassociative, it suffices to show that, at each point p, we have that

$$\psi(v_1, \ldots, v_4) = \|v_1 \wedge \ldots \wedge v_4\| \tag{5.8}$$

where $v_1, \ldots, v_4$ is any oriented basis of the tangent space to C_a at p. By the invariance of ψ under Sp_1, it suffices to check this condition at points p of the form $t(ai, bi)$ for $t > 0$. By the invariance of ψ under parallel translation, we may also assume that $t = 1$.

Let $p = (ai, bi)$ and consider the tangent space V_p to the orbit of Sp_1 through p. Then

$$V_p = \{(a(xi - ix), -bix) : x \in \mathrm{Im}(\mathbb{H})\}$$

To see this, note that $Sp_1 = \{e^x \in \mathbb{H} : x \in \mathrm{Im}(\mathbb{H})\}$. Set $q_t = e^{tx}$ for $t \in R$ and differentiate the action at $t = 0$. It follows that an oriented orthogonal basis of the tangent space to C_a at p is given by

$$v_1 = (0,1)$$

$$v_2 = -(2ak, bk)$$

$$v_3 = (2aj, bj)$$

$$v_4 = (ai, bi)$$

A direct computation shows that

$$[v_1,v_2,v_3] = (4abi,(4a^2 - b^2)i)$$

Hence, since $\psi(v_1,\ldots,v_4) = \langle v_4,[v_1,v_2,v_3]\rangle$, we have that

$$\psi(v_1,\ldots,v_4) = b(8a^2 - b^2)$$

Since the v_j are orthogonal, we have

$$\|v_1\wedge\ldots\wedge v_4\| = (4a^2 + b^2)$$

One can easily check that (5.8) holds if $a = \sqrt{5}/3$ (and $b = 2/3$). This completes the proof.

We now observe that the cone

$$C_{\sqrt{5}/3} \subset \mathrm{Im}(\mathbb{H}) \times \mathbb{H}$$

projects onto $\mathbb{H}$ in a one-to-one, nonsingular manner outside the origin. In fact, it is just the graph of the map $f:\mathbb{H} \to \mathrm{Im}(\mathbb{H})$ given by

$$f(x) = \frac{\sqrt{5}}{2\|x\|}\,\bar{x}ix \tag{5.9}$$

for $x \in \mathbb{H}$. Restricting f to the sphere $\|x\| = 1$ gives a map $S^3 \to S^2$. This is the classical Hopf map.

The cone

$$C_{\sqrt{5}/3}$$

and the associated mapping f in (5.9) were first discovered by the second author and R. Osserman (1977). They observed that f represents a Lipschitz solution to the minimal surface system which is not C^1. We have now shown that the graph of f is absolutely area minimizing. Furthermore, we have the following.

THEOREM 5.2. The function f given by (5.9) is a Lipschitz solution to the system (4.16) which is not C^1.

This makes sharp the regularity result (Corollary 4.4) that C^1 solutions are real analytic.

REFERENCES

Bindschadler, D. Absolutely area minimizing singular cones of arbitrary codimension. (To appear.)

Bombieri, E, diGiorgi, E., and Giusti, E. Minimal cones and the Bernstein Problem. Invent. Math., 7 (1969), pp. 243-268.

Federer, H. Some theorems on integral currents. Trans. A.M.S., 117 (1965), pp. 43-67.

Federer, H. Geometric Measure Theory. Springer Verlag, New York (1969).

Federer, H., and Fleming, W. H. Normal and integral currents. Ann. of Math., 72 (1960), pp. 458-520.

Harvey, R., and Shiffman, B. A characterization of holomorphic chains. Ann. of Math.(2), 99 (1974), pp. 553-587.

King, J. The currents defined by analytic varieties. Acta Math., 127 (1971), pp. 185-220.

Lawson, H. B., Jr. The equivariant Plateau problem and interior regularity. Trans. A.M.S., 173 (1972), pp. 231-250.

Lawson, H. B., Jr. Minimal Varieties in Real and Complex Geometry. Univ. of Montreal Press, Montreal (1973).

Lawson, H. B., Jr., and Osserman, R. Non-existence, non-uniqueness and irregularity of solutions to the minimal surface system. Acta Math., 134 (1977).

Morrey, C. B. Second order elliptic systems of differential equations. Ann. of Math Studies, No. 33, Princeton Univ. Press, Princeton (1954), pp. 101-160.

Schafer, R. An Introduction to Nonassociative Algebra. Academic Press, New York (1966).

Rice University
Houston, Texas
University of California
Berkeley, California

A GEOMETRIC VERSION OF SCATTERING THEORY AND BACKLUND TRANSFORMATIONS FOR SECOND ORDER LINEAR ORDINARY DIFFERENTIAL OPERATORS

Robert Hermann*

Introduction

Recent work on "solitons" and nonlinear waves has interrelated three topics in an amazing way: scattering theory for linear ordinary differential operators, nonlinear partial differential equations (e.g., the Korteweg-de Vries equation) and Bäcklund transformations. As a byproduct of my work (1966, 1971, 1977) on a geometric approach (due to H. Walquist and F. Estabrook, 1975 [see also, R. Hermann, ed, 1977]) to these matters, I have tried to find more direct ways of seeing these interrelations and understanding their geometric meaning. In this note, I shall investigate one of these approaches.

Scattering Theory for the Riccati Equation

Let $\mathcal{G}$ be the Lie algebra of 2×2 real matrices of trace zero. Let $G \equiv SL(2,R)$ be the Lie group of 2×2 real matrices of determinant one. (As the notation indicates, $\mathcal{G}$ is the Lie algebra of G.) Let x be a real variable $-\infty < x < \infty$.

Suppose

$$x \to A(x) = \begin{pmatrix} A_{11}(x) & A_{12}(x) \\ A_{21}(x) & -A_{11}(x) \end{pmatrix}$$

is a curve in $\mathcal{G}$. Let

$$z = \begin{pmatrix} z_1 \\ z_2 \end{pmatrix}$$

*Supported by a grant from the National Aeronautics and Space Administration, Grant No. NSG-2148.

denote a vector in R^2. Consider the linear differential equations

$$z_x = A(x)z \tag{2.1}$$

They can be solved as

$$z(x) = g(x)z(0) \tag{2.2}$$

with

$$g_x = Ag, \quad g(0) = 1 \tag{2.3}$$

i.e., $x \to g(x)$ is a curve in the Lie group $SL(2,R)$ whose infinitesimal generator is A. (As usual in the theory of nonlinear waves, derivatives are denoted by letter subscripts.)

We can convert (2.1) into a Riccati equation for the scalar y in the following way: set

$$y(x) = \frac{z_1(x)}{z_2(x)} \tag{2.4}$$

Then,

$$\begin{aligned} y_x &= \frac{z_2 z_{1,x} - z_1 z_{2,x}}{z_2^2} \\ &= \frac{z_2(A_{11}z_1 + A_{12}z_2) - z_1(A_{21}z_1 - A_{11}z_2)}{z_2^2} \\ &= 2A_{11}y + A_{12} - A_{21}y^2 \end{aligned}$$

Thus, y satisfies the following Riccati equation:

$$y_x = a + by + cy^2 \tag{2.5}$$

with

$$a = A_{12} \qquad b = 2A_{11} \qquad c = -A_{21} \tag{2.6}$$

Now, suppose that

$$\lim_{x\to\pm\infty} A(x) = A^0 = \begin{pmatrix} A^0_{11} & A^0_{12} \\ A^0_{21} & -A^0_{11} \end{pmatrix} \tag{2.7}$$

Set

$$a^0 = A^0_{12} \qquad b^0 = 2A^0_{11} \qquad c = -A^0_{21} \tag{2.8}$$

Let us *suppose* that the equation

$$a^0 + b^0 r + c^0 r^2 = 0$$

has two roots r_1, r_2. Let us also suppose that there are four solutions y_1, y_2, y_3, y_4 of (2.5) such that

$$\lim_{x\to\infty}\begin{pmatrix} y_1 \\ y_3 \end{pmatrix} = \begin{pmatrix} r_1 \\ r_2 \end{pmatrix}$$
$$\lim_{x\to-\infty}\begin{pmatrix} y_2 \\ y_4 \end{pmatrix} = \begin{pmatrix} r_1 \\ r_2 \end{pmatrix} \tag{2.9}$$

Set

$$S = \frac{y_1-y_2}{y_1-y_3}\frac{y_3-y_4}{y_2-y_4} \tag{2.10}$$

It is called the invariant scattering function. (It is a function of the curve $x \to A(x)$.) Notice the classical property of the cross-ratio appearing on the right hand side of (2.10) is that it is invariant under linear fractional transformations on the y's. In particular, S is independent of x. We shall see later that S is the absolute value squared of the "reflection coefficient", in the sense of traditional scattering theory. What is desired is an understanding of how S changes as the curve $x \to A(x)$ is changed. Suppose, then, that

$$x \to A'(x)$$

is another curve in G. Use it to form the Riccati equation

$$y'_x = a' + b'y' + c'y'^2 \tag{2.11}$$

THEOREM 2.1. Suppose that there is a curve

$$x \to g(x) = \begin{pmatrix} g_{11} & g_{12} \\ g_{21} & g_{22} \end{pmatrix}$$

in SL(2,R) with the following property. The following formula

$$y'(x) = \frac{g_{11}(x)y(x) + g_{12}(x)}{g_{21}(x)y(x) = g_{22}(x)} \tag{2.12}$$

maps a curve $x \to y(x)$ which satisfies (2.5) into a curve $x \to y'(x)$ which satisfies (2.11). Suppose further that

$$\lim_{x\to+\infty} \begin{pmatrix} y_1 \\ y_3 \end{pmatrix} = \begin{pmatrix} r_1 \\ r_2 \end{pmatrix}$$

$$\lim_{x\to-\infty} \begin{pmatrix} y_2 \\ y_4 \end{pmatrix} = \begin{pmatrix} r_1 \\ r_2 \end{pmatrix}$$

Then

$$S = S' \tag{2.13}$$

Proof. The proof of Theorem 2.1 is a "trivial" consequence of the invariance of the cross-ratio function under linear fractional transformations of the type (2.12). It then transfers the question to one of deciding when the "general solutions" of the Riccati equations (2.5) and (2.11) are related via (2.12). This, in turn, is reflected in a statement about the "orbit" of the "gauge group" of all "curves" $x \to g(x)$ in SL(2.R) acting on the "space" of Ricatti equations. What seems to happen in certain cases is that the conditions for the existence of such a curve $x \to g(x)$ can be described in terms of a system of ordinary differential equations between functions (a,b,c,a',b',c'). These differential equations define the Bäcklund transformation. Let us now see how these general ideas work in the main case.

Scattering Theory for the Schrödinger-Sturn-Liouville Equation and the Bäcklund Transformation for the Korteweg-de Vries Equation

Consider the following differential equation.

$$\psi_{xx} + u\psi = \lambda\psi \tag{3.1}$$

$x \to u(x)$ is a smooth function which vanishes sufficiently rapidly as $x \to \pm\infty$. Define the associated Riccati equation

$$\begin{aligned} y &= \psi_x/\psi \\ y_x &= \lambda - u - y^2 \end{aligned} \tag{3.2}$$

Let us also be given another equation of this type

$$\begin{aligned} \psi'_{xx} + u'\psi' &= \lambda\psi' \\ y' &= \psi'_x/\psi \end{aligned} \tag{3.3}$$

Thus, we have

$$a = \lambda - u \qquad b = 0 \qquad c = -1$$

$$a^0 = \lambda \qquad b^0 = 0 \qquad c^0 = -1$$

$$r_i = \pm\sqrt{\lambda} \quad \text{for} \quad i = 1,2,$$

Let us see how the "invariant scattering function" S is defined in terms of the usual "scattering data" (see Gardner, Greene, Kruskal and Muira, 1967, and Zakharov and Faddeev, 1971). Fix $\lambda < 0$. Let ψ_+, ψ_- be solutions of (3.1) such that

$$\lim_{x\to\pm\infty} \psi_\pm(x) = e^{\sqrt{\lambda}x} = 0 \tag{3.4}$$

Let $\overline{\psi_\pm(x)}$ denote their complex conjugate. They are also solutions of (3.1) and

$$\lim_{x\to\pm\infty} \overline{\psi_{\pm}(x)} - e^{-\sqrt{\lambda}x} = 0 \tag{3.5}$$

(This type of solution - determined by "boundary conditions at infinity" - is generally called a Jost function in the physics literature.) The relation between the Jost function at $+\infty$ and $-\infty$ is then determined by a relation of the following form

$$\psi_+ = a\psi_- + b\overline{\psi}_- \tag{3.6}$$

where a,b are functions of λ. If ψ,ψ' are solutions of (3.1), set

$$W(\psi,\psi') = \psi_x\psi' - \psi\psi'_x \equiv \text{the Wronskian} \tag{3.7}$$

Then,

$$W(\psi_+,\overline{\psi}_+) = 2\sqrt{\lambda} = W(\psi_-,\overline{\psi}_-) \tag{3.8}$$

Hence,

$$a = W\frac{(\psi_+,\overline{\psi}_-)}{2\sqrt{\lambda}}$$

$$b = \frac{(\psi_+,\psi_-)}{2\sqrt{\lambda}} \tag{3.9}$$

DEFINITION. Set

$$R(\lambda) = \frac{b(\lambda)}{a(\lambda)} \tag{3.10}$$

This is the reflection coefficient for the linear differential equation (3.1).

Now, set

$$y_1 = \frac{\psi_{+,x}}{\psi_+} \qquad y_2 = \frac{\psi_{-,x}}{\psi_-} \qquad y_3 = \frac{\overline{\psi}_{-,x}}{\overline{\psi}_-} \qquad y_4 = \frac{\overline{\psi}_{+,x}}{\overline{\psi}_+} \tag{3.11}$$

y_1,y_2,y_3,y_4 are solutions of the Riccati equation (3.2). They have the asymptotic behavior as $x \to \pm\infty$ needed (following the ideas of section 2) to define the invariant scattering function

$$S(\lambda) = \frac{y_1-y_2}{y_1-y_3}\frac{y_4-y_3}{y_4-y_2} \tag{3.12}$$

THEOREM 3.1

$$S(\lambda) = |R(\lambda)|^2 \tag{3.13}$$

Proof. Using (3.10)-(3.11), we have

$$R = \frac{b}{a} = \frac{W(\psi_+,\psi_-)}{W(\psi_+,\overline{\psi}_-)} = \frac{\psi_{+,x}\psi_- - \psi_+\psi_{-,x}}{\psi_{+,x}\overline{\psi}_- - \psi_+\overline{\psi}_{-,x}} = \frac{y_1-y_2}{y_1-y_3}\,\frac{\psi_+\overline{\psi}_-}{\psi_+\psi_-}$$

(using (3.11)).

Hence,

$$\overline{R} = \frac{y_4-y_3}{y_4-y_2}\,\frac{\psi_-}{\overline{\psi}_-}$$

hence,

$R\overline{R}$ = right hand side of (3.13)

DEFINITION. The Riccati differential equations

$$\begin{aligned} y_x &= \lambda - u - y^2 \\ y'_x &= \lambda - u' - y'^2 \end{aligned} \tag{3.14}$$

are related via a <u>Bäcklund transformation of Wahlquist-Estabrook type</u> if there are a pair (w,w') of function of x, and a constant λ' such that the following conditions are satisfied

$$u = -w_x \qquad u' = -w'_x \qquad (w + w')_x = (w - w')^2 - \lambda \tag{3.15}$$

We recognize (3.15) as part of the formulas associated with the Korteweg-de Vries Bäcklund transformation (see Hermann (ed.), 1977). That the solutions of (3.14) are connected via relations of the form (2.12) is now a consequence of the Bianchi-type superposition formula, first proved by Wahlquist and Estabrook (see Hermann (ed.), 1977).

Thus we see that, as a consequence of general principles, the absolute value of the reflection coefficient is invariant under Bäcklund transformations. In fact, it is known that the reflection coefficient itself is invariant. I do not know if this is also a "general" fact, or is more specially tied to the specific dynamical situation. I suspect it is the former.

Changes in the Scattering Function Induced by a Lie Algebra-valued One-form of Curvature Zero

Let us return to the general setting of section 2. Introduce a "time" variable t. Suppose, given a pair $(A(x,t),B(x,t))$ of maps $R^2 \to G$. It defines a G-valued one-form

$$\eta = A \otimes dx + B \otimes dt \tag{4.1}$$

on R^2. Its curvature is then the following G-valued two-form

$$\Omega = dA \wedge dx + dB \wedge dt + \frac{1}{2}[\eta,\eta] = (A_t - B_x - [A,B]) \otimes dt \wedge dx \tag{4.2}$$

Suppose that

$$\Omega = 0 \tag{4.3}$$

Then the following vectorial equations

$$\begin{aligned} z_x &= A(x,t)z \\ z_t &= B(x,t)z \end{aligned} \tag{4.4}$$

are completely integrable in the classical sense. Their solution can be written in the following form.

$$z(x,t) = g(x,t)(z(0,0)) \tag{4.5}$$

where (x,t) $g(x,t)$ is a surface in G, such that

$$\begin{aligned} g_x &= gA \\ g_t &= gB \end{aligned} \tag{4.6}$$

Now, set

$$y = \frac{z_1}{z_2} \tag{4.7}$$

It satisfies the following Riccati equations

$$y_x = a + by + cy^2 \tag{4.8}$$

$$y_t = \alpha + \beta y + \gamma y^2 \tag{4.9}$$

where the coefficients $(a,b,c,\alpha,\beta,\gamma)$ are functions of x and t constructed from A and B.

We can now form the invariant scattering function S using equation (4.8). It will depend on t. Its evolution in t is via a linear fractional transformation under which S (as a cross-ratio) is invariant. Hence,

$$S \text{ is independent of } t \tag{4.10}$$

Thus, we see that "deformations" determined by (4.3) and (4.10) leave the invariant scattering function unchanged. Thus, we see that there are two ways of changing the curve $x \to A(x)$ in G so as to leave unchanged the invariant scattering function - via the "discrete" Bäcklund transformation and via the "continuous" deformations generated by equations of type (4.3)-(4.4).

REMARK. I have been sloppy about the precise asymptotic conditions required for (4.4) in order to leave S unchanged. What must happen is that the equations (4.4) must be such as to imply that formally

$$\frac{\partial}{\partial t}(y(\pm\infty,t)) = 0 \tag{4.11}$$

In the examples with which I am familiar, this is accomplished by requiring that

$$\lim_{x\to\pm\infty} B(x,t) = \text{constant} \times \left(\lim_{x\to\pm\infty} A(x,t)\right) \tag{4.12}$$

At this point is established a close relation to the standard material concerning Korteweg-de Vries-Schrödinger obtained via the standard inverse scattering technique. If the potential function $x \to u(x)$ changes via a solution $x \to u(x,t)$ of the Korteweg-de Vries equation, it is known (see Dodd and Gibbon, 1977 and preprint) that the reflection coefficient $\lambda \to R(\lambda)$

changes via the rule

$$R(\lambda,t) = e^{\sqrt{\lambda^3}t} R(\lambda) \qquad (4.13)$$

In particular, $(R(\lambda,t)) = (R(\lambda))$. Again, the rule (4.12) is a more precise result which may follow from a sharpening of the tools developed here.

Prolongations and Generalized Conservation Laws for Exterior Differential Systems

Where do relations like (4.3) come from? Geometrically, (4.3) says that a curvature is zero. A comprehensive geometric genesis for relations of this sort has been provided by Estabrook and Wahlquist (1975) and I will now, briefly, develop their ideas in the form presented in my work (1961, 1976, 1977).

Let X be a manifold. An exterior differential system for X - abbreviated to ED - is a collection of differential forms with the following properties

ED is an ideal in the Grassman algebra formed by the differentials forms in X

$d(ED) \subset ED$

(ED is also called a differential ideal.)

Let X be a manifold with such an ED. A submanifold map $\alpha: Z \to X$ is an integral manifold of

$\alpha^*(ED) = 0$

DEFINITION. Let ED,ED' be exterior differential systems on manifolds X, X'. A map $\pi: X \to X$ is called a prolongation if

$\pi^*(ED) \subset ED$

Thus, such a prolongation maps integral manifolds of ED into integral submanifolds of ED'. In view of the standard relations between exterior differential systems and partial differential equations, this is one way of defining the notion of "geometric homomorphism" between solutions of different differential equations.

DEFINITION. Let (ED,X), (ED',X') be exterior differential systems. A Bäcklund transformation between them is another system (ED",X"), together with maps

$$\beta : X'' \to X \qquad \beta' : X'' \to X'$$

such that

$$\beta^*(ED) \subset ED'' \supset \beta'^*(ED')$$

Such a Bäcklund transformation provides a "correspondence" between integral submanifolds of ED and ED' (thus, a "correspondence" or "relation" between solutions of the underlying differential equations).

Estabrook-Wahlquist Prolongations and Generalized Conservation Laws

Start off with ED as an exterior differential system on the manifold X. Let (ED',X') be another exterior system, and let

$$\pi : X' \to X$$

be a prolongation map. It is said to be an Estabrook-Wahlquist prolongation if the following conditions are satisfied.

a) π is a submersion map, i.e., $\pi_* : T(X') \to T(X)$ is onto

b) There is a set of one-forms $\mathcal{P}$ on X' with the following properties:

   i) $\mathcal{P}$ is a module over the ring F(X') of C^∞ functions on X'.

   ii) A tangent vector $v \in T(X')$ is tangent to the fibers of π if and only if $\mathcal{P}(v) = 0$. (In other words, $\mathcal{P}$ defines an Ehresmann connection for the fiber space $\pi : X' \to X$.)

   iii) ED' is the Grassmann algebra ideal generated by $\pi^*(ED)$ and $\mathcal{P}$.

Condition (iii) is the geometric kicker. It says that, over the submanifolds

$$\phi : Z \to X$$

such that $\phi^*(ED) = 0$, the connection defined by $\mathcal{P}$ is flat. Another way of putting this is to say that the curvature form of the connection defined by $\mathcal{P}$ lie in the ED.

In view of the relations between "connections" and "Lie-algebra valued one-forms", it is convenient to make further definitions. Let G be a real Lie algebra (possibly infinite dimensional). A G-valued one-form on X is a linear mapping

$$\eta : T(X) \to G$$

Its curvature is a G-valued two-form on Ω defined by the following formula.

$$\Omega = d\eta + \frac{1}{2}[\eta,\eta] \tag{6.1}$$

DEFINITION. η is called a <u>generalized conservation law</u> for the system ED if Ω lies in ED $\otimes G$. (Identify G-valued differential forms in X with elements of $\mathcal{D}(X) \otimes G$, where $\mathcal{D}(X)$ = <u>all</u> differential forms on X and the tensor product $\otimes$ is that of real vector spaces.

Such an η defines connections. Let Y be a space on which G acts as a Lie algebra of vector fields. Set

$$X' = X \times Y$$

We can use η to define a connection in the following way. Suppose (y^a), $1 \leqslant a,b \leqslant m$, is a coordinate system for Y, and (x^i), $1 \leqslant i,j \leqslant n$, is a coordinate system for X. Suppose

$$\frac{\partial}{\partial x^i} = \Gamma^a_i(y)\frac{\partial}{\partial y^a} \tag{6.2}$$

Set

$$\theta^a = dy^a - \Gamma^a_i dx^i \tag{6.3}$$

These are one-forms on $X \times Y = X'$. They generate P, and the connection. It is readily seen that the connection is flat, i.e., the Pfaffian system P is completely integrable, if and only if the curvature form Ω vanishes. If η is a generalized conservation law for ED, it is readily verified that

$$X' = X \times Y \to X$$

the Cartesian projection map, is an Estabrook-Wahlquist prolongation in the sense defined above.

Let us now specialize to the case of two-dimensional integral submanifolds of ED, i.e., $Z = R^2$ parametrized by x and t, and G the algebra of 2×2 real matrices of trace zero. Let

$$\phi:Z \to X$$

be an integral submanifold. Suppose $\eta:T(X) \to G$ is a generalized conservation law. Then

$$\phi^*(\eta) = A(x,t)dx + B(x,t)dt$$

where $A(x,t)$, $B(x,t)$ are 2×2 matrix valued functions of (x,t). The condition $\phi^*(\Omega) = 0$ then gives the relation

$$A_t - B_x = [A,B]$$

that we encountered in the previous section in terms of scattering theory.

For the Korteweg-de Vries equation, X is the space of variables (x,t,u,u_x,u_{xx}). Estabrook and Wahlquist have determined (1975) all "generalized conservation laws" associated with one way of writing Korteweg-de Vries as an exterior differential system. This leads to an interesting infinite dimensional Lie algebra; it is generated (as a Lie algebra) by seven elements, satisfying structure relations given by formula (3.7) of their paper (1975). The standard "inverse scattering" equations associated with the Korteweg-de Vries equation is obtained by looking for Lie algebra homomorphisms from this Lie algebra to the Lie algebra of $SL(2,R)$. It is also clear, from the work of Wahlquist and Estabrook, that the notions of "Bäcklund transformation" and "superposition formulas" are closely tied with the Lie algebra. We would very much like to know a more intrinsic definition of this Lie algebra, we well as general techniques for computing it for other partial differential equations! (More extensive work toward generalization has been done by H. Morris, 1976, 1976 and 1977, J. Corones, 1976 and 1977, R. Dodd and J. Gibbon, 1977 and preprint.) Another interesting question is the possible physical meaning of these "generalized conservation laws". Further work is in progress.

REFERENCES

Corones, J. Solitons and Simple Pseudopotentials. J. Math. Phys., 17 (1976), pp. 756-759

Corones. J. Solitons, Pseudopotentials and Certain Lie Algebras. J. Math. Phys., 18 (1977), pp, 163-164.

Dodd, R. K., and Gibbon, J. D. The Prolongation Structure of a Class of

Nonlinear Evolution Equations. Preprint, 1977.

Dodd, R. K., and Gibbon, J. D. The Prolongation Structure of Some Higher Order Korteweg-de Vries Equations. Preprint.

Gardner, C. S., Greene, J. M., Kruskal, M. D. and Muira, R. Phys. Rev. Lett., 19 (1967), pp. 1095-1097.

Hermann, R. The Inverse Scattering Technique of Soliton Theory, Lie Algebras, the Quantum Mechanical Poisson-Moyal Bracket, and the Rigid Rotating Body. Phys. Rev. Lett., 37 (1961), p. 1591.

Hermann, R. The Pseudopotentials of Estabrook and Wahlquist, the Geometry of Solitons and the Theory of Connections. Phys. Rev. Lett., 36 (1976), p. 835.

Hermann, R. The Geometry of Nonlinear Differential Equations, Bäcklund Transformations, and Solitons. Interdisciplinary Mathematics, Math. Sci. Press, Brookline, Mass. Part A (1976), Part B (1977) and Part C (in preparation).

Hermann, R. Toda Lattices, Cosymplectic Manifolds, Bäcklund Transformations and Kinks. Interdisciplinary Mathematics, Vol. 15, Math. Sci. Press, Brookline, Mass. Part A (1977).

Hermann, R. (ed.) The Ames Research Center (NASA) 1976 Conference on the Geometry of Nonlinear Waves, Lie Groups. History, Frontiers and Applications, Vol. 6, Math. Sci. Press, Brookline, Mass. (1977).

Morris, H. C. Prolongation Structures and a Generalized Inverse Scattering Problem. J. Math. Phys., 17 (1976), pp. 1867-1869.

Morris, H. C. Prolongation Structures and Nonlinear Evolution Equations in Two Spatial Dimensions. J. Math. Phys., 17 (1976), pp. 1870-1872.

Morris, H. C. A Prolongation Structure for the AKNS System and its Generalizations. J. Math. Phys., 18 (1977), pp. 533-536.

Morris, H. C. Articles in Hermann, R. (ed.) above.

Wahlquist, H. D. and Estabrook, F. B. Prolongation Structures of Nonlinear Evolution Equations. J. Math. Phys., 16 (1975), pp. 1-7.

Zakharov, V. E., and Faddeev, L. D. Korteweg-de Vries Equation: A Completely Integrable Hamiltonian System. Func. Anal. and its Appli., 5 (1971), pp. 280-287.

This paper is reprinted from Interdisciplinary Mathematics, Volume 20, Chapter 29, Math. Sci. Press, Brookline, Mass. 1979.

Rutgers University
New Brunswick, New Jersey

STOCHASTIC JACOBI FIELDS

Paul Malliavin

Introduction

Given an elliptic operator Δ on a manifold, there is now a classical way, via an associated wave operator, to associate a flow. Then, the asymptotic properties of the flow reflect the asymptotic properties of the spectrum Δ.

Another point of view is to associate to Δ its Ito stochastic differential system. Then, the asymptotics of the generic solution of this system will yield some information on the lowest bound of the spectrum of Δ. As an analog of the two points of view in commutative harmonic analysis, it is well known that the Fourier transform gives some information on the behavior of the singularities of a measure when the Laplace transform gives the upper bound of the support of a measure with right limited support. By a general correspondence principle, a stochastic differential system can be viewed as the limit of an ordinary differential system. Then, the second point of view will lead, as the first, to the study of asymptotics of some ordinary differential system.

A classicial Jacobi field is a vector field realizing an infinitesimal transformation preserving the geodesic flow. In our setting, we shall give the components of a moving frame $(\xi_1(t),\ldots,\xi_n(t))$. To this moving frame corresponds a horizontal flow on the bundle of orthonormal frames of the Riemannian manifold. Then, an ordinary Jacobi field will be an infinitesimal transformation preserving this flow. The stochastic analog will be the stochastic Jacobi field. As the classical Jacobi fields are developed in classical calculus of variations, we shall have to work in stochastic calculus of variations.

Correspondence Principles Between Stochastic and Ordinary Differential Equations

We shall denote by u_1 a C^∞-real valued function, having its support in $[0,1]$ and satisfying $\int u dt = 1$. We define $u_\varepsilon(t) = \varepsilon^{-1} u_1(t\varepsilon^{-1})$. Now, let $b^k_\omega(t)$ be a sample path of Brownian motion on R^n $(k = 1,\dots,n)$ and we denote by N^t the increasing family of σ-fields defined by the past $t' \leqslant t$. We define

$$b^k_{\varepsilon(\omega)}(t_0) = \int b^k_\omega(t + t_0) u_\varepsilon(t) dt$$

Then, $b_{\varepsilon(\omega)}(\cdot)$ is a C^∞-function, and $b_{\varepsilon(\omega)}(t - \varepsilon)$ is N^t-adapted.

LIMIT THEOREM 2.1. Consider the stochastic differential system (with C^∞ coefficients)

$$dx^i_\omega(t) = a^i_k(x_\omega(t)) db^k_\omega(t) + C^i_\omega(x(t)) dt, \quad 1 \leqslant i \leqslant m, \quad 1 \leqslant k \leqslant n \tag{2.1}$$

and the ordinary differential system

$$dx^i_{\varepsilon(\omega)}(t) = a^i_k(x_{\varepsilon(\omega)}(t)) db^k_{\varepsilon(\omega)}(t) + \tilde{C}^i(x_{\varepsilon(\omega)}(t)) dt \tag{2.2}$$

where

$$\tilde{C}^i = C^i - \frac{1}{2} \sum_k a^s_k \frac{\partial}{\partial x^s} (a^i_k) \tag{2.3}$$

We shall denote by $x(\omega,v_0,t)$, $x(\omega,\varepsilon,v_\varepsilon,t)$ the solutions of (2.1) and (2.3) which satisfy Cauchy's condition

$$x(\omega,v_0,0) = v_0 \qquad x(\omega,\varepsilon,v_\varepsilon,2\varepsilon) = v_\varepsilon \tag{2.4}$$

Then it is possible to find a universal numerical sequence ε_j such that

$$v_j \to v_0 \quad (v_j = v_{\varepsilon_j}) \tag{2.5}$$

Then, for almost all ω, $x(\omega,\varepsilon_j,v_j,t) \to x(\omega,v_0,t)$ locally uniformly in t. Furthermore, if the convergence holds uniformly in v_0, we have, for almost all ω,

$$x(\omega,\varepsilon_{j'},v_j,t) \text{ converges locally uniformly in } (v_0,t) \tag{2.6}$$

Proof. See Malliavin (1976).

COROLLARY 2.2. Almost surely in ω, $x(\omega,v_0,t)$ exists for all v_0 and depends continuously upon v_0.

Proof. We define $x(\omega, v_0, t)$ by the limit (2.6). (See Malliavin, 1976.)

Define $U_{\omega,t}$ by $U_{\omega,t}(v_0) = x(\omega, v_0, t)$. Then $U_{\omega,t}$ is a pseudogroup of local C^∞-diffeomorphisms.

Proof. See Malliavin (1976).

We will now consider the stochastic multiplicative integral (see Ibero, 1976, and McKean, 1969).

Let G be a matrix group, $\mathcal{G}$ its Lie algebra. Given nonanticipating functions $E_1(t,\omega), \ldots, E_n(t,\omega)$, $C(t,\omega)$ with values in $\mathcal{G}$, we consider the ordinary matrix differential equation

$$dg_{\varepsilon(\omega)}(t) = g_{\varepsilon(\omega)}(t)\{E_k(t,\omega)db^k_{\varepsilon(\omega)}(t) + C(t,\omega)dt\} \tag{2.7}$$

Then, by a generalization of Theorem 2.1, this system has as a limit the stochastic matrix equation

$$dg_\omega(t) = g_\omega(t)(E_k(t,\omega)db^k_\omega(t) + \hat{C}(t,\omega)dt) \tag{2.8}$$

where

$$\hat{C} = C + \frac{1}{2}\sum_k E_k^2(t,\omega) \tag{2.9}$$

We can also consider the McKean stochastic multiplicative integral defined by the limit of products of the form

$$\ldots\exp(E_k(t_s,\omega)(b^k_\omega(t_{s+1}) - b^k_\omega(t_s)) + C(t_s,\omega)(t_{s+1} - t_s))\ldots$$

where $t_{s+1} > t_s$ and the terms are in increasing order of indices from left to right. We shall denote by

$$\exp\left(* \int_{[0,t]} (E_k db^k_\omega + Cdt)\right)$$

the McKean's stochastic multiplicative integral. Then, we have that the general solution of (2.8) is

$$g_\omega(t) = g_\omega(0)\exp\left(* \int_{[0,t]} E_k db^k_\omega + Cdt\right) \tag{2.10}$$

Furthermore, if $g_0 \in G$, then $g_\omega(t) \in G$ for every $t > 0$.

Ordinary Calculus of Variations on a Parallelized Manifold

The Darboux's point of view of the moving frames reduces some problems in differential geometry to integration of matrix linear differential systems.

We shall denote by V a C^3 manifold. We shall suppose that, at every point on V, is given a differential form θ with values in R^m, realizing for every $v_0 \in V$ an isomorphism of

$$T_{x_0}(V)$$

on R^m. We shall call θ the parallelism differential form. We denote $\theta = (\theta^1,\ldots,\theta^m)$. Then, for every differential form of degree k, λ can be uniquely written as

$$\lambda = f_{i_1,i_2,i_k}\theta^{i_1} \wedge \theta^{i_2} \wedge \ldots \wedge \theta^{i_k}$$

In particular, this is the case for the coboundary $d\theta^i$, which means that there exist functions $a^i_{k,r}$ such that we have the structural equation

$$d\theta^i = a^i_{k,r}\theta^k \wedge \theta^r \qquad k < r \tag{3.1}$$

We shall also write this in matrix form

$$(d\theta)(z \wedge \tilde{z}) = A(\theta(z),\theta(\tilde{z})) \tag{3.2}$$

where A is a bilinear antisymmetric map $R^n \times R^m \to R^m$ defined by the $a^i_{k,r}$. Let $C^2(R^m)$ be the vector space of C^2 maps of R^+ in R^m. If $f \in C^2(R^m)$, let z^f denote the vector field defined on $V \times R^+$ by

$$z^f_{\tau,v} = \theta^{-1}_v(\frac{df}{d\tau}) \tag{3.3}$$

Then we have a flow on V. It defines a semigroup of transformations on $V \times R^+$. Suppose $U_{f,\tau}$ solves Cauchy's problem

$$\frac{d}{d\tau}(U_{f,\tau}v_0) = z^f_{\tau,v(\tau)} \tag{3.4}$$

where $v(t) = U_{f,\tau}(v_0)$, $U_{f,0}(v_0) = v_0$. We fix τ in τ_0 and we wish to compute the differential U'_{f,τ_0} of the map

$$v \to U_{f,\tau_0}v$$

Then, using the parallelism, this differential can be express by a matrix

$$J(v_0,\tau_0,f) = \theta^{-1}_{v(\tau_0)}\ (U'_{f,\tau_0}(v_0))\ \theta^{-1}_{v_0} \tag{3.5}$$

Let Φ be any smooth map of $(R^+)^2 \to R^m$. Consider the pullback $\Phi^*\theta$ of θ by this map. Then

$$\Phi^*\theta = p(\tau,t)d\tau + q(\tau,t)dt \tag{3.6}$$

where p,q take their values in R^m.

LEMMA 3.1.

$$\frac{\partial q}{\partial \tau} = \frac{\partial p}{\partial t} + A_\Phi(p,q) \tag{3.7}$$

where A comes from the structural equation (3.2).

Proof. We shall compute $d(\Phi^*\theta)$ in two ways

$$d(\Phi^*\theta) = \left(\frac{\partial q}{\partial \tau} - \frac{\partial p}{\partial t}\right) d\tau \wedge dt$$

Now use (3.2), which gives us

$$d(\Phi^*\theta) = \Phi^*(d\theta) = A_\Phi(p,q)d\tau \wedge dt$$

which proves the lemma.

PROPOSITION 3.2. Let $z_o \in R^m$ and denote

$$z(\tau) = J(v_o,\tau,f)\cdot z_o \tag{3.8}$$

Then, $z(\tau)$ is determined as the solution of the Cauchy problem

$$\frac{dz(\tau)}{d\tau} = A_{v(\tau)}(f'(\tau),z(\tau)) \tag{3.9}$$

(3.9)

$$z(0) = z_o$$

Proof. Construct a C^2 curve on $V: t \to g(t)$ satisfying $\frac{dg}{dt}(0) = z_o$. Define a map Φ of $(R^+)^2$ into V by

$$\Phi(\tau,t) = U_{f,\tau}(g(t)) \tag{3.10}$$

Then

$$\Phi^*\theta = pd\tau + qdt \tag{3.11}$$

$\frac{\partial \Phi}{\partial \tau} = \frac{\partial}{\partial \tau} U_{f,\tau}(g(t)) = z^f_{\tau,\Phi}$. Therefore,

$$\Phi^*\theta = f'(\tau)d\tau + q(\tau,t)dt \tag{3.12}$$

Now, using equation (3.7), we get (3.9).

REMARK. As the differential equation (3.9) is linear in z, we can, therefore, put this solution in a matrix calculus setting. For ξ fixed, we denote by $i(\xi)A$ the map of $R^m \to R^m$ defined by $\eta \to A(\xi,\eta)$. We use the notation of the multiplicative integral for the solution of the matrix differential equation (see Ibero, 1976), then (3.9) reads

$$J(v_0,\tau_0,f) = \exp\left({}^* \int_{[\tau_0,0]} i(f'(\tau))A_{v(\tau)}d\tau \right) \tag{3.13}$$

We shall now suppose that the control function, f, will be modified by a small variation δf, and that the origin of the arc will be fixed in v_0. This will correspond to the equation

$$\frac{\partial p}{\partial t} = \frac{\delta f}{dt} \qquad q(o,t) = 0 \tag{3.14}$$

We want to compute the variation

$$(\tilde{\delta} v)(\tau) = U_{\delta f,\tau}(v_0) \tag{3.15}$$

We shall substitute this variation in the parallelism

$$(\delta v)(\tau) = \theta_{v(\tau)}(\tilde{\delta} v(\tau)) \tag{3.16}$$

Then we have

PROPOSITION 3.2. Let J be given by (2.10). Then

$$(\delta v)(\tau_0) = \int_0^{\tau_0} J(v(\tau),\tau_0 - \tau,f)\cdot(\delta f')(\tau)d\tau \tag{3.17}$$

$$(J(v_0,\tau_0,f))^{-1}\delta v(\tau_0) = \int_0^{\tau_0} (J(v_0, \ ,f))^{-1}(\delta f')(\tau)d\tau \tag{3.18}$$

Proof. We remark that $(\delta v)(\tau) = q(\tau,o)$. Then, (2.7) and (2.2) give

$$\frac{d}{d\tau}\delta v = \delta f' + i(f')A\cdot\delta v$$

This gives a linear differential equation in δv which is solved by Lagrange's method, using $(\delta v)(0) = o$, and then getting (3.17).

REMARK. Suppose that, instead of making a smooth variation, δf of f, we proceed by a finite number of jumps. More precisely, we shall define a <u>discrete variation</u>, δ_D, by a finite partition, $0 \leqslant \tau_1 < \tau_2 < \ldots < \tau_n < \tau_o$. At each point τ_k, we associate $c_k \in R^m$ and we define

$$(f + \delta_D f)(\tau) = f(\tau) + \delta t \sum_{\tau_k < \tau} c_k \tag{3.19}$$

Then we deduce from (3.9) that

$$(\delta_D v)(\tau_o) = \delta t \sum_{o \leqslant \tau_k < \tau_o} J(v(\tau_k), \tau_o - \tau_k, f) c_k \tag{3.20}$$

In this form, (3.17) is nothing but a continuous version of (3.20), and, in fact, it would be possible to deduce (3.17) from (3.20) using a discretization technique. In conclusion, (3.9) could be used as a basic formula.

Given a parallelism on V we shall introduce the <u>normal chart</u> at v_o as the map $n_{v_o} : R^m \to V$ defined by

$$n_{v_o}(\xi) = U_{f_\xi, 1}(v_o) \tag{3.21}$$

where $f_\xi(\tau) = \tau\xi$. As the Jacobian of n_{v_o} at v_o is the identity, we then get a local chart at v_o.

PROPOSITION 3.3. Denote

$$\phi = n^{-1}_{v(\tau_o)} \circ U_{f,\tau_o} \circ n_{v_o} \tag{3.22}$$

Then ϕ is a map of R^m into itself. The first two terms of its Taylor expansion are

$$\phi(\xi) = J(v_o, \tau_o, f) + \frac{1}{2}(D_\xi J)\cdot\xi + (0(|\xi|^3)) \tag{3.23}$$

where, for $\xi = (\xi' \ldots \xi^m)$, we denote

$$(D_\xi J)\cdot\xi = \sum_k \left(\frac{\partial J}{\partial \xi^k} \left(n_{v_o}(\xi) \right) \right)_{\xi = o} \cdot \xi \tag{3.24}$$

Proof. Consider the pullback of θ by U^*_{f,τ_o}. Then

$$(U^*_{f,\tau_o}\theta)_v = J(v,\tau_o,f)o\theta_v \tag{3.25}$$

and we have two parallelisms on the neighborhood of v_o - one given by θ and the other by $U^*_{\tau_o}\theta = \tilde{\theta}$. Denote by $n_{v_o}, \tilde{n}_{v_o}$ the associated normal charts. Then,

$$U_{f,\tau_o} o\tilde{n}_{v_o} = n_{v(\tau_o)} = \tilde{n}^{-1}_{v_o} on_{v_o} \tag{3.26}$$

We shall forget the indices for these normal charts. Denote $J_0 = J(v_o,\tau_o f)$. J_1 will be the first order term in the Taylor expansion $\xi \to J(n(\xi),\tau_o,f)$. Finally, we shall denote $\tilde{\xi} = J_o\xi$, $\phi(\xi) - \tilde{\xi} = r$. By (3.8)

$$r = 0(|\xi|^2) \quad \xi \to 0 \tag{3.27}$$

Now write (3.26) in the form

$$\tilde{n}(\tilde{\xi} + r) = n(\xi) \tag{3.28}$$

The differential system defining the two normal charts are

$$\begin{aligned} &\frac{d}{dt} n(t(\tilde{\xi} + r)) = \theta^{-1}_{\tilde{n}(t(\tilde{\xi}+r))}(\tilde{\xi} + r) \\ &\frac{d}{dt} n(t\xi) = \theta^{-1}_{n(t\xi)}(\xi) \end{aligned} \tag{3.29}$$

By (3.27) and (3.28), we have

$$\tilde{n}(t\tilde{\xi}) = n(t\xi) + 0(|\xi|^2) \tag{3.30}$$

Therefore, the first equation of (3.29) leads to

$$\tilde{n}(\tilde{\xi} + r) = \int_0^1 \theta^{-1}_{n(t\xi)}(\tilde{\xi} + r)dt + 0(|\xi|^3)$$

Now use (3.25), which gives $\tilde{\theta} = (J_0 + J_1 + 0(|\xi|^2))$, and, therefore,

$$\tilde{n}(\tilde{\xi} + r) = \int_0^1 \theta^{-1}_{n(t\xi)}(1 - J_0^{-1}J_1(t\xi))J_0^{-1}(\tilde{\xi} + r)dt + 0(|\xi|^3)$$

Now, use the second equation of (3.29), with $J_0^{-1}\tilde{\xi} = \xi$, to obtain

$$\tilde{n}(\tilde{\xi} + r) = n(\xi) + \theta_0^{-1}J_0^{-1}r - \tfrac{1}{2}\theta_0^{-1}J_0^{-1}J_1(\xi)\cdot\xi + 0(|\xi|^3)$$

Again using (3.28), we obtain

$$r = \frac{1}{2}J_1(\xi)\cdot\xi + O(|\xi|^3)$$

which proves (3.23)

The above has reduced the problem of computing the second derivative of $U_{f,\tau}$ to the problem of computing the first derivative of the matrix J.

This last problem is a problem of calculus of variation of the first order. In fact, the matrix J is expressed by (3.13) as a path integral along $U_{f,\cdot}(v_0)$. Therefore, its variation can be computed in terms of the variation of the path, and, finally, in the variation of the path from v_0, as in (3.17).

LEMMA 3.4. Let $\tau \to A(\tau)$ be a continuous map of R in m×m matrices. Denote

$$\Phi(A) = \exp\left(* \int_{[\tau_o,0]} A(\tau)d\tau\right)$$

Let $A + \delta A$ be a variation of $A(\cdot)$. Denote by $\delta\Phi$ the first order variation of Φ. Then,

$$(\Phi(\tau_o))^{-1}(\delta\Phi(\tau_o)) = \int_0^{\tau_o} ((\delta A)(\tau))^{\Phi(\tau)}d\tau \qquad (3.31)$$

where $B^{\Psi} = \Psi^{-1}B\Psi$.

Proof. Use the associated differential equation $\frac{\partial\Phi}{\partial\tau} = A\Phi$. We have the usual variational equation

$$(\delta\Phi)' = (\delta A) + A(\delta\Phi) \qquad (3.32)$$

and Lagrange's method leads to (3.31).

PROPOSITION 3.5. Let $D_\xi J_{\tau_0}$ be defined as in (3.24). Then it satisfies

$$J_{\tau_0}^{-1}(D_\xi J_{\tau_o}) = \int_0^{\tau_o} (i(f'(\tau))D_{J_\tau \xi}A)^J d\tau$$

Proof. It follows directly from (3.13) and (3.31).

Stochastic Calculus of Variation on a Parallelized Manifold

First, we will consider semielliptic operators and parallelisms. Given a parallelized manifold (V,θ), we have the canonical vector fields $A_1 \ldots A_m$ defined as $\langle A_1,\theta\rangle = (1,0\ldots0)$ and so on. Choosing $n \leqslant m$, we consider the semielliptic operator

$$L = \frac{1}{2}\sum_{k=1}^{n} \mathcal{L}^2_{A_k} \tag{4.1}$$

Conversely, let us give a hypoelliptic operator on a manifold W as

$$\tilde{L} = \frac{1}{2}\sum_{k}^{n} \mathcal{L}^2_{\tilde{A}_k} \tag{4.2}$$

Suppose the $\tilde{A}_k(w_0)$ are linearly independent for every $w_0 \in W$. Then, it is always possible to find locally a parallelism such that $\tilde{L}$ takes the form (3.2). Therefore, from a local point of view, (4.1) brings no restriction other than the above.

Ito's stochastic system for (4.2) reads in a local chart, ϕ,

$$dx^i_{\phi,\omega} = A^i_s db^s_\omega + \frac{1}{2}\left(\sum_s \mathcal{L}_{A^s} A^i_s\right) dt \tag{4.3}$$

Now, given a sample of the Brownian motion b_ω on R^n, $\omega \in \Omega(R^n)$, we recover, integrating (4.3) with the given initial condition $a \in V$, a sample $x(\tau,\phi,a,\omega)$ of $\Omega(L)$. Therefore, we have defined a mapping

$$u_{\phi,a}:\Omega(R^n) \to \Omega(L) \tag{4.4}$$

PROPOSITION 4.1. The mapping $u_{\phi,a}$ is intrinsic, i.e., for two local charts ϕ, $\tilde{\phi}$, we have

$$x(\tau,\tilde{\phi}, a,\omega) = (\tilde{\phi}o\phi^{-1})(x(\tau,\phi,a,\omega))$$

where τ runs in the common domain of the definition of the two members.

Proof. Denote $\tilde{\phi}o\phi^{-1} = f$. Then

$$A^i_k = \frac{\partial f^i}{\partial x^q} A^q_k$$

Thus, $\tilde{L}$ can be expressed in terms of the A^q. On the other hand, Ito's stochastic calculus gives the differential of $d(f(s))$. Making these two straightforward computations, we get the proof.

NOTATION. We shall denote the trajectory associated to b_ω by

$$dv = A_k d_s b^k_\omega \quad \text{(Stratanovitch's stochastic calculus)} \tag{4.5}$$

LEMMA 4.2. Suppose the lifetime of the diffusion v_ω is infinite. Then, for every $a \in V$, there is a canonical map

$$\Omega(R^n) \to \Omega_a(L)$$

$$\omega \to v_{\omega,a}(\cdot)$$

Proof. We use a locally finite covering of V by local charts. Then, all of the paths of $\Omega(L)$ on $[0,\tau_o]$ are compact. It is sufficient for a given path to use a finite number of charts.

The correspondence between paths of $\Omega(L)$ and $\Omega(R^n)$ will be established step by step in each different chart, using (4.3). But, thanks to Proposition 4.1, this correspondence will be independent of the chart used.

PROPOSITION 4.3. Suppose (3.17). Then, for every $\omega \in \Omega(R^n)$, $\tau_o > 0$, there exists a mapping

$$U_{\omega,\tau_o} : V \to V$$

defined by

$$U_{\omega,\tau_o}(a) = v_{\omega,a}(\tau_o)$$

Proof. It follows from (3.15) and Corollary 2.2.

THEOREM 4.4. Under the conditions that V is a compact manifold of class C^4 without boundary, the map $v \to U_{\omega,\tau}(v)$ is almost surely of class C^2. Furthermore, its first differential is read in the parallelism as

$$J(v_o,\tau_o,\omega) = \exp\left(* \int_{[\tau_o,0]} (B_i(v_\omega(\tau))db^i_\omega(\tau) + Cd\tau \right) \tag{4.6}$$

where

$$C^i = \sum_k A^i_k \frac{\partial}{\partial v^i} B_k \qquad B_i(\xi) = A(e_i,\xi)$$

(A is defined in (4.2), $e_1 \ldots e_n$ being the canonical basis of R^n) and where the multiplicative integral appearing in (4.6) is a right multiplicative integral in the sense of Section II. A second differential, read in the normal charts n_{v_0}, $n_{v(\tau_0)}$, is given by (3.23) and the analog of Proposition 3.5, which follows.

$$J_{\tau_0}^{-1}(D_\xi J_{\tau_o}) = \int_0^{\tau_o} (D_{J_\tau \xi} B_i)^{J_\tau} d_s b_\omega^i \tag{4.7}$$

where J_τ is given by (4.6) and where the notation d_s means the Stratanovitch stochastic integral.

Proof. The fact that $U_{\omega,\tau}$ is of class C^2 follows from Section II. To obtain the results stated, we use the correspondence principle. Consider $U_{\varepsilon(\omega),\tau}$ defined in (3.3). Then, we can apply the results of Section III. This map is twice differentiable; its first differential is, according to (3.13)

$$U'_{\varepsilon(\omega),\tau_0} = \exp\left(* \int_{[\tau_0,0]} B_i(v_{\varepsilon(\omega)}(\tau)) db^i_{\varepsilon(\omega)}(\tau) \right) \tag{4.8}$$

We shall write (4.8) as the linear differential system

$$\frac{dz_\varepsilon(\tau)}{d\tau} = A_{v(\tau)}(db_{\varepsilon(\omega)}, z_\varepsilon(\tau)) = B_{k,v}(z_\varepsilon(\tau)) db^k_{\varepsilon(\omega)}$$

$$dv_\varepsilon = A_k(v) db^k_{\varepsilon(\omega)}$$

By the transfer principle

$$dz = B_{k,v}(z(\tau)) db^k_\omega + (\partial_i B_{\underline{k},v}) A^k_k + B^2_{\underline{k},v}(z(\tau)) \frac{d\tau}{2}$$

$$dv + A_k db^k + \frac{1}{2}(\partial_i A_{\underline{k}}) A^i_{\underline{k}} d\tau$$

Define V_t, $\tilde{z}_t$ by

$$V_t = \exp\left(* \int_{[\tau_0,0]} B_k db^k_\omega \right) \qquad \tilde{z}_t = V_t^{-1} z_t$$

Then

$$B_k db^k_\omega d\tilde{z}_t + V_t d\tilde{z}_t = (\partial_i b_{\underline{k},v}) A^i_{\underline{k}} d\tau$$

Then, dz_t is given by an ordinary differential equation, and we get

$$z = \exp\left(* \int_{[\tau,0]} B_k db^k_\omega + \sum_K \frac{1}{2}(\partial_i B_{\underline{k},v}) A^i_{\underline{k}} d\tau \right) z_0$$

Therefore, the family of C^1 maps $U_{\varepsilon(\omega),\tau_0}$, a.s. converges uniformly with its differential. Thus the limit U_{ω,τ_0} is a.s. of class C^1 with (4.6) as first differential. Similar reasoning gives the C^2-regularity and (4.7)

We now express the semielliptic operator L in a normal chart

$$(Lf)(v_0) = \left[\frac{1}{2} \sum_{s=1}^{n} \frac{\partial^2}{\partial \xi_s^2} (f \circ n_{v_0})(\xi) \right]_{\xi=0} \tag{4.9}$$

Proof. Denote by $\hat{\theta}$ the parallelism form read in the normal chart. We shall write $\hat{\theta}_\xi$ = Identity + R_ξ. Then, by the normality of the chart

$$R_\xi(\xi) = 0 \tag{4.10}$$

Differentiating this on the first coordinate vector field e_1, we deduce

$$\frac{\partial}{\partial \xi_1} R_\xi \quad (\xi) + R_\xi(e_1) = 0$$

Apply this identity to $\xi = e_1$. Then, according to (4.10)

$$\left(\frac{\partial}{\partial \xi_1} R_\xi\right)_{\xi=e_1} (e_1) = 0 \tag{4.11}$$

On the other hand,

$$(A_1)_\xi = (I + R_\xi)^{-1}(e_1) = e_1 - R_\xi(e_1) + 0(|\xi|^2)$$

$$\mathcal{L}^2_{A_1} = \frac{\partial^2}{\partial \xi_1^2} - \frac{\partial}{\partial \xi_1} R_\xi(e_1) + 0(|\xi|)$$

We then get (4.9). A consequence of (4.9) is that the normal charts are intimately linked to the stochastic calculus of variation for L.

We shall follow the approach of (3.19) and (3.20). Given $0 \leqslant \tau_1 < \tau_2 \ldots < \tau_q$, we shall denote by $c_1, \ldots, c_q$ <u>independent copies</u> of the normal Gaussian variable on R^n. We then define the discrete variation of the Brownian motion. Define $\delta_D\omega$ as the curve

$$b_{\omega+\delta_D\omega}(t) = b_\omega(t) + \sum_{\tau_k \leqslant \tau} (\delta t)^{1/2} (\tau_k - \tau_{k-1})^{1/2} c_k \tag{4.12}$$

where (δt) must be understood as a parameter which will tend to zero. We define a corresponding variation $v_{\omega+\delta_D\omega}$ by the recursion formula

$$v_{\omega+\delta_D\omega}(\tau) = U_{\omega,\tau-\tau_j}(v_{\omega+\delta_D\omega}(\tau_j)) \qquad \tau_j \leqslant \tau < \tau_{j+1} \tag{4.13}$$

At the point τ_j, we will have a jump. Denoting the normal chart at the point $v_{\omega+\delta_D\omega}(\tau_j^-)$ by n_j, we define the value on the right by

$$n_j(v_{\omega+\delta_D\omega}(\tau_j^+)) = ((\tau_j - \tau_{j-1})\delta t)^{1/2} c_j \tag{4.14}$$

where $c_j \in R^m$ by the natural imbedding of $R^n \to R^m$.

PROPOSITION 4.5. τ_j and c_j being fixed, n_τ denoting the normal chart at the point $v_\omega(\tau)$, we have

$$\begin{aligned} n(v_{\omega+\delta_D\omega}(\tau)) &= (\delta t)^{1/2} \sum_{\tau_j<\tau} J(v_\omega(\tau-\tau_j),\tau-\tau_j,\omega^{\tau_j})(\tau_j - \tau_{j-1})^{1/2} c_j \\ &+ \delta t \sum_{\tau_j<\tau} K(v_\omega(\tau_j),\tau - \tau_j,\omega^{\tau_j})(\tau_j - \tau_{j-1}) + O((\delta t)^{3/2-\varepsilon}) \end{aligned} \tag{4.15}$$

where $\varepsilon > 0$, ω^λ denotes the shift of the time λ on the probability space and where

$$K(v_0,\tau,\omega) = \frac{1}{2} \sum_{k=1}^{n} U''_{\omega,\tau}(v_0)(e_k,e_k) \tag{4.16}$$

U'' being computed by (4.7) and (3.23).

Proof. This follows from (4.12) and from the well known Ito's change of variables formula.

Given a path v_ω, we shall define the infinitesimal variation as the continuous analog of the discrete sum appearing in (4.16), to which we could add the drift coming from the $-x\cdot\nabla$ in the Ornstein-Uhlenbeck infinitesimal generator (see Malliavin, 1976).

LEMMA 4.6. Let b_ω fixed, $b_{\varepsilon(\omega)}$ defined in (4.8) and consider the variation

$$b_{\varepsilon(\omega)} = -\frac{\delta t}{2} b_{\varepsilon(\omega)} \qquad \delta v_0 = 0$$

Then, when $\varepsilon \to 0$, the variation of v has for its limit $\delta v = -D_1 \delta t$, where

$$D_1(\tau_0,v_0,\omega) = \frac{1}{2}(J(v_o,\tau_o,\omega)) \int_0^{\tau_o} (J(v_0,\tau,\omega)^{-1}\cdot e_k) d_s b_\omega^k(\tau) \tag{4.17}$$

Proof. This follows from (3.18) and the principle of transfer I.

REMARK. It is important to use (3.18) instead of (3.17), which will lead to anticipating stochastic integration.

DEFINITION. We call a variation $\delta\omega$ of the Brownian motion b_ω defined on $[0,1]$ the data of an independent new Brownian motion $b_{\delta\omega}$, defined on the same interval. We then denote

$$b_{\omega+\delta\omega}(\tau) = b_\omega(\tau) + (\delta t)^{1/2} b_{\delta\omega}(\tau)$$

DEFINITION. We shall define $v_{\omega+\delta\omega}(\tau_0)$, $\tau_0 \in [0,1]$, by

$$n_{v_\omega(\tau_0)}(v_{\omega+\delta\omega}(\tau_0)) = (\delta t)^{1/2} Q + (\delta t)(D_1 + D_2) \tag{4.18}$$

where D_1 is defined in (4.17).

$$Q(v_o, \tau_o, \omega, \delta\omega) = \int_0^{\tau_o} J(v_\omega(\tau), \tau_0 - \tau, \omega^\tau) e_k \, db_{\delta\omega}^k(\tau) \tag{4.19}$$

$$D_2(v_o, \tau_o, \omega) = \int_0^{\tau_o} K(v_\omega(\tau), \tau_0 - \tau, \omega^\tau) d\tau \tag{4.20}$$

K is as defined in (4.16).

REMARK. The integrand which appears in (4.16) is anticipatory, but this does not matter, because b_ω and $b_{\delta\omega}$ are independent.

THEOREM 4.7. Fix $v_0 \in V$, $\tau_0 \in [0,1]$. Let g_0 be the map of $\Omega(R^n)$ into V defined by $\omega \to U_{\omega,\tau_0}(v_0)$. Then, g_{τ_0} has, with respect to the Ornstein-Uhlenbeck process, a stochastic differential dg_{τ_0}. dg_{τ_0} is expressed in the chart $n_{v_\omega(t_0)}$ by (4.18). The covariance matrix σ_{ij} of dg_{τ_0} is given by

$$\sigma_{ij}(\tau_0, \omega) \ell^i \ell^j = \int_0^{\tau_0} \sum_{k=1}^{n} \ell^2 (\tilde{J}(\tau_0, \tau, \omega) e_k) d\tau \tag{4.21}$$

where $\tilde{J}$ is

$$\tilde{J}(\tau_0, \tau, \omega) = J(v_\omega(\tau), \tau_0 - \tau, \omega^\tau) \tag{4.22}$$

Finally, the infinitesimal generator Λ of the process is computed on g_{τ_0} by

$$\Lambda(n_{v_\omega(\tau_o)} \circ g_{\tau_o}) = D_1 + D_2 \tag{4.23}$$

Proof. We first remark that the increase during the time $t, t + \delta t$ of the Ornstein-Uhlenbeck process is

$$b_{\omega+\delta\omega} - \frac{\delta t}{2} b_\omega \tag{4.24}$$

where $b_{\omega+\delta\omega}$ is defined in Lemma 4.6. The second term will contribute to δv by $D_1 \delta t$. We now deal with $b_{\omega+\delta\omega}$. We consider $v_{\omega+\delta\omega}$ defined by the solution of the Cauchy problem

$$\begin{aligned} &dv_{\omega+\delta\omega}(\tau) + A_k(v_{\omega+\delta\omega}(\tau))d_s b^k_{\omega+\delta\omega}(\tau) \\ &v_{\omega+\delta\omega}(0) = v_o \end{aligned} \tag{4.25}$$

The stochastic differential of $v_{\omega+\delta\omega}(\tau_o)$, ω being fixed, $\delta\omega$ varying now in the Brownian space, will give the contribution of $b_{\omega+\delta\omega}$. To take advantage of the semigroup property, we introduce a more general system than (4.25) in the following lemma.

LEMMA 4.8. Let $a \in V$ be given, and let ω and δt be fixed. Then $\delta\omega$ will be considered as a martingale. Now consider the two stochastic systems; the first in $\delta\omega$, the second in ω

$$\begin{aligned} du_{\omega+\delta\omega} &= A_k(u_{\omega+\delta\omega})(d_s b^k_\omega + d_s b^k_{\delta\omega}) \\ du_\omega &= A_k(u_\omega) d_s b^k_\omega \end{aligned} \tag{4.26}$$

with the same initial condition

$$u_\omega(0) = u_{\omega+\delta\omega}(0) = a \tag{4.27}$$

Let

$$Y_{\delta\omega}(\tau) = n_a(u_{\omega+\delta\omega}(\tau)) - n_a(u_\omega(\tau)) \tag{4.28}$$

Now, almost surely with respect to ω, $Y_{\delta\omega}(\tau)$ has a stochastic differential in $\tau = 0$, with respect to $\delta\omega$, which is given by

$$\lim_{\tau\to 0} \tau^{-1} E(Y_{\delta\omega}(\tau)) = 0 \tag{4.29}$$

$$\lim_{\tau \to 0} \tau^{-1} E(\ell^2(Y_{\delta\omega}(\tau))) = (\delta t) \sum_{k=1}^{n} \ell_k^2 \tag{4.30}$$

where $\ell = (\ell_1 \ldots \ell_m)$.

Proof. We shall work with Ito's stochastic differential equation, written in the fixed normal chart n_a. By abuse of notation, u will stand for $n_a ou$ and so on. Then, according to (3.23)

$$u_\omega(\tau) = \int_0^\tau A_k(u_\omega(\lambda))db_\omega(\lambda) + o(\tau) \tag{4.31}$$

and the same is true for $u_{\omega+\delta\omega}$. Now, use a first order expansion for A_k, $A_k(\xi_0+\xi) = A_k(\xi_0) + \xi^s(\partial_s A_k)(\xi_0) + G_k(\xi)$, where $G_k(\xi) = 0(|\xi|^2)$. Now, construct the difference of (4.31) and the corresponding equation for $b_{\omega+\delta\omega}$, $u_{\omega+\delta\omega}(\tau) - u_\omega(\tau) = r(\tau) = r_1(\tau) + r_2(\tau)$ with

$$r_1(\tau) = A_k(0)b^k_{\delta\omega}(\tau) + \int_0^\tau (A_k(u_\omega) - A_k(0))db^k_{\delta\omega} + \int_0^\tau (A_k(u_\omega+r) - A_k(u_\omega))db^k_{\delta\omega} \tag{4.32}$$

$$r_2(\tau) = \int_0^\tau \{A_k(u_\omega+r) - A_k(u_\omega)\}db^k_\omega \tag{4.33}$$

Define $M(\omega,\delta\omega,\tau) = \max|r(\tau')|$, $0 \leqslant \tau' \leqslant \tau$. Now, almost surely, in ω,

$$|r_2(\tau)| \leqslant (\tau M^2)^{1/2-\varepsilon} \tag{4.34}$$

Furthermore,

$$|r_1(\tau)| = 0(\tau^{1/2-\varepsilon}) \tag{4.35}$$

almost surely in $\delta\omega$, therefore $M^{1/2-\varepsilon} \leqslant M^{1-2\varepsilon} + \tau^{1/2-\varepsilon}$. In particular, M tends to zero. Therefore $\tau^{1/2-\varepsilon}$ is the leading term in the right member of the last inequality, and we have

$$M \leqslant \tau^{1/2-\varepsilon} \tag{4.36}$$

Now, using (4.34), we get

$$|r_2(\tau)| \leqslant \tau^{3/4-\varepsilon} \tag{4.37}$$

Now, the second integral of (4.32) is almost surely, in $\delta\omega$, bounded by $\tau^{1-\varepsilon}$, so

$$|r_1(\tau) - A_1(0)b^k_{\delta\omega}(\tau)| = 0(\tau^{1-\varepsilon}) \tag{4.38}$$

We have $E^{\delta\omega}(r_1) = 0$ and

$$E^{\delta\omega}(r_2) = \int_0^{\tau} \partial_s A_k(u_\omega) E^{\delta\omega}(r^s) db^k + 0(\tau^{5/2-\varepsilon}) \tag{4.39}$$

Define

$$\Psi(\omega,\tau) = E^{\delta\omega}(r(t)) = E(r(t))$$

$$\Psi^*(\omega,\tau) = \max_{0\leqslant\tau'\leqslant\tau} |\Psi(\omega,\tau')|$$

Then, from (4.39), we have

$$\Psi(\omega,\tau) \leqslant ([\Psi^*(\omega,\tau)]^2\tau)^{1/2-\varepsilon} + 0(\tau^{5/2-\varepsilon})$$

which implies

$$\Psi^*(\omega,\tau) \leqslant (\Psi^*(\omega,\tau))^{1-2\varepsilon}\, \tau^{1/2-\varepsilon} + 0(\tau^{5/2-\varepsilon}) \tag{4.40}$$

In conclusion, we have

(i) either $\Psi^*(\tau) \leqslant 2(\Psi^*(\tau))^{1-2\varepsilon}\,\tau^{1/2-\varepsilon}$ (then we deduce $(\Psi^*(\tau))^{2\varepsilon} \leqslant 2\tau^{1/2-\varepsilon}$ and therefore $\Psi^*(\tau) \leqslant 2\tau^{1/2\varepsilon-1/2})$

(ii) or $\Psi^*(\tau) > 2(\Psi^*(\tau))^{1/2-\varepsilon}\,\tau^{1/2-\varepsilon}$ (and then we deduce from (4.40)

$$\frac{1}{2}\Psi^* = 0(\tau^{5/2-\varepsilon}) \tag{4.41}$$

Then, in any case, the weakest inequality, i.e., (4.41), holds - which proves (4.29). In (4.30) it is clear that (4.37) eliminates r_2, and r_1 appears only in its first term.

LEMMA 4.9. Fixed ω, v_0, τ_0, and δt, define

$$Z_{\delta\omega}(\tau) = n_{v_\omega(\tau_0)} \circ U_{\omega,\tau_0-\tau}(v_{\omega+\delta\omega}(\tau)) \tag{4.42}$$

where $v_{\omega+\delta\omega}$ has been defined in (4.25). Then, $Z_{\delta\omega}(\tau)$ has, for every τ, a stochastic differential in $\delta\omega$ which is the image by the map

$$g_{\omega,\delta\omega}(\tau) = n_{v_\omega(\tau_0)} \circ U_{\omega,\tau_0-\tau} \circ n^{-1}_{\omega+\delta\omega}(\tau)$$

of $dY_{\delta\omega}$ defined by (4.29) and (4.31) with $a = v_{\omega+\delta\omega}(\tau)$.

Proof. We have the following semigroup property which corresponds to the Markov property for the diffusion v_ω

$$U_{\omega,\tau_1+\tau_2} = U_{\omega^{\tau_1},\tau_2} \circ U_{\omega,\tau_1}$$

We have

$$Z_{\delta\omega}(\tau+\varepsilon) = g_{\omega,\varepsilon}(\tau) n_{v_{\omega+\delta\omega}(\tau)} \quad v_{\omega+\delta\omega}(\tau+\varepsilon)$$

where

$$g_{\omega,\varepsilon}(\tau) = n_{v_\omega(\tau_0)} \circ U_{\omega^\bullet,\tau-\tau_0-\varepsilon} \circ n^{-1}_{v_{\omega+\delta\omega}(\tau)}$$

By the semigroup property

$$U_{\omega^\bullet,\tau_0-\tau-\varepsilon} \circ U_{\omega,\varepsilon} = U_{\omega,\tau_0-\tau}$$

Therefore

$$Z_{\delta\omega}(\tau) = g_{\omega,\varepsilon}(\tau) \circ n_{v_{\omega+\delta\omega}(\tau)} \circ U_{\omega^\bullet,\varepsilon} \circ v_{\omega+\delta\omega}(\tau)$$

Taking, in (4.28), $a = v_{\omega+\delta\omega}(\tau)$, we get

$$Z_{\delta\omega}(\tau+\varepsilon) - Z_{\delta\omega}(\tau) = g_{\omega,\varepsilon}(\tau)(\xi+Y(\varepsilon)) - g_{\omega,\varepsilon}(\tau)(\xi)$$

where

$$\xi = n_{v_{\omega+\delta\omega}(\tau)} \circ v_{\omega+\delta\omega}(\tau)$$

Now, ξ is N^τ measurable. Therefore, we use the Taylor expansion for $g_{\omega,\varepsilon}(t)$, starting from ξ. We get

$$Z_{\delta\omega}(\tau+\varepsilon) - Z_{\delta\omega}(\tau) = g'_{\omega,\varepsilon}(\tau,\xi)Y_{\delta\omega}(\varepsilon) + \frac{1}{2} g''_{\omega,\varepsilon}(\tau,\xi)(Y_{\delta\omega}(\varepsilon),Y_{\delta\omega}(\varepsilon))$$

$$+ 0((Y(\varepsilon))^2)$$

Now, as $\varepsilon \to 0$, $\xi \to 0$ and

$$g'_{\omega,\varepsilon}(\tau,\xi) \to g'_\omega(\tau), g''_{\omega,\varepsilon}(\tau,\xi) \to g''_\omega(\tau)$$

we get

$$E^{N^\tau}(\ell^2(Z_{\delta\omega}(\tau+\varepsilon) - Z_{\delta\omega}(\tau))) = E^{N^\tau}(\tilde{\ell}^2(Y_{\delta\omega}(\varepsilon)))$$

where $\ell = (g'_{\omega,\varepsilon}(\tau,\xi))*\tilde{\ell}$. Now $\tilde{\ell}$ tends to a limit as $\varepsilon \to 0$ and (4.30) gives us the announced limit when we divide by ε. Now

$$\varepsilon^{-1}E^{N^\tau}((Z_{\delta\omega}(\tau+\varepsilon) - Z_{\delta\omega}(\tau))) = \frac{1}{2}\,\partial_r\partial_s g\; E\left(\frac{Y^s_{\delta\omega}Y^s_{\delta\omega}}{\varepsilon}\right)$$

When $\varepsilon \to 0$, all the terms of this quadratic form converge and this finishes the proof.

Proof of Theorem 4.7.

We fix δt and ω. Now, $Z_{\delta\omega}(\tau)$ being defined by (4.42), by Lemma 4.9, we have an exact expression for its stochastic differential. Therefore,

$$Z_{\delta\omega}(\tau_o) = (\delta t)^{1/2}\int_0^{\tau_o} g'_{\omega,\delta\omega}(\tau)e_k\,db^k_{\delta\omega}(\tau) + \delta t\int_0^{\tau_o}\sum_{k=1}^{n} g''_{\omega,\delta\omega}(\tau)(e_k,e_k)\,d\tau$$

Let δt tend to zero. Then $v_{\omega+\delta\omega} \to v_\omega$, and the two integrands tend to $U'_{\omega,\tau_o-\tau}$ and $U''_{\omega,\tau_o-\tau}$, which proves the theorem. (For another approach, see Michel, 1977.)

Stochastic Moving Frames on the Frame Bundle of a Riemannian Manifold and Harmonic Differential 1-forms

Given a Riemannian manifold M of dimension n, call a frame an isometry

$$r: R^n \to T_{x_o}(M)$$

Then the orthogonal group $O(n)$ operates on the frames on the left

$$r \to rg = r \circ g$$

(We use a right action of $O(n)$ instead of a left action, as in Kobayashi-Nomizu, 1963.) $O(M)$ is a principal $O(n)$ bundle. The Riemannian connection defines a parallelism on $O(M)$, associated to a 1-form π, defined on $O(M)$, with values in $D(n)$, the Lie algebra of the group of Euclidean motion of R^n. We can also define π by its components, π^i, π^i_j, with $\pi^i_j + \pi^j_i = 0$. Then, the structure equations can be read (see Kobayashi-Nomizu, 1963, p. 129)

$$\begin{aligned}\langle z\wedge\tilde z, d\pi^i\rangle &= -(\pi^i_j(z)\pi^j(\tilde z) - \pi^i_j(\tilde z)\pi^j(z))\\ \langle z\wedge\tilde z, d\pi^i_j\rangle &= -(\pi^i_q(z)\pi^q_j(\tilde z) - \pi^j_q(\tilde z)\pi^q_j(z)) + R^i_{j,k,\ell}\pi^k(z)\pi^\ell(\tilde z)\end{aligned} \tag{5.1}$$

Where R denotes the curvature tensor, read in the frame r.

The stochastic moving frame is the limit of the differential system

$$\left\langle \frac{dr}{d\tau}, \pi^i \right\rangle = \frac{db^i_{\varepsilon(\omega)}}{d\tau}$$

$$\left\langle \frac{dr}{dt}, \pi^i_j \right\rangle = 0 \tag{5.2}$$

(5.2) generates a C^∞ flow, $U_{\varepsilon(\omega),\tau}$. We introduce

$$J(\varepsilon(\omega),r_0,\tau_0) = \pi_{U_{\varepsilon(\omega),\tau_0}(v_0)} \circ U'_{\varepsilon(\omega),\tau_0}(v_0) \circ \pi^{-1}_{v_0}$$

and

$$z(\tau_0) = J(\varepsilon(\omega),r_0,\tau_0)z(0)$$

Then, by Lemma 3.1 and (5.1), we obtain

$$\frac{dz^i}{d\tau} = z^i_j \frac{db^j_{\varepsilon(\omega)}}{d\tau}$$

$$\frac{dz^i}{d\tau} = R^i_{j,k,\ell} \frac{db^k_{\varepsilon(\omega)}}{d\tau} z^\ell \tag{5.3}$$

We introduce the matrix

$u_j \in \mathrm{End}(D(n))$ by the formula

$$(u_j(z))^i = z^i_j$$

$$(u_j(z))^i_q = R^i_{q,j,\ell} z^\Omega \tag{5.4}$$

Finally, define the matrix $M \in \mathrm{End}(D(n))$ by

$$M^i_{q,\ell} = \sum_j \nabla_j R^i_{q,j,\ell} \quad \text{(where } \nabla \text{ is the covariant derivative)}$$

$$M\, z = (0, M^i_{q,\ell} z^\ell) \tag{5.5}$$

REMARK. By Bianchi's identity, M can be expressed in terms of covariant derivatives of the Ricci tensor. In particular, for a Ricci flat manifold, we have $M = 0$.

THEOREM 5.1.

$$J(\omega,r_0,\tau_0) = \exp\left(* \int_{[\tau_0,0]} (u_j db^j + \frac{1}{2} M\, d\tau) \right) \tag{5.6}$$

Proof. (5.2) can be written

$$dz_{\varepsilon(\omega)} = u_j z_{\varepsilon(\omega)} db^j_{\varepsilon(\omega)} \tag{5.7}$$

Now, using the correspondence principle, the differential system (5.7) will tend to the stochastic system

$$dz = u_j z_\omega db^j_\omega + \frac{1}{2} \sum_j (u_j^2 + \nabla_j u_j) z_\omega d\tau \tag{5.8}$$

Expressing this in the form of a multiplicative stochastic integral has the effect of erasing the first term in $d\tau$.

REMARKS.

$$\begin{aligned} \sum_j (u_j^2(z))^i &= \sum_j R^i_{j,j,\ell} z^\ell = -(\mathrm{Ricci}(z^\bullet))^i \\ \Big(\sum_j u_j^2(z)\Big)^i_q &= \sum_j R^i_{q,j,\ell} z^\ell_j \end{aligned} \tag{5.9}$$

$$\det(J(\omega,\tau_o,\tau_o)) = 1 \tag{5.10}$$

In fact, we have $\mathrm{trace}(u_j) = 0$. Therefore, the resolvant of (5.7) is of determinant 1. Then (5.10) follows by making $\varepsilon \to 0$. We can interpret (5.10) as saying that the <u>stochastic flow</u> $U_{\omega,\tau}$ <u>conserves the volume element of</u> $O(M)$. The computation of $U''_{\omega,\tau}(\tau_o)$ will be given by (4.7). The Stratanovitch integral, written in Ito's notation, introduces terms of the form $\nabla_i \nabla_j R^k_{\ell,m,n}$.

We consider a form, ρ, defined on M, satisfying

$$\begin{aligned} &d\rho = 0 \\ &d\delta\rho = 0 \quad \text{(where } \delta \text{ is the adjoint of } d) \end{aligned} \tag{5.11}$$

We denote by $\sigma = p^*\rho$ the pullback of ρ on $O(M)$. Then, if λ is a 1-differential form on $O(M)$, it can be read as a map $O(M) \to \lambda_r \in (D(n))^*$. Given λ, a differential 1-form on $O(M)$, we define

$$\lambda_{\omega,\tau} = (U_{\omega,\tau})^* \lambda \tag{5.12}$$

Then

$$, (r_o) = \exp \; * \int_{[0,\ _o]} ({}^t u_j db^j + \frac{1}{2}\, {}^t Mf\ )\ (r\ (\ _o) \tag{5.12}$$

$$\lambda_{\omega,\tau}(r_o) = \exp\left(* \int_{[0,\tau_o]} ({}^t u_j db^j + \frac{1}{2}\, {}^t M\, d\tau)\lambda(r_\omega(\tau_o))\right) \tag{5.13}$$

The statement of the following theorem was suggested to me by D. Sullivan.

THEOREM 5.2. Let ρ be a 1-form on the compact Riemannian manifold M, satisfying (5.11). Then, for all positive τ

$$\lambda = E(U^*_{\omega,\tau}\lambda) \tag{5.14}$$

REMARK. The hypothesis of compactness on M can be replaced by the hypothesis that the diffusion on M has an infinite lifetime and the expectation appearing in (5.14) is locally uniformly absolutely convergent in t.

Proof. Define

$$\lambda^\tau = E(U^*_{\omega,\tau}\lambda)$$

Now,

$$\frac{d}{d\tau}\lambda^\tau = \lim_{\varepsilon\to 0} E\left(\frac{1}{\varepsilon}(U^*_{\omega,\tau+\varepsilon} - U^*_{\omega,\tau})\lambda\right)$$

$$U_{\omega,\tau+\varepsilon} = U_{\omega^\tau,\varepsilon} \circ U_{\omega,\tau}$$

Therefore,

$$\frac{d}{d\tau}\lambda^\tau = \lim_{\varepsilon\to 0} E(U^*_{\omega,\tau}\varepsilon^{-1}(U^*_{\omega^\tau,\varepsilon} - I)\lambda)$$

Let N^τ be the σ-field adapted to b_ω. Then,

$$\frac{d}{d\tau}\lambda^\tau = \lim_{\varepsilon\to 0} E(U^*_{\omega,\tau}E^{N^\tau}\varepsilon^{-1}(U^*_{\omega^\tau,\varepsilon}\lambda-\lambda))$$

$$\langle z, U^*_{\omega^\tau,\varepsilon}\lambda\rangle_{r_o} = \langle J(\omega^\tau, r_o, \varepsilon)z, \lambda_{r_\omega(t)}\rangle = (Jz)^i(\lambda_{r_\omega(t)})_i$$

We have

$$Jz = \exp(u_j\delta b^j + \frac{\varepsilon}{2}M + o(\varepsilon))z + \nabla_k u_j(z)\int_0^t b^k db^j$$

$$= z + u_j(z)b^j + \frac{\varepsilon}{2}\left(\sum_j u_j^2 + M\right)z + o(\varepsilon) + \nabla_k u_j(z)\int_0^t b^k db^j$$

We denote by $f_\rho(r)$ the function with values in R^n which associates to the covector $\rho_{p(r)}$ its components in the frame r. Then

$$f_\rho(r) = (\lambda_r)_i$$

Therefore,

$$\langle z, U^*_{\omega,\varepsilon}\lambda\rangle = \langle z, f_\rho(r_\omega(\varepsilon))\rangle + \sum_j \langle u_j(z), f_\rho(r_\omega(\varepsilon))\rangle \delta b^j$$
$$+ \frac{\varepsilon}{2}\Big(\sum_j u_j^2 + M\Big)\, z, f_\rho(r_\omega(\varepsilon))) + \langle \nabla_k u_j(z), f_\rho(r_\omega(\varepsilon))\rangle \int_0^t b^k db^j$$

Using the fact that $(\nabla u_j(z))^i = 0$, the last terms disappear. Now, making $\varepsilon \to 0$

$$\lim_{\varepsilon\to 0} \langle z, U^*_{\omega,\varepsilon}\lambda - \lambda\rangle = \frac{1}{2}\langle z, \Delta_{0(M)} f_\rho\rangle + \sum_j \langle u_j(z), \partial_j f_\rho\rangle + \frac{1}{2}\langle \sum_j u_j^2(z), f_\rho\rangle \quad (5.15)$$

The first sum can be written as

$$\sum_{j,i} z_j^i \partial_j f_\rho^i = \sum_{i,j} a_j^i(\partial_j f_\rho^i - \partial_i f_\rho^j) = 0 \quad (5.16)$$

according to the first equation of (5.11). (5.11) implies

$$\Box\rho = 0 \quad (5.17)$$

(where $\Box$ is the deRham-Hodge Laplacian. It is known by the Weitzenböck formula (see Malliavin, 1974) that (5.17) is equivalent to

$$-\Delta_{0(M)} f_\rho + \mathrm{Ricci}(f_\rho) = 0 \quad (5.18)$$

Now, using (5.9), we get that (5.15) is equal to zero and the theorem is proved.

As the sign appearing in (5.18) and in $\sum u_j^2$ is crucial, we shall make some well known computations below in full.

We denote f_ρ by f, and by f_i its components. ξ^i denotes the canonical linear forms on R^n. Then

$$f = f_i \xi^i$$

$$df = \partial_j f_i \xi^j \wedge \xi^i$$

$$\delta f = -\partial_{\underline{i}} f_{\underline{i}}$$

$$\delta df = \sum_{i \neq j} \partial_{\underline{i}} \partial_j f_{\underline{i}} \xi^j - \partial_j^2 f_i \xi^i$$

$$d\delta f = -\partial_k \partial_{\underline{i}} f_i \xi^k$$

$$(\delta df + d\delta f)_q = -\Delta_{O(M)} f_q + \sum_{i \neq q} (\partial_i \partial_q - \partial_q \partial_i) f_i$$

$$(\partial_i \partial_q - \partial_q \partial_i) f_i = ([\partial_i, \partial_q] f)_i$$

$$\langle \partial_i \wedge \partial_q, d\pi \rangle + \langle [\partial_i, \partial_q], \pi \rangle = 0$$

Therefore, $[\partial_i, \partial_q] = -R^{\bullet}_{\bullet,i,q}$ and $([\partial_i, \partial_q]f)_i = -R^s_{i,i,q} f_s$. On the other hand, by (5.9)

$$\langle \sum_j u_j^2(z), f \rangle = \sum_i R^s_{i,i,q} z^q f_s = \langle z^q, \sum_i R^s_{i,i,q} f_s \rangle$$

Finally,

$$-\Delta_{O(M)} f + \sum [\partial_i, \partial_q] f = -\Delta_{O(M)} f + \sum_j ({}^t u_j)^2 f$$

REMARK. In (5.16) we have again used the fact that $\Box\rho = 0$. This point of view can be developed in an abstract way to get exotic mean values formulas (see Malliavin, 1977). The extension of Theorem 5.2 to degree > 1 does not seem completely immediate.

Minoration of Equivariant Spectrum Above a Homogeneous Riemannian Manifold

Let M be a Riemannian manifold. We consider an irreducible representation χ of the orthogonal group $O(n)$ and F_χ the space of this representation. (If the bundle $O(M)$ can be reduced to a principal subbundle K, we shall take for χ a representation of K.) We consider L^2_χ, the space of L^2 functions, f, defined on $O(M)$, vector valued in F_χ, and χ-equivariant. That is,

$$f(rg) = \chi(g^{-1}) f(r) \tag{6.1}$$

If $f \in L^2_\chi$ and f is smooth, then $\Delta_{O(M)} f \in L^2_\chi$. Therefore, we can define

$$m(\chi) = \inf(\text{spectrum of } -\Delta_{O(M)} \text{ on } L^2_\chi) \tag{6.2}$$

where we have chosen a selfadjoint extension of $\Delta_{O(M)}$ on L^2_χ.

We shall suppose that

$$\nabla_i R^s_{r,\ell,m} = 0 \tag{6.3}$$

Then $M = 0$ and

$$J(\omega,r_o,\chi) = \exp\left(* \int_{[\tau,0]} u_j db^j_\omega\right) \tag{6.4}$$

It can also be remarked that $J(\omega,r_o,\chi)$ is independent of r_o. Furthermore, we have, by (4.7), that

$$J^{-1}_{\tau_o} D_\xi J_{\tau_o} = \int_0^{\tau_o} (D_{J_\tau \xi} u_i)^{J_\tau} d_s b^i_\omega = 0$$

as the u_i are constant by (5.11). Therefore,

$$U''_{\omega,\tau}(r_o) = 0 \quad \text{(in the normal charts)} \tag{6.5}$$

We can also introduce the group G of all isometries of M. Then, $M = G/K$ and the frame bundle $0(M)$ contains G as a reducing subbundle of structural group K.

The vector field $\hat{e}_1$, defined as

$$\pi^i(\hat{e}_1) = 1 \quad \text{for } i = 1, \; 0 \text{ otherwise}$$

and

$$\pi^i_j(\hat{e}_1) = 0$$

can be identified with an element, e_1, of G. The process on $0(M)$ reduces on G by

$$g_\omega(\tau) = \exp\left(* \int_{[0,\tau]} e_q db^q_\omega\right) \tag{6.6}$$

Then

$$J(\omega,\tau) = \mathrm{Ad}(g^{-1}_\omega(\tau)) \tag{6.7}$$

Ad is a representation of G.

$$\mathrm{Ad}(g^{-1}_\omega(\tau)) = \exp\left(* \int_{[\tau,0]} - \mathrm{ad}(e_q) db^q_\omega\right) \tag{6.8}$$

and therefore $-\mathrm{ad}(e_k)$ is the reduction of u_k. Then, the variation of $g_\omega(\tau_o)$ is given by

$$g_{\omega+\delta\omega}(\tau_o) = g_\omega(\tau_o)\exp\left(\int_0^{\tau_o} \mathrm{Ad}(g_\omega^{-1}(\tau_o)g_\omega(\tau))\ e_q db^q_{\delta\omega}(\tau)\right) \tag{6.9}$$

We can take $b_{\delta\omega}$ composed of two discrete jumps, one at $\tau = 0$, and the other at $\tau = \tau_o$. More precisely, we shall define a map of $\Omega(R^{2n}) \to G$ by the stochastic differential system

$$dg_{\tilde{\omega}}(\tau) = \frac{1}{\sqrt{2}}\,[\mathrm{Ad}(g_{\tilde{\omega}}^{-1}(\tau))e_q db^q_{\tilde{\omega}}(\tau) + e_q db^{q+n}_{\tilde{\omega}}(\tau)] \tag{6.10}$$

Then we have (see Malliavin, 1976)

THEOREM 6.1. The law of $g_{\tilde{\omega}}(\tau)$ and $g_\omega(\tau)$ are equal. The law $p_\tau(g)dg$ of $g_\omega(\tau)$ satisfies the parabolic equation

$$4\,\frac{\partial}{\partial\tau}\,p_\tau = (\Delta + \tilde{\Delta})p_\tau \tag{6.11}$$

where

$$(\Delta\phi)(g_o) = \sum_q \frac{d^2}{dt^2}\,\phi(g_o\ \exp(te_q))$$

$$(\tilde{\Delta}\phi)(g_o) = \sum_q \frac{d^2}{dt^2}\,\phi(g_o\ \exp(t\ \mathrm{Ad}(g_o^{-1})e_q))$$

REMARK. In some cases, $\Delta + \tilde{\Delta}$ can be a strictly elliptic operator (see Malliavin, 1976).

In considering the estimation of the equivariant spectrum, we shall limit ourselves to the case where M is homogeneous, and we shall use (6.10) and (6.11) of the simplified calculus of variations. The method used here does not seem immediately extendable to a more general case. We shall deal with the following situation. G will be a Lie group, K a compact subgroup of G such that there exists a complementary subspace V of $\underline{K}$ into $\mathcal{G}$, stable under the action of $\mathrm{Ad}(K)$. A Euclidean metric on V, invariant under the action of $\mathrm{Ad}(K)$, will define a Riemannian structure on G/K. Now, given χ, a representation of K on the finite dimension space F, we want to evaluate

$$m(\chi) = \inf(\text{spectrum of } -\Delta_G \text{ on } L^2_\chi) \tag{6.12}$$

with

$$\Delta_G = \frac{1}{2}\sum_q L^2_{e_q} \qquad (e_1 \ldots e_n \text{ an orthonormal basis of } V.)$$

We shall associate to the representation χ its vector bundle F_χ over G/K. Let us introduce the following equivalent relation over $G \times F$. $(g,\xi) \sim (g',\xi') \Leftrightarrow \exists k$ such that $g' = gk^{-1}$ and then $\xi' = \chi(k)\xi$. The quotient of $G \times F$ by this equivalence relation is the vector bundle F_χ. Therefore, we have a natural map

$$q: G \times F \to F_\chi$$

PROPOSITION 6.2. Define on $G \times F$ the diffusion process

$$z_{\tilde{\omega}}(t) = (g_{\tilde{\omega}}(t),\xi)$$

where $g_{\tilde{\omega}}$ is defined in (6.10) and ξ stays constant. Then, the projection $q(z_{\tilde{\omega}}(t))$ is a diffusion on F_χ.

Proof. We introduce the Ito invariants of $q(z_{\tilde{\omega}}(t))$, i.e., for every C^2 function ϕ defined on F^χ, we consider

$$\lim_{\varepsilon\to 0} \frac{1}{\varepsilon} E^{N^t}[\phi(q(z_{\tilde{\omega}}(t+\varepsilon)) - \phi(z_{\tilde{\omega}}(t)] = L^{N^t}\phi$$

$$\lim_{\varepsilon\to 0} \frac{1}{\varepsilon} E^{N^t}[(\phi(q(z_\omega(t+\varepsilon)) - \phi(z_\omega(t)))^2] = \|\nabla^{N^t}\phi\|^2$$

Then, L^{N^t}, ∇^{N^t} are two operators defined on the C^2 function on F^χ. These operators are known in terms of $z_\omega(t)$. It is well known that the proposition is equivalent to

L^{N^t}, ∇^{N^t} <u>dependent only on</u> $q(z_\omega(t))$

Consider z, z' such that $q(z) = q(z')$

$$z = (g,\xi)$$

$$z' = (gk^{-1},\chi(k)\xi)$$

We want to prove

$$L^z = L^{z'}$$

$$\nabla^z = \nabla^{z'}$$

Consider the diffeomorphism Ψ of $G \times F$ defined by

$$\Psi:(g'',\xi'') \to (g''k^{-1},\chi(k)\xi'')$$

Then $q \circ \Psi = q$ and Proposition 6.2 will result from

LEMMA 6.3. The diffusion $z_{\tilde{\omega}}(t)$ is invariant under the diffeomorphism Ψ.

Proof. Set $\delta b_{\tilde{\omega}} = b_{\tilde{\omega}}(\varepsilon) - b_{\omega}(0)$. Then

$$\exp(\mathrm{Ad}(g_o^{-1})e_r\delta b^{r+n} + e_r\delta b^r)$$

$$= \exp(\mathrm{Ad}(g_o^{-1})e_r\delta b^{r+n})\exp(e_r\delta b^r)\exp(-\tfrac{1}{2}[\mathrm{Ad}(g_o)e_q,e_{q'}]\delta b^{r+n}\delta b^{r'} + 0(\varepsilon))$$

As $\delta b^{r'}$ and $\delta b^{r''}$ are independent for $r' \neq r''$, we can neglect the last exponential in the computation of the stochastic differential. Furthermore, $g_o\exp(\mathrm{Ad}(g_o^{-1})e_r\delta b^{r+n}) = \exp(e_r\delta b^{r+n})g_o$ implies

$$g_{\tilde{\omega}}(t_o+\varepsilon) = \exp(e_q\delta b_{\tilde{\omega}}^{q+n})g_{\tilde{\omega}}(t_o)\exp(e_q\delta b_{\tilde{\omega}}^q) \tag{6.13}$$

Now, apply the diffeomorphism Ψ

$$\Psi(g_{\omega}(t_o+\varepsilon) = \exp(e_q\delta b_{\tilde{\omega}}^{q+n})\Psi(g_{\tilde{\omega}}(t_o))\exp(\mathrm{Ad}(k)e_q\delta b_{\tilde{\omega}}^q)$$

As $\mathrm{Ad}(k)$ is an orthogonal transformation on the $\{e_q\}$, we find that the last exponential has the same stochastic differential, up to an isomorphism $\tilde{\omega} \to \tilde{\omega}'$ of the probability space that $e_r\delta b^r$. Therefore, $d(\Psi(g_{\tilde{\omega}}(t_o+\cdot))) = dg_{\tilde{\omega}'}(t_o+\cdot)$ where $g^!_{\tilde{\omega}'}(t_o+\varepsilon) = g_{\tilde{\omega}}(t_o)k^{-1}\exp(\mathrm{Ad}(kg_{\tilde{\omega}}(t_o))e_r\delta b_{\tilde{\omega}'}^{r+n} + e_r\delta b_{\tilde{\omega}'}^r$. As the process $z_{\tilde{\omega}}$ is constant on the F component, we deduce

$$d(u(z_{\tilde{\omega}}(t+\cdot)) = d(z^!_{\tilde{\omega}'}(t+\cdot))$$

At each point, g_o, we consider the positive quadratic form defined on G', the real dual space of G, by

$$q_g(\ell) = \sum_{r=1}^{n} \ell^2(e_r) + \ell^2(\mathrm{Ad}(g^{-1})e_r) \tag{6.14}$$

Let P be the subspace of G' consisting of the linear forms vanishing on $\underline{K}$. Then, it is possible to split (not uniquely if q is semidefinite) G' as

$$G' = P \oplus V \quad \text{(orthogonal decomposition for } q_{g_o}) \tag{6.15}$$

There exists a Gaussian variable, U_g, supported by $\underline{K}$ and a Gaussian variable, W_g, independent of U_g and supported by the annihilator of V, such that

$$2\log(E(\exp(\ell(U_g))) = q_g(v) \qquad (\ell = p + v)$$
$$2\log(E(\exp(\ell(W_g))) = q_g(p) \tag{6.16}$$

We associate to U_g the semielliptic operator

$$(\Omega_g\phi)(g_0) = -\lim_{\varepsilon\to 0}\frac{1}{2\varepsilon}[\phi(g_0\exp(\varepsilon U_{g_o})) - \phi(g_0)] \tag{6.17}$$

(The minus sign is introduced in order to get a positive operator.)

We define a diffusion

$$g_w(t) = \exp\left(* \int_{[0,t]} (\delta t)^{1/2} W_{g_w(t)}\right) \tag{6.18}$$

In matrix form, this means that $g_w(t)$ satisfies the following stochastic differential equation

$$\delta g_w(t) = g_w(t)\left[(\delta t)^{1/2} W_{g_w(t)} + \frac{\delta t}{2} W^2_{g_w(t)}\right] \tag{6.19}$$

When $g_w(t)$ is known, we shall define the process

$$k_{w,u}(t) = \exp\left(* \int_{[0,t]} (\delta t)^{1/2} U_{g_w(t)}\right) \tag{6.20}$$

THEOREM 6.4. There exists an isomorphism between the probability space of $g_{\tilde{\omega}}(\cdot)$ and the probability space of $(g_w(\cdot), k_{w,u}(\cdot))$ such that

$$g_{\tilde{\omega}}(t) = g_w(t)k^{-1}_{w,u}(t) \tag{6.21}$$

REMARK. See Malliavin and Malliavin, Lecture Notes, for a result of this nature.

Proof. We shall compute, in a left invariant exponential chart, the stochastic differential D of the second member of (6.21)

$$D = k_{w,u}(t)g_w^{-1}(t)g_w(t + \delta t)k^{-1}_{w,u}(t + \delta t)$$
$$= k_{w,u}(t)g_w^{-1}(t)g_w(t + \delta t)[k^{-1}_{w,u}(t)k_{w,u}(t + \delta t)]^{-1}k^{-1}_{w,u}(t)$$

Now, we use the identity

$$\exp((\delta t)^{1/2}W')\exp((\delta t)^{1/2}U') = \exp((\delta t)^{1/2}(U'+W') + \tfrac{1}{2}\delta t[W',U'] + O(\delta t))$$

When U', W' are independent random variables in this identity, we can delete the term $[W',U']$ which does not give a stochastic contraction when we compute a stochastic differential. Therefore,

$$D = k_{w,u}(t)\exp\left[(\delta t)^{1/2}\left(W_{g_w} + U_{g_w}\right)\right]k^{-1}_{w,u}(t)$$

or

$$D = \exp\left((\delta t)^{1/2}\mathrm{Ad}(k_{w,u})\left(W_{g_w} + U_{g_w}\right)\right).$$

Now, by Proposition 6.2, D has the same law as

$$\tilde{D} = \exp\left((\delta t)^{1/2}\left(W_{g_w k^{-1}_{w,u}} + U_{g_w k^{-1}_{w,u}}\right)\right)$$

and, by the analysis of the variances, we find that the two members of (6.21) have the same stochastic differential at t_o if they take the same value at t_o. As for $t = 0$, the two members of (6.21) are equal to the identity. They are equal for all $t > 0$.

MAIN THEOREM 6.5. We have

$$m(\chi) \geqslant \lim_{t_o\to\infty} \frac{-1}{t_o} \log E_e\left(\left\| \exp\left(* \int_{[0,t_o]} -\underline{\chi}(\Omega_{g_w(t)}dt\right) \right\|\right)$$

where $m(\chi)$ is defined in (6.12), Ω in (6.17), g_w in (6.18) and we denote by χ the differential of the representation χ.

COROLLARY 6.6. Suppose that χ is a representation of degree 1. Then

$$(\chi) \geqslant \lim_{t_o\to\infty} \frac{1}{t_o}(\log E_e)\exp\left(-\int_0^{\tau_o}(\Omega_{g_w(t)})dt\right)$$

Proof of the theorem. For every $\varepsilon > 0$, there exists $f \in L^2_\chi(G)$ such that the solution f_t of the Cauchy problem

$$\frac{\partial f_t}{\partial t} = \Delta_{O(M)}f_t \qquad f_o = f$$

satisfies

$$\|f_{t_o}\|_{L^2} \geq e^{-(m(\chi)+\varepsilon)t_0}\|f\|_{L^2} \quad \text{for all} \quad t_o > 0 \tag{6.23}$$

On the other hand, using (6.21), we get

$$f_{t_o}(g_o) = E_{g_{o,w}}\{E_u(f(g_w(t)k^{-1}_{w,u}(t)))\} = E^w_{g_o}\{[E^u(\chi(k_{w,u}(t))]f(g_w(t))\}$$

$$= E^w_{g_o}\left\{\exp\left(*_{[0,t_o]} -\underline{\chi}(\Omega_{g_w(t)})dt\right)f(g_w(t))\right\}$$

Finally,

$$f_t(g_o) = \int f(g_o g)K_{t_o}(dg)$$

with

$$K_{t_o}(A) = E_e\left(\exp\left(*\int_{[0,t_o]} -\underline{\chi}(\Omega_{g_w(t)})dt\right) \; 1_A(g_w(t_o))\right)$$

Now, the inequality

$$\|f_t\|_{L^2} \leq \|f\|_{L^2}\|K_{t_o}\|_{L^1}$$

combined with (6.23) gives the theorem.

REMARK. When G/K is noncompact, we can also take advantage of the fact that the density of probability for the Riemannian diffusion on G/K can have an exponentially small norm.

This computation is related to the stochastic holonomy (see Malliavin and Malliavin, 1975, and Gaveau, 1977).

REFERENCES

Gaveau, B. Estimées hypoelliptiques sure les groupes nilpotents d'ordre 2. Acta Math. (1977).

Ibero, M. Intégrales stochastiques multiplicatives. Bull. des Sci. Math. (1976).

McKean, H. P. Stochastic Integrals. Academic Press, New York, New York (1969).

Kobayashi, S., and Nomizu, K. Foundations of Differential Geometry, Vol. I. Tracts in Math., 15, Interscience Publishers, New York, New York (1963).

Malliavin, Marie Paule and Malliavin, Paul. Factorizations et lois limites de la diffusion horizontale au-dessus d'un espace riemannien symetrique. Lecture Notes 404, pp. 161-210.

Malliavin, Marie Paule, and Malliavin, Paul. Holonomie stochastique au-dessus d'un espace riemannien symetrique. Compte Rendus, 280 (1975) pp. 793-795.

Malliavin, Paul. Stochastic calculus of variations and hypoelliptic operators. Proc., International Conf. on Stoch. Diff. Equa., Kyoto (1976).

Malliavin, Paul. Formule de la moyenne pour les formes harmoniques. J. of Func. Anal., 17 (1974), pp. 274-291.

Malliavin, Paul. Sur certaines formules exotiques de la moyenne. Comptes Rendus (1977)

Michel, D. Formules de Stokes stochastiques. Bull. des Sci. Math. (1977)

Institut Mittag-Leffler
Djursholm, Sweden

THETA FUNCTIONS, SOLITONS, AND SINGULAR CURVES

H. P. McKean*

Introduction

The purpose of this paper is to explain the application of theta functions of singular curves to the equation of Korteweg-de Vries: $\partial q/\partial t = X_2 q = 3q\,\partial q/\partial x - (1/2)\partial^3 q/\partial x^3$. The equations of translation, $\partial q/\partial t = X_1 q = \partial q/\partial x$ and $\partial q/\partial t = X_2 q$ comprise the first two of a series of commuting ∞-dimensional Hamiltonian flows in function space of similar form: $\partial q/\partial t = X_j q$ $(j = 1,2,3,\ldots)$. The common invariant manifolds of the latter foliate the function space in a geometrically interesting way. The typical invariant manifold Q may be viewed as the Jacobi variety J of a (singular or nonsingular) curve of genus $g \leqslant \infty$, and the remarkable fact is that, in the natural coordinate q of J, the flow $\partial q/\partial t = Xq$ is a straight-line motion at speed 1 in some direction x, tangent to $J[\dot{q} = x]$. The function $q \in Q$ is expressed in terms of the theta function ϑ of the variety by means of the formula $q(x) = -2[\ell g\vartheta(q + xx_1)]''$ in which q is the point of J corresponding to q, x_1 is the direction tangent to J corresponding to the infinitesimal translation $X_1 q = q'$, and $'$ means differentiation with regard to x. Q is specified by fixing certain spectral data pertaining to the operator $L = -d^2/dx^2 + q(x)$. The details of the recipe are explained in three special cases within the class of infinitely differentiable functions: a) q periodic, b) q rapidly vanishing at $\pm\infty$, c) q rational. The three cases are linked by means of the singularization of curves

The Periodic Case

Let q be of period 1. The spectral data employed to specify Q comprises the periodic (and antiperiodic) spectrum of L obtained by solving $Lf = \lambda f$ with $f(x+1) = +1(-1) \times f(x)$. This produces a spectrum

*The author gratefully acknowledges the support of the National Science Foundation under Grant No. NSF-MCS76-07039.

of $2g+1 \leqslant \infty$ simple eigenvalues $\lambda_0 < \lambda_1 \leqslant \lambda_2 < \ldots < \lambda_{2g-1} \leqslant \lambda_{2g}$ with additional (possibly interlacing) double eigenvalues. Q is a g-dimensional torus, which may be identified with the real part of the Jacobi variety J of the hyperelliptic curve $C:\ell^2(\lambda) = -(\lambda-\lambda_0)\ldots(\lambda-\lambda_{2g})$ with a suitable interpretation of the product if $g = \infty$. ϑ is the customary Riemann theta function:

$$\vartheta(x) = \sum e^{2\pi\sqrt{-1}\, x\cdot m - \pi C[m]}$$

in which the sum is extended over the points m of the dual of the real period lattice and $C[m]$ is the quadratic form based upon the period matrix of C. The (transitive) action of the commuting flows $\partial q/\partial t_j = X_j q$ ($j = 1,\ldots,g$) is expressed by

$$e^{t_1 X_1 + \ldots + t_g X_g} q(x) = -2[\ell g \vartheta(q + x x_1 + t_1 x_1 + \ldots + t_g x_g)]''$$

the direction $x_j = X_j q$ tangent to J being the image of the infinitesimal motion $X_j q$ ($j = 1,\ldots,g$) (see Its and Matveev, 1975, and/or McKean and Moerbeke, 1975, for $g \leqslant \infty$, and McKean and Trubowitz, 1976 and 1977, for $g = \infty$). For simplicity, only the case $g < \infty$ is discussed below.

The Rapidly Vanishing Case

Now let $0 < k_g < \ldots < k_1$ be fixed, let $\lambda_{2g} = 0$, and let the handles of C be pinched by making the segment $[\lambda_{2i-2}, \lambda_{2i-1}]$ shrink to the point $-k_i^2$ ($i = 1,\ldots,g$). The curve becomes singular $\ell^2(\lambda) = (\lambda+k_1^2)\ldots(\lambda+k_g^2)\sqrt{-\lambda}$, i.e., it becomes a sphere $k = \sqrt{-\lambda}$ with points $\pm\sqrt{-1}k_i$ ($i = 1,\ldots,g$) identified in pairs, the (real) periods of J tend to ∞ so that J becomes a cylinder, and the leading part of $\vartheta(q + t_1 x_1 + \ldots + t_g x_g)$ is a multiple of

$$\vartheta_1 = \sum_{n_1,\ldots,n_g = 0,1} e^{(q + t_1' k + t_2' k^3 + \ldots + t_g' k^{2g-1})\cdot n + K[n]}$$

with $k = (k_1,\ldots,k_g)$, $k^3 = (k_1^3,\ldots,k_g^3)$, etc.

$$K[n] = \sum_{i<j} n_i n_j \,\ell g \left(\frac{k_i - k_j}{k_i + k_j}\right)^2$$

and a trivial affine substitution $t \to t'$. The class Q is now the family of many-soliton functions arising from the bound states $-k_i^2$ ($i = 1,\ldots,g$).

The formula $q = -2(\ell g\vartheta_1)''$ is due to Hirota (1971), the only novelty being that ϑ_1 is now seen to be the theta function of the singular curve. The computation goes back to Poincaré (1895) (see also Fay, 1973, and Matveev, 1976.) The customary many-soliton formula for $t' = 0$:

$$\vartheta_1 = \det \left[1 + \frac{2\sqrt{k_i k_j}}{k_i + k_j} e^{q_i/2} e^{q_j/2} \right]$$

was derived before by Kay and Moses (1956) and Gardner, Greene, Kruskal and Muira (1974) by means of the scattering theory of Gelfand and Levitan (1951), Agranovich and Marcenko (1963) and Faddeev (1964). The adjective <u>many-soliton</u> refers to the fact that for $t \to \pm\infty$, the solution $q(x) = -2[\ell g\vartheta_1(q+xk+tk^3)]''$ of $\partial q/\partial t = X_2 q$ separates out into an approximate sum of solitary waves: $-(k_i^2/2)\cosh^{-2}(k_i x/2 = 2k_i^3 t)$ $(i = 1,\ldots,g)$. Now, for the general function q which vanishes rapidly at $\pm\infty$, the solution of $\partial q/\partial t = X_2 q$ consists of a many-soliton part arising from the bound states $-k_i^2$ $(i = 1,\ldots,g)$ of $L = -d^2/dx^2 + q(x)$, plus an oscillating part which dies away as $t \to \pm\infty$ arising from the continuous part of the spectrum of L. The latter is described by means of scattering data epitomized by the reflection coefficient* $s_{12}(k)$ $(-\infty < k < \infty)$ of L, and the flow $\partial q/\partial t = X_2 q$ is mirrored in a) a g-dimensional many-soliton motion $q \to q + tk^3$ as above, and b) an ∞-dimensional motion $q \to q + tk^3$ of $q(k)$ = phase $s_{12}(k)$; in particular, the invariant manifold Q is a kind of ∞-dimensional Jacobi variety (cylinder) specified by the energy levels $-k_i^2$ $(i = 1,\ldots,g)$ and by the gain $|s_{12}(k)|$ $(-\infty < k < \infty)$, and the recipe of Gelfand and Levitan (1951) and Agranovich and Marcenko (1963) for recovering q from the scattering data can be cast into the preferred form $q = -2(\ell g\vartheta_2)''$, ϑ_2 being (by some stretch of the imagination) the theta function of a singular curve of genus $g = \infty$ expressible as a Fredholm determinant; compare Dyson (1976) and the formula of Kay and Moses (1956) cited above.

The Rational Case

The special g-soliton function $q(x) = -g(g + 1)\cosh^{-2}(x)$ leads to the connection between many-soliton functions and the rational solutions of $\partial q/\partial t = X_2 q$ of Airault, McKean and Moser (1977) (see Adler and Moser, 1978, for an elementary and more satisfactory account.) ϑ_1 is formed with

*$s_{12}(k) = 0$ in the pure many-soliton case.

$k = (1,\dots,g)$ and a special choice of q. The substitution $q \quad q + \sqrt{-1}\,k$ changes $q(x)$ into $-g(g+1)\sinh^{-2}(x)$. The Jacobi cylinder is now blown up in the vicinity of $q + \sqrt{-1}\,k$ by the substitution $k \to k/B$ $(B \uparrow \infty)$. The curve becomes a sphere with a nasty singularity at $\lambda = 0$, $[\ell(\lambda) = \lambda^g\sqrt{-\lambda}]$, and for $n = (1/2)g(g+1)$

$$\vartheta_1(q + \sqrt{-1}\,k + xk/B + t_1k/B + t_2k^3/B^3 + \dots + t_gk^{2g-1}/B^{2g-1})$$

converges to a polynomial

$$\vartheta_5 = \prod_{i=1}^{n}(x_i(t) - x)$$

The roots x_i $(i = 1,\dots,n)$ of ϑ_5 fill out the part of the closure of the <u>complex</u> locus

$$x_i \neq x_j \quad (i \neq j), \quad \sum_{j\neq 1}(x_i - x_j)^{-3} = 0 \qquad (i = 1,\dots,n)$$

for which the corresponding functions $q(x) = -2(\ell g\vartheta_5)'' = 2[(x - x_1)^{-2} + 2[(x - x_1)^{-2} + \dots + (x - x_n)^{-2}]$ are <u>real</u>. The latter fill out <u>the</u> g-dimensional invariant manifold Q of the flows $\partial q/\partial t = Xq$ in the class of real rational functions of x. The spectral description of Q has to do with the spectra of powers of L; see Adler and Moser (1978). The same computation has been carried out by Ablowitz.

The Jacobi Variety in the Periodic Case

The following information may be found in Dubrovin and Novikov (1974), and/or McKean and Moerbeke (1975). Let $2g + 1$ numbers $\lambda_0<\lambda_1<\lambda_2<\dots<\lambda_{2g-1}<\lambda_{2g}$ be fixed and let Q be the manifold of real infinitely differentiable functions q of period A for which the <u>simple</u> part of the periodic and antiperiodic spectrum of $L = -d^2/dx^2 + q(x)$ coincides with λ_i $(i = 0,1,\dots,2g)$, the rest of its spectrum being <u>double</u>. Q is void in general; in fact, if C is the hyperelliptic curve $\ell^2(\lambda) = -(\lambda-\lambda_0)\dots(\lambda-\lambda_{2g})$, if $\omega = \ell^{-1}(\lambda)(c_0+c_1\lambda + \dots + c_{g-1}\lambda^{g-1})d\lambda$ is the general differential of the first kind upon it, if a_i is the simple closed curve passing once counter-clockwise about the segment $[\lambda_{2i-1},\lambda_{2i}]$, and if $m_i = 1 +$ the number of double (or pairs of simple) eigenvalues to the left of λ_{2i-1} $(i = 1,\dots,g)$, then Q is nonvoid only if

$$\sum_{i=1}^{g} m_i \int_{a_i}\omega = A\,c_{g-1}$$

Q is then a g-dimensional torus isomorphic to the real part of the Jacobi variety J of C. The identification is made as follows. Let $\mu_i \in [\lambda_{2i-1},\lambda_{2i}]$ (i = 1,...,g) be the so-called auxiliary spectrum obtained by solving $Lf = \mu f$ with $f(0) = f(1) = 0$. The map of $q \in Q$ to the divisor $p_i = (\mu_i, \ell(\mu_i))$ (i = 1,...g) of C is 1:1 onto the product of the circles produced by opening up the banks of the segments $[\lambda_{2i-1},\lambda_{2i}]$ (i = 1,...,g), as in Figure 1. The supplementary map*

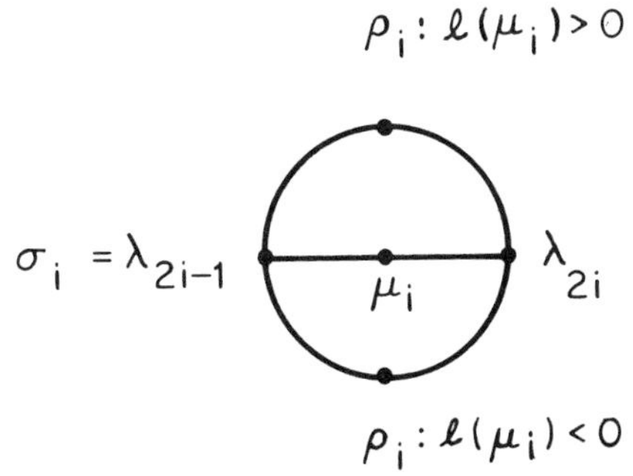

Figure 1

$$(p_1,\ldots,p_g) \to \sum_{i=1}^{g} \int_{o_i}^{p_i} \omega = x(\omega)$$

from the divisor p_i (i = 1,...,g) to the application x of differentials of the first kind is an isomorphism between divisors and the real part of the Jacobi variety J of C when considered modulo the lattice of (real) periods x arising from closed paths of integration in the sum. The inverse map $J \to Q$ has an elegant expression in terms of the theta function ϑ of $J: q(x) = -2[\ell g \vartheta(q + x x_1)]'' + \int_0^1 q(x)dx$, in which q is the point of J corresponding to the infinitesimal translation $X_1 q = q'$, and the prime stands for differentiation with regard to x; for simplicity, $\int_0^1 q(x)dx$ is made to vanish below. Now, $X_1 q = q'$ is only the first of a series of infinitesimal motions of Q expressed in Hamiltonian form by† $X_j q =$

*$o_i = (\lambda_{2i-1}, 0)$ (i = 1,...,g).

†$D = d/dx$. $\partial H/\partial q$ signifies the gradient in function space.

$D\,\partial H_j/\partial q$ $(j = 1,2,\dots)$. The Hamiltonian H_j is the integral over a period of a polynomial in $q(x), q'(x), q''(x), \dots$ without constant term, uniquely specified by a) that proviso, b) the rule $X_{j+1}q = K\,\partial H_j/\partial q$ with $K = qD + Dq - (1/2)D^3$, and c) $H_{-1} = 1$; for example 0) $H_0 = \int q\,dx$, $X_0 q = 0$, 1) $H_1 = \int (1/2)q^2 dx$, $X_1 q = q'$, 2) $H_2 = \int [(1/2)q^3 + (1/4)(q')^2]dx$, $X_2 q = 3qq' - (1/2)q'''$, etc., $X_2 q$ being the infinitesimal motion of Korteweg-de Vries. Now, the X_j $(j = 1,\dots,g)$ commute and span the tangent space of Q at every point, reexpressing the fact that Q is a g-dimensional torus; in particular, Q inherits from R^g a coordinate system obtained by fixing an origin of Q, e.g., the point o of Q with $\mu_i = \lambda_{2i-1}$ $(i = 1,\dots,g)$, and viewing the map $t \in R^g$, $e^X = \exp(t_1X_1 + \dots + t_gX_g)$ as an isomorphism between Q and the torus obtained from R^g by dividing out the lattice of periods such that e^X acts as the identity on Q. The remarkable fact about the connection between Q and J is that, up to a trivial substitution, the coordinate so obtained on Q is the same as that obtained from the coordinate x on J, i.e., if x_j is the direction tangent to J corresponding to the direction X_j on Q $(j = 1,\dots,g)$, then the point of J corresponding to $q = e^X(o)$ is $t_1'x_1 + \dots + t_g'x_g + o$ with an affine substitution $t' = at + b$, which is to say that the map $Q \to J$ converts the complicated nonlinear motions $\partial q/\partial t = X_j q$ $(j = 1,2,\dots)$ of Q into straight-line motions at constant speed: $q \to q + tx_j$ $(j = 1,2,\dots)$.

Singular Curves

The purpose of this article is to explain what happens when the curve $C: \ell^2(\lambda) = -(\lambda-\lambda_0)\dots(\lambda-\lambda_{2g})$ becomes singular. The computation goes back to Poincaré (1895); see Fay (1973, pg. 54) and Matveev (1976). Let $\lambda_{2g} = 0$, fix numbers $0 < k_g < \dots < k_1$, and pinch the handles of C to points by shrinking the segment $[\lambda_{2i-1}, \lambda_{2i}]$ to the point $-k_i^2$ $(i = 1,\dots,g)$ so that the curve becomes $\ell(\lambda) = (\lambda+k_1^2)\dots(\lambda+k_g^2)\sqrt{-\lambda}$, i.e., a sphere $k = \sqrt{-\lambda}$ with the points $\pm\sqrt{-1}\,k_i$ $(i = 1,\dots,g)$ identified in pairs*. Fix the homology basis a_i, b_i $(i = 1,\dots,g)$ as in Figure 2 and introduce the basis of the differentials of the first kind. The imaginary period matrix

---

* The Singularization requires letting $A \uparrow \infty$. By Marcenko and Ostroviskii (1975), this has the desired effect; in particular, any numbers k_i $(i = 1,\dots,g)$ achieved by proper choice of $\lambda_0,\dots,\lambda_{2g}$ for $A = 1$, say. During the deformation, the lengths $\lambda_{2i-1} - \lambda_{2i-2} = o(1)$ $(i = 1,\dots,g)$ are comparable; this is important for the appraisal of periods carried out below.

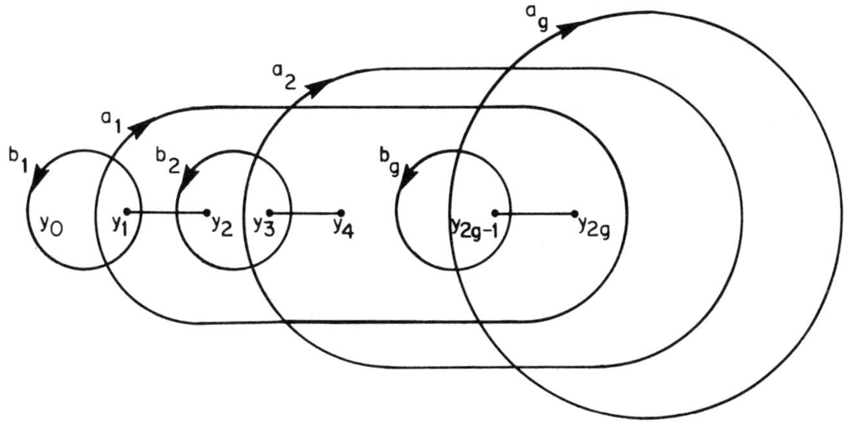

$$\omega_i = \ell^{-1}(\lambda) \prod_{g>j\neq 1} (\lambda-\lambda_{2j})\, d\lambda \times \sqrt{-\lambda_{2i}} \times (2\pi)^{-1} \quad (i = 1,\cdots,g)$$

Figure 2

$$B_{ij} = \sqrt{-1} \int_{b_i} \omega_j = \sqrt{-1}(2\pi)^{-1}\sqrt{-\lambda_{2j}} \int_{b_i} \frac{1}{\lambda-\lambda_{2j}} \frac{d\lambda}{\sqrt{-\lambda}} + o(1)$$

tends to the identity, while the real period matrix

$$A_{ij} = \int_{a_i} \omega_j$$

consists of two parts: a singular diagonal and a nonsingular off-diagonal. The 11 entry is typical of the diagonal:

$$A_{11} = \frac{\sqrt{-\lambda_1}}{\pi} \int_1^0 \frac{1}{\sqrt{(\lambda-\lambda_0)(\lambda-\lambda_1)}} \frac{d\lambda}{\sqrt{-\lambda}} + o(1)$$

$$= \frac{1}{\pi} \lg \frac{1}{\lambda_1-\lambda_0} + \frac{2}{\pi} \lg k_1 + c + o(1)$$

The 12 entry is typical of the off-diagonal:

$$A_{12} = \frac{\sqrt{-\lambda_2}}{2\pi} \int_{a_1} \frac{\lambda-\lambda_0}{\lambda-\lambda_1} \frac{1}{\sqrt{(\lambda-\lambda_2)(\lambda-\lambda_3)}} \frac{d\lambda}{\sqrt{-\lambda}} + o(1)$$

$$= \frac{\sqrt{-\lambda_2}}{\pi} \int_{\lambda_1}^{0} \frac{1}{\lambda-\lambda_2} \frac{d\lambda}{\sqrt{-\lambda}} + o(1) = \frac{1}{\pi} \ell g \frac{k_2+k_1}{k_2-k_1} + o(1)$$

the second integral being contrued as a principal value. The upshot is that the Riemann matrix $C = BA^\dagger$ can be estimated as

$$\frac{1}{\pi} \ell g \frac{1}{\lambda_{2i-1} - \lambda_{2i-2}} + c_i + o(1) \qquad \text{on diagonal}$$

$$\frac{1}{\pi} \lambda g \left| \frac{k_j+k_i}{k_j-k_i} \right| + o(1) \qquad \text{off diagonal}$$

The theta function of the curve is defined by

$$\vartheta(x) = \sum e^{2\pi\sqrt{-1}\, x\cdot m - \pi C[m]}$$

with m running over the dual of the real period lattice and $C[m]$ the quadratic form based upon $C = BA^\dagger$. The Jacobi transformation is now employed to clarify the behavior of ϑ as the curve becomes singular: up to a constant factor

$$\vartheta(x) = \sum e^{-\pi(x-\ell)(BA^\dagger)^{-1}(x-\ell)}$$

with ℓ running over the real period lattice itself. The formula is applied with $x + A(1/2)$ $[1/2 = (1/2,\ldots,1/2]$ in place of x, $\ell = An$, and n running over the g-dimensional integral lattice Z^g. The sum may now be expressed as*

$$\sum e^{-2\pi(1/2 - n)x - \pi(1/2 - n)B^{-1}A(1/2 - n)} + o(1)$$

and the appraisal of periods confirms that the principal part of the sum comes from terms with

$$\sum_{i=1}^{g} |\ell g(\lambda_{2i-1} - \lambda_{2i-2})|(n_i - 1/2)^2$$

as small as possible, i.e., with $n_i = 0$ or 1 $(i = 1,\ldots,g)$. The moral

*$BA^\dagger = AB^\dagger$ by the Riemann period relations.

is that, aside from an affine substitution in χ and an inessential factor of the form $e^{a\chi+b}$, the principal part of $\vartheta(\chi+A(1/2))$ is just

$$\vartheta_1(\chi) = e^{\chi\cdot n+K[n]}$$

$$K[n] = \sum_{i<j} n_i n_j \, \ell g \left(\frac{k_i-k_j}{k_i+k_j}\right)^2$$

with $n_i = 0$ or 1 $(i = 1,\ldots,g)$ in the sum, ϑ_1 being regarded as the theta function of the singular curve*.

The Jacobi Cylinder and Solitons

The meaning of the singular theta function is not far off. The singularization of the curve blows up the variety J into a cylinder with imaginary periods $2\pi\sqrt{-1}\,(n_1/k_1,\ldots,n_g/k_g):n \in Z^g$ with no real periods at all. The directions χ_j $(j = 1,\ldots,g)$ corresponding to the infinitesimal translation $X_1 q = q'$, etc., go over into directions tangent to the cylinder: for example†,

$$\chi_1:\omega_j \to \sum_{i=1}^{g} m_i \int_{a_i} \omega = A \times (2\pi)^{-1} \sqrt{-\lambda_{2j}} \qquad (j = 1,\ldots,g)$$

is proportional to $k + o(1)$, while a closer investigation confirms the fact that χ_j $(j = 1,\ldots,g)$ is proportional to $k^{2j-1} + o(1)$ $(j = 1,\ldots,g)$††. The upshot is that, for

$$\vartheta_0 = {}_1(q + xk + t_1 k + t_2 k^3 + \ldots + t_g k^{2g-1})$$

$$= \sum e^{(q + xk + t_1 k + t_2 k^3 + \ldots + t_g k^{2g-1})\cdot n - K[n]}$$

the function $q = -2(\ell g\vartheta_0)''$ satisfies $\partial q/\partial t_j = X_j q$ $(j = 1,\ldots,g)$ up to an inessential scaling; for example, if $g = 1$, then

$$\vartheta_1(xk - tk^3/4) = 1 + e^{xk - tk^3/4}$$

---

* $\left(\frac{a-b}{a+b}\right)^2$ is just the cross ratio $[a,-a;b,-b]$.

† a_i is now the customary loop passing once counterclockwise about the segment $[\lambda_{2i-1},\lambda_{2i-2}]$ $(i = 1,\ldots,g)$.

†† $k^j = (k_1^j,\ldots,k_g^j)$.

and $q(x) = k(k^2/2)\cosh^{-2}(x/2 - tk^3/4)$ is the familiar solitary wave for $\partial q/\partial t = X_2 q$. The general formula for ϑ_1 is due to Hirota (1971); in its dependence on x (or t_1) and $t_2 = -t/2$, it expresses the general many-soliton solution of $\partial q/\partial t = X_2 q$. The adjective <u>many-soliton</u> refers to the fact that for $t \to \pm\infty$, $q = -2[\ell g \vartheta_1(q + xk + tk^3/4)''$ separates out into a sum of solitary waves

$$q(x) = \sum_{i=1}^{g} -(k_i^2/2)\cosh^{-2}(q_i/2 + xk_i/2 = tk_i^3/4)$$

or approximately so. The more conventional formula

$$\vartheta_1(x) = \sum e^{x \cdot n - K[n]} = \det\left[1 + \frac{2\sqrt{k_i k_j}}{k_i + k_j} e^{x_i/2} e^{x_j/2}\right]$$

of Kay and Moses (1956) and Gardner, Greene, Kruskal and Muira (1974) is obtained by expansion of the determinant in the form $\det(1 + \Delta) = 1 + \mathrm{sp}\,\Delta + \ldots + \det\Delta$, but more of this below.

Scattering Theory

The singular theta function may be profitably viewed from the standpoint of scattering theory. Let $q(x)$ vanish rapidly at $x = \pm\infty$, as is the case for many-soliton functions, let $L = -d^2/dx^2 + q(x)$, and let $\psi_-(x,k)$ and $\psi_+(x,k)$ be the wave functions determined by the rule indicated in Table 1, with $e_\pm = \exp(\pm\sqrt{-1}\,kx)$.

TABLE 1

	$x \uparrow \infty$	$x \downarrow -\infty$
ψ_+	$s_{11}e_+$	$e_+ + s_{12}e_-$
ψ_-	$e_- + s_{21}e_+$	$s_{22}e_-$

The recipe specifies the (unitary) scattering matrix $s_{ij}(k)$ $(i,j = 1,2)$ as a function of the wave number $k \neq 0$; actually only the so-called reflection coefficient s_{12} is necessary, e.g.,

$$s_{11}(k) = \exp\left[\frac{1}{2\pi\sqrt{-1}} \int \ell g(1 - |s_{12}|^2(k'))\, \frac{dk'}{k' - k - \sqrt{-1}\,0+}\right]$$

The latter pertains to the continuous spectrum of L which may have an additional discrete spectrum comprising a number $g < \infty$ of bound states at energy levels $-k_1^2 < \ldots < -k_g^2 < 0$ with wave functions

$$\psi(x,\sqrt{-1}\,k_i) \sim \ell_i \exp(\mp k_i x) \qquad (\mp x \uparrow \infty) \qquad (i = 1,\ldots,g)$$

The map from q so $s_{12}(k)$ $(-\infty < k < \infty)$ and k_i, ℓ_i $(i = 1,\ldots,g)$ is 1:1, the inverse map being expressed as follows. Let K^+ denote the upper triangular part of the integral operator K, i.e., $K^+f(x) = \int_x^\infty K(x,y)f(y)dy$, and let $Af(x) = \int A(x+y)f(y)dy$ with

$$A(x) = \sum_{i=1}^{g} \ell_i e^{-k_i x/2} + (2\pi)^{-1} \int e^{\sqrt{-1}\,kx} s_{12}(k)dk$$

The inverse map is now effected by solving $A^+ + (BA)^+ + B = 0$ for the triangular operator B, the upshot being $q(x) = -2[B(x,x)]'$. The formal solution is $B = -A^+ + (A^+A)^+ - ((A^+A)^+A)^+ + \ldots$, and with the notation Δ for the restriction of A to $[x,\infty)$, it develops that

$$\begin{aligned} B(x,x) &= -\sum_{n=0}^{\infty} (-1)^n \int_{[x,\infty)^n} A(x+y_1)A(y_1+y_2)\ldots A(y_n+x)d^n y \\ &= \frac{d}{dx} \sum_{n=0}^{\infty} \frac{(-1)^n}{n+1} \int_{[x,\infty)^{n+1}} A(y_0+y_1)\ldots A(y_n+y_0)d^{n+1}y \\ &= [\mathrm{sp}\ \ell g(1+\Delta)]' \\ &= [\ell g \det(1+\Delta)]' \end{aligned}$$

The formula for q now takes the form $q = -2(\ell g\vartheta_2)''$ with $_2 = \det(1+\ ) =$

$$\vartheta_2 = \det(1+\Delta) = \sum_{n=0}^{\infty} \frac{1}{n!} \int_{[x,\infty)^n} \det[A(y_i + y_j):1 \leqslant i,j \leqslant n]d^n y$$

see Dyson (1976). The formula of Kay and Moses (1956) for the singular theta function arises from the present recipe in the reflectionless case $[s_{12} = 0]$; in fact, as an elementary computation confirms, if $s_{12} = 0$

$$\begin{aligned} \det(1+\Delta) &= \det\left[1 + \frac{2\sqrt{\ell_i\ell_j}}{k_i+k_j} e^{-k_i x/2} e^{-k_j x/2} :1 \leqslant i,j \leqslant g\right] \\ &= \sum e^{(q-xk)\cdot n} \prod_{i<j} \frac{k_i-k_j}{k_i+k_j}^{2n_i n_j} \end{aligned}$$

which is just the old formula for ϑ_1 with $-k$ in place of k and $e^q = (\ell_1/k_1,\ldots,\ell_g/k_g)$; in particular, the many-soliton functions are now seen to be characterized by $s_{12} = 0$, and the numbers $-k_i^2$ and

$$k_i e^{q_i} = \ell_i \qquad (i = 1,\ldots,g)$$

are recognized, repsectively, as the energy levels and norming constants comprising, in the absence of s_{12}, the spectral data of L. The same kind of formula is obtained in the absence of bound states $[g = 0]$; in fact, with $k^+ = k + \sqrt{-1}\ 0+$ and $q(k) = \text{phase } s_{12}(k)$,

$$\vartheta_2 = \sum_{n=0}^{\infty} \frac{1}{n!} \int_{[x,\infty)^n} \det\left[\frac{1}{2\pi}\int e^{\sqrt{-1}\,k(y_i+y_j)} s_{12}(k)dk\right] d^n y$$

$$= \sum_{n=0}^{\infty} \frac{1}{n!}(2\pi)^{-n}\int s_{12}(k_1)\ldots s_{12}(k_n)d^n k \int_{[x,\infty)^n}\det\left[e^{\sqrt{-1}(k_i^+ + k_j^+)y_j}\right] d^n y$$

$$= \sum_{n=0}^{\infty} (-2\pi\sqrt{-1})^{-n} \int_{k_1<\ldots<k_n} s_{12}(k_1)\ldots s_{12}(k_n)d^n k \det\left[\frac{e^{\sqrt{-1}(k_i^+ + k_j^+)x}}{k_i^+ + k_j^+}\right]$$

$$= \sum_{n=0}^{\infty} (-\sqrt{-1}\pi)^{-n} \int_{k_1<\ldots<k_n} \frac{|s_{12}(k_1)\ldots s_{12}(k_n)|}{2^n k_1^+\ldots k_n^+} d^n k$$

$$\times\ e^{\sqrt{-1}[q(k_1)+\ldots+q(k_n)+2x(k_1^+ +\ldots+k_n^+)]} \prod_{i<j}\left(\frac{k_i - k_j}{k_i^+ + k_j^+}\right)^2$$

the limit indicated by k^+ being performed outside the sum. ϑ_2 now appears as some kind of singular (theta) function of the phase $q(k) + 2xk$. Now, let q be general. The flows $\partial q/\partial t = X_j q$ $(j = 1,2,\ldots)$ commute, the action of the jth flow on the scattering data being to advance the phase of $s_{12}(k)$ by k^{2j-1} and $\ell g(\ell_i/k_i)$ by $k_i^{2j-1}t$ $(i = 1,\ldots,g)$. Thus, $|s_{12}(k)|$ and k_i $(i = 1,\ldots,g)$ determine an invariant manifold Q of g or $g + \infty$ dimensions according as $s_{12} = 0$ or not; especially, Q is a kind of Jacobi cylinder ϑ relative to which the flows $\partial q/\partial t = X_j q$ $(j = 1,2,3,\ldots)$ are straight-line motions at constant speed: schematically,

$$e^{tX_j}q(x) = -2[\ell g\vartheta_2(q + xk + tk^{2j-1})]'' \qquad (j = 1,2,\ldots)$$

The theta function aspect of the scattering theory is still to be fully investigated and is presented only in a provisional and formal way.

Rational Functions of e^x

The special many-soliton case with $k = (k_1,\ldots,k_g)$ integral is of particular interest. $\vartheta_1(q + xk)$ is a polynomial in $y = e^x$ of degree $n = k_1 + \ldots + k_g$ with roots

$$y_i = e^{x_i} \neq 0 \qquad (i = 1,\ldots,n)$$

in the closure of the locus (see Airault, McKean and Moser, 1977).

$$y_i \neq y_j \quad (i \neq j)$$

$$\sum_{j\neq i} \frac{y_i+y_j}{(y_i-y_j)^3} = 0 \qquad (i = 1,\ldots,n)$$

and

$$q(x) = -2(\lg\vartheta_1)'' = (1/2)\sum_{i=1}^{n} \sinh^{-2}\left(\frac{x-x_i}{2}\right)$$

is a rational function of $y = e^x$; in fact, for fixed $n = 1,2,3,\ldots,$ the closed locus breaks up into subloci identifiable via the map

$$(y_1,\ldots,y_n) \to q = -2(\lg\vartheta_1)''$$

as the many-soliton families Q arising from the solutions of $k_1+\ldots+k_g = n$ in unequal integers k_i $(i = 1,\ldots,g)$, such Q being the most general invariant manifolds of the flows $\partial q/\partial t = Xq$ in the class of rational functions of $y = e^x$ vanishing at $x = \pm\infty$. The proof proceeds in two steps. To begin with, if q is a rational function of e^x vanishing at $x = \pm\infty$ and retaining that rational form under the flows $\partial q/\partial t = Xq$, then (see Airault, McKean and Moser, 1977) it can be expressed as $-2(\lg\vartheta_3)''$ with

$$\vartheta_3 = \prod(e^x - e^{x_i})$$

and

$$y_i = e^{x_i} \qquad (i = 1,\ldots,n)$$

from the closed locus for some $n - 1,2,3,\ldots.$ The proof is finished by confirming the cited break-up of the locus into many-soliton subloci. Let

e^{x_i} $(i = 1,\dots,n)$ belong to the open locus and form $q = -2(\lg\vartheta_3)''$ with ϑ_3 as before. Then (see Airault, McKean and Moser, 1977) q retains this form under the flows $\partial q/\partial t = Xq$ for short times and the (open) manifold Q so produced is of dimension $g < \infty$. But, then, $s_{12} = 0$, q is a many-soliton function $-2(\lg\vartheta_1)''$, k is integral since e^x is of period $2\pi\sqrt{-1}$, and $k_1 + \dots + k_g = n$ by a degree count.

EXAMPLE. The sublocus with smallest n for fixed g is associated with $k = (1,2,\dots,g)$ $[n = (1/2)g(g+1)]$. The corresponding class Q contains the function $q_g(x) = -(1/4)g(g+1)\cosh^{-2}(x/2)$. This is plain for $g = 1: q_1(x) = -2[\lg(1 + e^x)]''$. The general proof follows Deift and Trubowitz (1978). Let $g = 2,3,\dots$. Then $\psi_0(x) = \cosh^{g+1}(x/2)$ is an eigenfunction of $L_g = -d^2/dx^2 + q_g(x)$ and $q_{g+1} = q_g - 2(\lg\psi_0)''$.

Let ψ be a wave function of L_g with wave number $k \neq 0$. Then, $\psi_+ = \psi_0^{-1}(\psi_-'\psi_0 - \psi_-\psi_0')$ is a wave function of L_{g+1} with the same wave number. The reflectionless property of q_g is now seen to be inherited by q_{g+1}; also, L_g is seen to have at least g bound states: $\psi(x) = \cosh(x/2)$ is one such with energy level $-g^2$ and $g - 1$ more at energy levels $-1^2,-2^2,\dots,-(g-1)^2$ are inherited from L_{g-1}. The formula $q_g = -2(\lg\vartheta_1)''$ is now applied: k is integral since e^x is of period $2\pi\sqrt{-1}$ and $n = k_1 + \dots + k_g$ is identified as $(1/2)g(g+1)$ by integrating $(\lg\vartheta_1)'' = (1/2)g(g+1)[\lg(1 + e^x)]''$ to obtain $\vartheta_1 = (1 + e^x)^n$. The identity $k_1 + \dots + k_g = (1/2)g(g+1)$ now forces $k = (1,2,\dots,g)$. The connection with the periodic case is noted: Let $p(x)$ be the Weierstrassian elliptic function of real period A and imaginary period $2\pi\sqrt{-1}\,B$. Ince (1940) proved that the periodic and antiperiodic spectrum of $q(x) = g(g+1)[p(x + \pi\sqrt{-1}\,B) - 1/3]$ begins with $2g + 1$ simple eigenvalues, the rest of the spectrum being double. Let $A \uparrow \infty$. Then, $q(x) \sim -g(g+1)$, $[2B\cosh(x/2B)]^{-2} = -2(\lg\vartheta_1)''$ with $k = B^{-1}(1,2,\dots,g)$.

Second Singularization: Rational Functions of x

The purpose of this section is to make a second singularization of the curve $\ell(\lambda) = (\lambda+1^2)\dots(\lambda+g^2)\sqrt{-\lambda}$ of the previous section by making the imaginary periods of the Jacobi cylinder tend to ∞. This lead to maifolds Q of rational functions of x invariant under the flows $\partial q/\partial t = Xq$; see Airault, McKean and Moser (1977) and, for an elementary and more satisfactory account not employing theta functions, Adler and Moser (1978). Let

q and $k = (1,2,\ldots,g)$ be as in the example of the last section so that $\vartheta_1(q + xk) = (1 + e^x)^n$ with $n = (1/2)g(g + 1)$. The singularization is effected by blowing up the cylinder in the vicinity of the point $q + \sqrt{-1}\, k$ by letting $B \uparrow \infty$ in the singular theta function

$$\vartheta_4 = \vartheta_1(q + \sqrt{-1}\, k + xk/B + t_1 k/B + t_2 k^3/B^3 + \ldots + t_g k^{2g-1}/B^{2g-1}$$

The statement is that for $n = (1/2)g(g + 1)$ and $B \uparrow \infty$, $B^n\vartheta_4$ approaches a polynomial

$$\vartheta_5 = \prod_{i=1}^{n}(x_i(t) - x)$$

This is clear for $t = 0$: $B^n\vartheta_4 = B^n(1 - e^{x/B})^n \sim \vartheta_5 = (-x)^n$, the associated function $q = -2(\ell g\vartheta_5)''$ being $2n\, x^{-2}$; it is also clear that, for general $t_1,\ldots,t_g$, $q = -2(\ell g\vartheta_5)''$ will be a rational function of x and t and will satisfy $\partial q/\partial t_j = X_j q$ $(j = 1,2,\ldots)$ up to an inessential scaling*. The g-dimensional manifold Q of rational functions $q = -2(\ell g\vartheta_5)'' = 2[(x - x_1)^2 + \ldots + (x - x_n)^2]$ so produced may be described in another way : it is the class of real rational functions of the stated form with poles x_i $(i = 1,\ldots,n)$ from the closure of the complex locus

$$x_i \neq x_j \qquad (i \neq j)$$

$$\sum_{j\neq i} (x_i - x_j)^3 = 0 \qquad (i = 1,\ldots,n)$$

For $g = 1,2,3,\ldots,$ (see Airault, McKean and Moser, 1977) this provides a complete list of the manifolds Q of real rational functions which retain a rational form under the flows $\partial q/\partial t = Xq$. The proof of the existence of

$$\vartheta_5 = \lim_{B\uparrow\infty} B^n\vartheta_4$$

is equivalent to the verification of

THE VANISHING THEOREM† . $(\partial^m/\partial t^m)\vartheta_4$ vanishes at $t = 0$ if

$$|m| = m_1 + 3m_2 + \ldots + (2g - 1)m_g < n = (1/2)g(g + 1)$$

---

*$X_j q$ is an isobaric polynomial in q and $' = D$, counting q as of degree 2 and $'$ as of degree 1; for instance, $X_2 q = 2qq' - (1/2)q'''$ is of degree 5.

† $\partial^m/\partial t^m = (\partial/\partial t_1)^{m_1} \ldots (\partial/\partial t_g)^{m_g}$.

The idea of the proof is as follows:

$$\vartheta_1(q + \sqrt{-1}\,k + t_1 k) = (1 - e^{t_1})^n \sim (-t_1)^n$$

as $t_1 \to 0$, so the statement is plain if $t_2 = \ldots = t_g = 0$. Now

$$\vartheta_4 = \vartheta_1(q + \sqrt{-1}\,k + t_1 k + t_2 k^3)$$

is a polynomial of degree $1 + \ldots + g = n$ in e^{t_1} and of degree $1^3 + \ldots + g^3$ in e^{t_2}; also*, $q = -2(\ell g\vartheta_4)''$ satisfies $\partial q/\partial t_2 = X_2 q = 3qq' - (1/2)q'''$ up to an inessential scaling, whence, for $t_2 = \ldots = t_g = 0$ and $t_1 \neq 0^\dagger$, $\vartheta_4 \sim (-t_1)^n$ and $-2(\vartheta_{42}/\vartheta_4)'' = X_2 q = X_2 O(t_1^{-2}) = O(t_1^{-5})$, with the result that $\vartheta_{42} = O(t_1^{n-3})$, i.e., $(\partial/\partial t_1)^{m_1} \partial/\partial t_2 \vartheta_4 = 0$ at $t = 0$ if $m_1 + 3 \cdot 1 < n$. Now, let $|m| = m_1 + 3m_2 + \ldots + (2g - 1)m_g$. Then**, for $t_2 = \ldots = t_g = 0$ and $t_1 \neq 0$, $\partial^m q/\partial t^m = X^m q$ implies

$$(\partial^m/\partial t^m) - 2(\ell g\vartheta_4)'' = X^m O(t_1^{-2}) = O(t_1^{-2-|m|})$$

and, by induction, $\partial^m \vartheta_4/\partial t^m = O(t_1^{n-|m|})$ vanishes at $t = 0$ if $|m| < n$. The proof of the vanishing theorem is finished, assuring the existence of††

$$\lim_{B\uparrow\infty} B^n \vartheta_4 = \vartheta_5 = \sum t^m (k\cdot n)^{m_1} (k^3\cdot n)^{m_2} \ldots (k^{2g-1}\cdot n)^{m_g}$$
$$\times (-1)^{k\cdot n} e^{q\cdot n + \sum_{i\neq j} n_i n_j \,\ell g|(i-j)/(i+j)|}$$

the sum being taken over $|m| = \frac{1}{2}g(g + 1)$ and $n = (n_1,\ldots,n_g)$ with coordinates 0 or 1.

See Adler and Moser (1978) for a more elementary and satisfactory expression.

REMARK. The vanishing theorem holds only for the present choice of q.

REFERENCES

Adler, M., and Moser, J. On a class of polynomials connected with the Korteweg-de Vries equation. To appear (1978).

* $' = \partial/\partial t_1$

† $\vartheta_{42} = \partial\vartheta_4/\partial t_2$

** $X^m = X_1^{m_1} \ldots X_g^{m_g}$

†† $t^m = t_1^{m_1} \ldots t_g^{m_g}$

Agranovich, Z. S. and Marcenko, V. A. *The Inverse Problem of Scattering Theory*. Gordon and Beach, New York (1963).

Airault, H., McKean, H. P., and Moser, J. Rational and elliptic solutions of the Korteweg-de Vries equation and a related many-body problem. *Comm. Pure Appl. Math.*, *30* (1977), pp. 95-148.

Deift, P., and Trubowitz, E. Inverse scattering on the line. *Comm. Pure Appl. Math.*, to appear.

Dubrovin, B. A., and Novikov, S. P. A periodicity problem for the Korteweg-de Vries and Sturm-Liouville equations. *Sov. Math. Dokl.*, *15* (1974) pp. 1597-1601.

Dyson, F. J. Fredholm determinants and inverse scattering problems. *Comm. Math. Phys.*, *47* (1976), pp. 171-183.

Faddeev, L. Properties of the S-matrix of the one-dimensional Schroedinger equation. *Trudy Mat. Inst. Steklov*, *73* (1964), pp. 314-333; *AMST*, *65* pp. 139-166.

Fay, J. Theta Functions on Riemann Surfaces. *Lect. Notes in Math.*, *352* Springer-Verlag, New York (1973).

Gardner, C., Greene, J. M., Kruskal, M., and Miura, R. Korteweg-de Vries equation and its generalizations 6. *Comm. Pure Appl. Math.*, *27* (1974) pp. 97-133.

Gelfand, I. M., and Levitan, B. M. On the determination of a differential equation from its spectral function. *Izvest. Akad. Nauk*, *15* (1951) pp. 309-360; *AMST*, *1* (1955), pp. 253-304.

Hirota, R. Exact solutions of the Korteweg-de Vries equation for multiple collisions of solitons. *Phys. Rev. Lett.*, *27* (1971), pp. 1192-1194.

Ince, E. L. The periodic Lamé functions. *Proc. Roy. Soc. Edinburgh*, *60* (1940), pp. 47-63.

Its, A. R., and Matveev, V. B. Hill's operator with finitely many gaps. *Funk. Anal. Appl.*, *9* (1975), pp. 65-66.

Kay, I., and Moses, H. Reflectionless transmission through dielectrics and scattering potentials. *J. Appl. Phys.*, *27* (1956), pp. 1503-1508.

Marcenko, V. A., and Ostrovskii, I. V. A characterization of the spectrum of Hill's operator. *Mat. Sbornik SSSR*, *139* (1975), pp. 540-606.

Matveev, V. B. Abelian functions and solitons. *Instytut Fizyki Teoretycznej Uniwersytetu Wroclawskiego*, preprint no. 373, Wroclaw (1976).

McKean, H. P., and van Meorbeke, P. The spectrum of Hill's equation. *Invent. Math.*, *30* (1975), pp. 217-274.

McKean, H. P., and Trubowitz, E. Hill's operator and hyperelliptic function theory in the presence of infinitely many branch points. Comm. Pure Math., 29 (1976), pp. 143-226.

McKean, H. P., and Trubowitz, E. Hill's surfaces and their theta functions Bull. Amer. Math. Soc., to appear.

Poincaré, H. Remarques diverses sur les fonctions abeliennes. J. de Math., 1 (1895).

Courant Institute of Mathematical Sciences
New York, New York

PARAMETRICES WITH C^∞ ERROR FOR $\Box_b$ AND OPERATORS OF HORMANDER TYPE

Linda Preiss Rothschild* and David Tartakoff†

Introduction

We construct here parametrices for certain second order hypoelliptic differential operators including the boundary Laplacian $\Box_b$ of the Cauchy-Riemann operator. These classes of operators have been studied in Folland and Stein (1974), and Rothschild and Stein (1976), where parametrices were constructed to invert the operator modulo an error which is "smoothing" (see Section 2) of any preassigned finite order. Starting with the approximate inverses defined in Rothschild and Stein (1976) via singular integrals on nilpotent Lie groups, we obtain operators which are inverses modulo an error which is infinitely smoothing. In fact, the error will be given by an operator with a smooth kernel.

A differential operator D is called <u>hypoelliptic</u> if $Df = g$ with $g \in C^\infty(U)$ implies $f \in C^\infty(U)$ for any open set U. We shall consider two classes of hypoelliptic differential operators. First, let M be a partially complex (or C-R) manifold of dimension $2\ell + 1 = m$ with a fixed Riemannian metric. (See, for example, Folland and Kohn, 1972, for relevant definitions.) $\Box_b$ is defined as the Laplacian $\theta_b\bar{\partial}_b + \bar{\partial}_b\theta_b$, where θ_b is the formal adjoint of $\bar{\partial}_b$, the tangential Cauchy-Riemann operator. $\Box_b$ operates on (p,q) forms on M. Kohn (1964) has proved that the following conditions on the Levi form ρ on M imply that $\Box_b$ is hypoelliptic on (p,q)-forms.

Y(q) The Levi form ρ has at least $\min(q + 1, \ell - q + 1)$ pairs of eigenvalues of opposite sign or $\max(q + 1, \ell - q + 1)$ eigenvalues of the same sign.

We shall construct a two-sided inverse, modulo an infinitely smoothing error error, S, for $\Box_b$ acting on (p,q)-forms, provided the condition Y(q)

*Supported in part by N. S. F. Grant MCS77-01155.

†Supported in part by N. S. F. Grant MCS72-05055A04.

is satisfied. Parametrices for $\square_b$ have previously been given by Boutet de Monvel (1974) and by Sjöstrand (1974). However, our operators are proved to be bounded on the appropriate L^p Sobolev spaces for $1 < p < \infty$, while the methods used to construct the parametrices of Boutet de Monvel and Sjöstrand do not seem to lend themselves to such a proof.

A key step in the proof of hypoellipticity of $\square_b$ is the observation that the highest order term is a negative definite quadratic expression in real vector fields which, together with their commutators, span the tangent space. Hörmander (1967) generalized this idea as follows. Let $X_1, X_2, \ldots, X_n$ be real vector fields on a manifold M, $m = \dim M$, such that $X_1, X_2, \ldots, X_n$, together with their commutators

$$[X_{i_1}, [X_{i_2}, \ldots [X_{i_1}, X_{i_j}] \ldots],] \qquad j \leqslant r$$

up to length r span the tangent space at each point. Then

$$\mathcal{L} = \sum_{i=1}^{n} X_i^2$$

is hypoelliptic. We shall construct an inverse, modulo an infinitely smoothing operator, for $\mathcal{L}$. To our knowledge, there is no other method known for constructing such an operator. For the special case $r = 2$, however, the pseudodifferential operators defined by Beals (preprint), Boutet de Monvel (1974) and Sjöstrand (1974) contain parametrices for many operators of the form $\mathcal{L}$. Furthermore, Grusin (1970) constructs parametrices for certain operators of this type with arbitrary r. None of these classes of parametrices is shown to preserve L^p for all $1 < p < \infty$.

We remark here that our methods could also be applied to the more general operators

$$X_0 + \sum_{i=1}^{n} X_i^2$$

considered in Hörmander (1967). For simplicity, we restrict out attention to sums of squares.

We wish to express our thanks to E. M. Stein and J. Rauch for several helpful conversations on this work.

Main Results

Let $X_1, X_2, \ldots, X_n$ be real vector fields which, together with their commutators of length $\leqslant r$, span the tangent space, TM. Since we are interested in local results, we may replace M by a relatively compact subset and define the following function spaces in $\mathbb{R}^m$:

$$L^{\infty} = \{f: f \text{ essentially bounded}\}$$
$$L^p = \{f: \|f\|_{L^p} = (\int |f|^p dx)^{1/p} < \infty\}$$

The weighted Sobolev spaces S_k^p are defined for nonnegative integers k:

$$S_k^p = \{f \in L^p : X_{i_1} X_{i_2}, \ldots, X_{i_j} f \in L^p, \quad j \leq k$$

$$\text{with norm } \|f\|^p_{S_n^p} = \sum \|X_{i_1} \ldots X_{i_j} f\|^p_{L^p}$$

The sum is taken over all indices $(i_1, \ldots, i_j)$ $j \leq k$, $0 \leq i_\ell \leq n$, with

$$X_{i_0} = I$$

We say $f \in S^p_{k,loc}$ if $\phi f \in S_k^p$ for any $\phi \in C_0^{\infty}$. The classical L^p Sobolev spaces are defined as usual and denoted L_α^p, $L^p_{\alpha,loc}$ (cf. Stein, 1970). The Lipschitz spaces are also defined with reference to $\mathbb{R}^m$ ($\|\ \|$ denoting Euclidean length): For $0 < \beta < 1$

$$\Lambda_{\beta,loc} = \{f: \sup |\phi f(x+y) - \phi f(x)|/\|y\|^{\beta} < \infty, \text{ all } \phi \in C_0^{\infty}\}$$

with norm

$$\|\phi f\|_{\Lambda_\beta} = \|\phi f\|_{\infty} + \sup_{\|y\|>0} |\phi f(x+y) - \phi f(x)|/\|y\|^{\beta}$$

$$\Lambda_{1,loc} = \{f: \phi f \in L^{\infty}; \sup_{\|y\|>0} |\phi f(x+y) + \phi f(x-y) - 2\phi f(y)|/\|y\| < \infty \quad \forall \phi \in C_0^{\infty}\}$$

Inductively, for $\alpha > 1$,

$$\Lambda_{\alpha,loc} = \{f: \phi f \in L^{\infty} \text{ and } \frac{\partial}{\partial x_j}(\phi f) \in \Lambda_{\alpha-1}, \quad \phi \in C_0^{\infty}\}$$

with norm

$$\|\phi f\|_{\Lambda_\alpha} = \|\phi f\|_{L^{\infty}} + \sum_{j=1}^{m} \left\| \frac{\partial}{\partial x_j}(\phi f) \right\|_{\Lambda_{\alpha-1}}$$

Let L, L' be any two of the spaces defined above. A mapping $T: L \to L'$ will be said to be bounded from L_{loc} to L'_{loc} if $\phi_1 T \phi_2$ is bounded in the appropriate norms for any $\phi_1, \phi_2 \in C_0^{\infty}$. T will be said to be smoothing of order λ if T is bounded from $L^p_{\alpha,loc}$ to $L^p_{\alpha+\lambda/r,loc}$, for all real α, from $S^p_{k,loc}$ to $S^p_{k+\lambda,loc}$, all nonnegative k, from $\Lambda_{\alpha,loc}$ to $\Lambda_{\alpha+\lambda/r,loc}$ all $\alpha > 0$, and L^∞_{loc} to $\Lambda_{\lambda/r,loc}$. T is smoothing of infinite order if it is smoothing of order λ for all $\lambda > 0$.

THEOREM 1. Let

$$\mathcal{L} = \sum_{j=1}^{n} X_j^2$$

and $\phi \in C_0^\infty(M)$ be given. Then there exist operators P and P', smoothing of order 2, and S_∞, S'_∞, smoothing of infinite order, such that for $f \in L^p(M)$, $1 < p < \infty$,

$$P\mathcal{L}f = \phi f + S_\infty f$$

$$\mathcal{L}P'f = \phi f + S'_\infty f$$

Furthermore, $P\phi = \phi P'$. The errors S_∞ and S'_∞ are integral operators with infinitely smoothing kernels.

THEOREM 2. Let $\Box_b^{(q)}$ be the boundary Laplacian acting on (p,q)-forms on a CR-manifold M. Suppose $Y(q)$ is satisfied. Then, for any $\phi \in C_0^\infty(M)$, there exist $P^{(')}$ and $S_\infty^{(')}$ as in Theorem 1 such that for $f \in C_0^\infty(M)$,

$$P\Box_b f = \phi f + S_\infty f$$

$$\Box_b P'f = \phi f + S'_\infty f$$

(All operators here are vector valued, and f is a (p,q)form.)

In Folland and Stein (1974), Theorem 2 is proved (in the case of a definite Levi form) with $P = P_k$ and the error, S_k, is smoothing of any preassigned order k. In Rothschild and Stein (1976), Theorem 1 and the general case of Theorem 2 are proved, also with an error smoothing of any finite order.

Approximation by Operators on a Nilpotent Lie Group

We shall construct P and P' for $\mathcal{L}$ only; the construction for $\Box_b$ is similar. As in Rothschild and Stein (1976), we begin by extending each X_i to a smooth vector field $\tilde{X}_i$ on a product manifold $\tilde{M} = M \times \mathbb{R}^q$. We mention here some of the important features of this extension and refer to Rothschild and Stein (1976) for details

(i) $\tilde{X}_1, \tilde{X}_2, \ldots, \tilde{X}_n$ and their commutators of length at most r span $T\tilde{M}$.

(ii) $\tilde{X}_j = X_j$ on functions independent of the new variables.

(iii) The commutators of length j, $1 \leqslant j \leqslant r$, satsify as few linear relations as possible, i.e., the only linear relations are those generated by the Jacobi identity and antisymmetry.

As a consequence of (iii), it is possible to identify M with a nilpotent Lie group as follows. Let N be the "free" nilpotent group of step r on n generators. Then, there is a 1-1 correspondence

$$\tilde{X}_{jk} \leftrightarrow Y_{jk} \qquad 1 \leqslant j \leqslant r \tag{3.1}$$

between a basis $\{\tilde{X}_{jk}\}$ for $T\tilde{M}$, and a basis $\{Y_{jk}\}$ of the Lie algebra $\mathfrak{n}$ of N. Here, each $\tilde{X}_{jk}$ is a commutator of the $\tilde{X}_\ell$ of length j, and each Y_{jk} is the commutator of length j of generators $Y_1, Y_2, \ldots, Y_n$ of $\mathfrak{n}$. We make the convention $Y_{1k} \equiv Y_k$, $\tilde{X}_{1k} = \tilde{X}_k$, $1 \leqslant k \leqslant n$.

For $\tilde{\xi} \in \tilde{M}$ fixed, the basis $\{\tilde{X}_{jk}\}$ provides a coordinate system around $\tilde{\xi}$ via the exponential map

$$\tilde{\eta} = \exp(\textstyle\sum u_{jk}\tilde{X}_{jk})\tilde{\xi} \leftrightarrow (u_{jk}) \tag{3.2}$$

The correspondences (3.1) and (3.2) give common coordinate systems for a neighborhood of $\tilde{\xi}$ in $\tilde{M}$ and a neighborhood of 0 in N:

$$\tilde{\eta} = \exp(\textstyle\sum u_{jk}\tilde{X}_{jk})\tilde{\xi} \leftrightarrow \operatorname{Exp}(\sum u_{jk}Y_{jk}) \leftrightarrow (u_{jk})$$

We define the important map $\Theta: \tilde{M} \times \tilde{M} \to N$ by

$$\Theta(\tilde{\xi}, \tilde{\eta}) = \operatorname{Exp}(\textstyle\sum u_{jk}Y_{jk}) \tag{3.3}$$

Integral Operators of Type λ

Following Rothschild and Stein (1976), we shall define the parametrix of $\sum\tilde{X}_j^2$ on $\tilde{M}$ as an integral operator obtained from "homogeneous" kernels on N. We first review homogeneity on N.

The dilations $\delta_s: Y_j \to sY_j$, $s > 0$ extend to automorphism of $\mathfrak{n}$, and via the exponential map to automorphisms of the group N. We denote the automorphisms again by $x \to \delta_s(x)$, $x \in N$. A function, f, is <u>homogeneous of degree</u> α if

$$f(\delta_s(x)) = s^\alpha f(x) \quad \text{for all } x$$

A differential operator D is called homogeneous of degree λ if $D(f\circ\delta_s) = s^{\lambda}(Df)\circ\delta_s$, $s > 0$, and of local degree $\leqslant \lambda$ if its Taylor expansion is a formal sum of homogeneous differential operators of degrees $\leqslant \lambda$. In this sense, the differential operator

$$\mathcal{D} = \sum_{j=1}^{n} Y_j^2$$

is homogeneous of degree 2, as is, for instance, $[Y_j, Y_k]$. By a careful choice of (3.1), one can prove (see Rothschild and Stein, 1976), that for each $\tilde{\xi}$, in the coordinates (u_{jk}), we have the important correspondence

$$\tilde{X}_j = Y_j + R_j \tag{4.1}$$

where R_j is of local degree $\leqslant 0$. It easily follows that

$$\sum_{j=1}^{n} \tilde{X}_j^2 = \sum_{j=1}^{n} Y_j^2 + D \tag{4.2}$$

where D is of local degree $\leqslant 1$.

A norm function on N is a mapping $x \to \langle x\rangle$, smooth away from $x = 0$ and homogeneous of degree 1 such that

(i) $\langle x\rangle = \langle x^{-1}\rangle$

(ii) $\langle x\rangle \geqslant 0$ for all x, and $\langle x\rangle = 0$ only if $x = 0$.

We define

$$\rho(\tilde{\xi},\tilde{\eta}) = \langle\Theta(\tilde{\xi},\tilde{\eta})\rangle \tag{4.3}$$

ρ is a pseudometric in the sense that we have the "triangle" inequality

$$\rho(\tilde{\xi},\tilde{\eta}) \leqslant c(\rho(\tilde{\xi},\tilde{\zeta}) + \rho(\tilde{\zeta},\tilde{\eta})) \tag{4.4}$$

The homogeneous dimension Q of N is

$$Q = \sum_{\alpha=1}^{r} \alpha \dim n^{\alpha}$$

where n^{α} is the eigenspace of δ_s with eigenvalue s^{α}. Any function g, smooth except at 0 and homogeneous of degree $-Q + \beta$, $0 < \beta$ defines a distribution by group convolution: $f \to f*g$. A homogeneous function g of degree $-Q$ smooth away from the origin whose mean value, $\int_{a\leqslant\langle u\rangle\leqslant b} g(u)\,du$, is zero for all a,b defines a distribution in principal value:

$$f \to \lim_{\varepsilon\to 0} \int_{\langle u\rangle > \varepsilon} g(u^{-1}v) f(u)\,du$$

g will then be called a kernel of type β, for $0 \leqslant \beta < Q$. By the map Θ, these kernels define operators on $\tilde{M}$ as follows. A kernel of type on $\tilde{M}$ is a function $K(\tilde{\xi},\tilde{\eta})$ on $\tilde{M} \times \tilde{M}$ such that, for every integer $s > 0$, we can write

$$K(\tilde{\xi},\tilde{\eta}) = \sum_{i=1}^{\ell(s)} a_i(\tilde{\xi}) k_{\tilde{\xi}}^{(i)}(\Theta(\tilde{\eta},\tilde{\xi})) b_i(\tilde{\eta})) + E_s(\tilde{\xi},\tilde{\eta})$$

such that

(a) $E_s \in C_0^s(\tilde{M}\times\tilde{M})$

(b) $a_i, b_i \in C_0^\infty(\tilde{M})$, $i = 1,2,\ldots,\ell$

(c) the functions $u \to k_{\tilde{\xi}}^{(i)}(u)$ are kernels of type $\geqslant s$, depending smoothly on $\tilde{\xi}$.

An operator of type β, $\beta > 0$ is one given by

$$T_K f(\tilde{\xi}) = \int K(\tilde{\xi},\tilde{\eta}) f(\tilde{\eta})\,d\tilde{\eta}$$

where K is a kernel of type β; for $\beta = 0$, an operator of type β is given by a pair (K,a), where $a \in C_0^\infty$ and

$$T_K f(\tilde{\xi}) = \lim_{\varepsilon\to 0} \int_{\langle\Theta(\tilde{\xi},\tilde{\eta})\rangle > \varepsilon} K(\tilde{\xi},\tilde{\eta}) f(\tilde{\eta})\,d\tilde{\eta} + a(\tilde{\xi}) f(\tilde{\xi})$$

We review here some important facts concerning kernels K_λ of type λ and the corresponding operators T_{K_λ}.

The important inequality

$$|K_\lambda(\tilde{\xi},\tilde{\eta})| \leqslant C\rho(\tilde{\xi},\tilde{\eta})^{-Q+\lambda} \tag{4.5}$$

is an almost immediate consequence of the definitions. Furthermore, since $\int_{a\leqslant|u|\leqslant b} |u|^{-Q+\lambda} \leqslant C(b^\lambda - a^\lambda)$ (see Folland and Stein, 1974), it can be shown that

$$\begin{aligned} &\int |K_\lambda(\tilde{\xi},\tilde{\eta})|\,d\tilde{\eta} \leqslant CC_{K,\lambda} \\ &\int |K_\lambda(\tilde{\xi},\tilde{\eta})|\,d\tilde{\xi} \leqslant CC_{K,\lambda} \end{aligned} \tag{4.6}$$

where

$$C_{K,\lambda} = \sup\{\rho(\tilde{\xi},\tilde{\eta}^{\lambda}) : K_\lambda(\tilde{\xi},\tilde{\eta}) \neq 0\}$$

PROPOSITION 4.1. For $\lambda > 0$, T_{K_λ} extends to a bounded operator on L^p, with norm $\leq CC_{K,\lambda}$.

The proposition follows from (4.6); see, for example, Rothschild and Stein (1976). The following are fundamental properties of kernels K_λ of type λ. (See Rothschild and Stein, 1976.)

If D is a differential operator of local degree $\leq d$, then $D^{\tilde{\eta}}K_\lambda$ and $D^{\tilde{\xi}}K_\lambda$ are kernels of type $\lambda - d$ for $d \leq \lambda$. (4.7)

T_{K_λ} is smoothing of order λ, $\lambda = 0,1,2,\ldots$ (4.8)

$X_i^{\xi}K = \sum X_j^{\eta}K_j + K_0$ where each K_j and K_0 is a kernel of type λ. (4.9)

Reduction to the Case $M = \tilde{M}$

Before proceeding to the construction we shall show that it suffices to assume $\tilde{M} = M$. For, suppose there exist $\tilde{P}$ smoothing of order 2 on $\tilde{M}$ and $\tilde{S}_\infty^{(1)}$ smoothing of infinite order as in Theorem 1. Let E be the extension operator (from functions on M to functions on $\tilde{M}$) and R the restriction operator (vice versa) defined in Rothschild and Stein (1976). Then, if T is an operator which is smoothing of order α on $\tilde{M}$, RTE is smoothing of order α on M. Let $P = R\tilde{P}E$ and $S_\infty^{(1)} = R\tilde{S}_\infty^{(1)}E$. Then $R\tilde{P}\tilde{\phi}E = R\tilde{\phi}IE + R\tilde{S}_\infty E$. However, since $\tilde{\mathcal{L}}E = E\mathcal{L}$, $P\mathcal{L} = \phi I + R\tilde{S}_\infty E$ where $P = R\tilde{P}E$ and $\tilde{\phi}$ is chosen so that $R\tilde{\phi}E = \phi$.

To construct the right parametrix, let $\mathcal{L}^t$ be the transpose of $\mathcal{L}$. Then

$$\mathcal{L}^t = \sum_{j=1}^{n} X_j^{t^2}$$

and so the above construction gives a left parametrix P' for $\mathcal{L}^t$ with error operator S'_∞. We shall show that P'^t and S'^t_∞ also have the desired smoothing properties, which will complete the proof.

Construction of P

As in Folland and Stein (1974) and Rothschild and Stein (1976), we begin with the existence of a kernel k of type 2 on N such that

$\mathcal{D}k = \delta$, the delta function (cf. Folland, 1975). Then define P_1 by

$$P_1 f = \int \phi_1(\xi) k(\Theta(\eta,\xi))\phi_2(\eta) f(\eta) d\eta$$

where $\phi_1, \phi_2 \in C_0^\infty(M)$ and $\phi_2 \equiv 1$ on the support of ϕ_1. To compute the error which results from using P_1 as a first approximation to the parametrix, note that $\mathcal{L}^\eta k(\Theta(\eta,\xi)) = \delta_{\eta=\xi} - S_1'$, where S_1' is a kernel of type 1. In fact, $S_1' = -D^\eta k(\Theta(\eta,\xi))$ by (4.2). Also, $\mathcal{L}$ is equal to its transpose modulo a differential operator R of local degree $\leqslant 1$. Thus, using (4.2)

$$\begin{aligned} P_1\mathcal{L}f &= \int \phi_1(\xi) k(\Theta(\eta,\xi))\phi_2(\eta)\mathcal{L}f(\eta) d\eta \\ &= \int \phi_1(\xi)(\mathcal{L} + R)^\eta (k(\Theta(\eta,\xi))\phi_2(\eta)) f(\eta) d\eta \\ &= \phi_1(\xi) f(\xi) + T_1 f(\xi) + T_2 f(\xi) - S_1\phi_2 f \end{aligned} \tag{6.1}$$

where

$$\begin{aligned} T_1 f &= \int \phi_1(\xi) k(\Theta(\eta,\xi))\phi_2''(\eta) f(\eta) d\eta \\ T_2 f &= \int \phi_1(\xi) K_1(\Theta(\eta,\xi))\phi_2'(\eta) f(\eta) d\eta \\ S_1\phi_2 f &= \int \phi_1(\xi) K_1^{(1)}(\Theta(\eta,\xi))\phi_2(\eta) f(\eta) d\eta \end{aligned} \tag{6.2}$$

Here ϕ_2', ϕ_2'' denote first and second order differential operators, applied to ϕ_2, $K_1^{(1)} = (R + D)^\eta k$, which is, therefore, of type 1, and K_1 is a finite sum of kernels of the form $(X_j)k$, which are of type 1. Since $\phi_1(\xi)\phi_2'(\eta)$ and $\phi_1(\xi)\phi_2''(\eta)$ have support bounded away from the diagonal $\xi = \eta$, and $k(\Theta(\eta,\xi))$ is smooth away from the diagonal, T_1 and T_2 are infinitely smoothing operators.

Now, let $\xi_0 \in M$ be fixed and suppose a sequence $\{\phi_i\}$ of functions in $C_0^\infty(M)$ is chosen such that, for suitable ε,

$$\begin{aligned} &\text{(i)} \quad \sup\{\rho(\xi_0,\xi) : \xi \in \operatorname{supp}\phi_i, \ \text{some } i \leqslant \varepsilon\} \\ &\text{(ii)} \quad \phi_{i+1} \equiv 1 \ \text{on} \ \operatorname{supp}\phi_i \end{aligned} \tag{6.3}$$

Define P_i, S_i, and R_i by

$$P_i f(\xi) = \int \phi_i(\xi) k(\Theta(\eta,\xi))\phi_{i+1}(\eta) f(\eta) d\eta$$

$$S_i\phi_{i+1}f(\xi) = -\int\phi_i(\xi)K_1^{(1)}(\Theta(\eta,\xi))\phi_{i+1}(\eta)f(\eta)d\eta$$

$$R_if(\xi) = -\int\phi_i(\xi)[k(\Theta(\eta,\xi))\phi''_{i+1}(\eta) + K_1(\Theta(\eta,\xi))\phi'_{i+1}(\eta)]f(\eta)d\eta$$

where the ϕ derivatives are as in (6.2). Then, for any N

$$[P_1 + S_1P_2 + S_1S_2P_3 + \ldots + S_1S_2 \ldots S_{N-1}P_N]\mathcal{L}f$$

$$= [\phi_1 f + (R_1+S_1R_2+S_1S_2R_3+\ldots+S_1\ldots S_{N-1}R_N)]f + (S_1S_2\ldots S_N\phi_{N+1})f$$

In passing to the limit, we must choose the ϕ_j with care. All will have support in a fixed small ball. Since $\phi_j \equiv 1$ near $\text{supp}\ \phi_{j-1}$,

$$\|\phi'_j\|_{L^\infty} \to \infty$$

as $j \to \infty$, and, therefore, the norm of $R_j \to \infty$ as $j \to \infty$. In fact, the norm of R_1 may be greater than one. This explains, also, why, in general, it is not possible to take the whole error $(S_1 + R_1$ at the first stage) and iterate it to obtain a fundamental solution with no error: the series will not converge. We choose the ϕ_j so that $\phi_i(\xi) = 1$ when

$$\rho(\xi,\xi_0) \leqslant \varepsilon \sum_{\ell=1}^{j} \frac{1}{\ell^2}$$

and

$$\text{supp}\ \phi_j \subset \left\{\xi:\rho(\xi,\xi_0) \leqslant \varepsilon \sum_{\ell=1}^{j+1} \frac{1}{\ell^2}\right\}$$

with

$$|\phi'_j| \leqslant C(j+1)^{2r} \qquad |\phi''_j| \leqslant C(j+1)^{4r} \tag{6.4}$$

ε will be chosen later. (We could have chosen $\sum(1/\ell^s)$ for another $s > 1$ just as well.) To construct the ϕ_j, we begin by constructing auxilliary functions ψ_j on the group N. Choose $\psi_0 \in C_0^\infty(N)$ with support in $\{x:\langle x\rangle \leqslant \varepsilon\}$, the ball of radius ε around 0, such that $\psi_0 = 1$ for $\langle x\rangle \leqslant \varepsilon/2$. Define $\overline{\psi}_j$ as the dilation of ψ_0 by a factor of $2/(j+1)^2$, i.e., $\overline{\psi}_j(x) = \psi_0(\delta_{s_j}(x))$, where $s_j = (j+1)^2/2$. Then,

$$|\overline{\psi}'_j(x)| \leqslant ((j+1)^2/2)^r|\psi'_0(\delta_{s_j}(x))|$$

$$\overline{\psi}''_j(x) \leqslant ((j+1)^2/2)^{2r} |\psi''_0(\delta_{s_j}(x))|$$

so that $\overline{\psi}$ satisfies the inequalities (6.4) with ϕ_i replaced by $\overline{\psi}_i$. Then put

$$\psi_j = \begin{cases} 1 & \langle x\rangle < \varepsilon \sum_{\ell=1}^{j+1} \frac{1}{\ell^2} \\ \overline{\psi}_j(\delta_{c_j}(x)) & \varepsilon \sum_{\ell=1}^{j} \frac{1}{\ell^2} \leqslant \langle x\rangle \leqslant \varepsilon \sum_{\ell=1}^{j+1} \frac{1}{\ell^2} \\ 0 & \varepsilon \sum_{\ell=1}^{j+1} \frac{1}{\ell^2} < \langle x\rangle \end{cases}$$

where

$$c_j = 1 - \frac{\varepsilon}{\langle x\rangle}\left(\sum_1^j \frac{1}{\ell^2} - \frac{1}{(j+1)^2}\right)$$

Since the c_j have a common bound, (6.4) is again satisfied. We put

$$\phi_j(\xi) = \psi_j(\Theta(\xi_0,\xi)) \tag{6.5}$$

Now, by Proposition 4.1, the L^p norm of the operator S_i is $\leqslant d(\varepsilon)$, where $d(\varepsilon) \to 0$ as the diameter of the fixed ball containing the supports of all the ϕ_i goes to 0. Hence

$$\| S_1 \dots S_N \phi_{N+1} f\|_{L^p} \leqslant d(\varepsilon)^N C^N \| \phi_{N+1} f\|_{L^p} \tag{6.6}$$

for $1 \leqslant p \leqslant \infty$, where C depends only on K. $d(\varepsilon)$ may be made arbitrarily small by shrinking supports. Thus, we may choose ε and the ϕ_i so that $d(\varepsilon)C < 1$ in (6.6). For such ε, the terms in (6.6) tend to zero as $N \to \infty$.

Next, we examine the limit of $E_N = R_1 + S_1R_1 + \dots + S_1S_2 \dots S_{N-1}R_N$ as $N \to \infty$.

Since the norm of R_{j+1} is $\leqslant C(j+1)^{4r}$, the norm of E_N on L^p is bounded by

$$\sum_{j=1}^{N} d(\varepsilon)^j C^j C(j+1)^{4r}$$

for $1 \leqslant p \leqslant \infty$, which converges as $N \to \infty$. Also, the L^p norm of the operator $S_1S_2\ldots S_NP_{N+1}$ is bounded by $cd(\varepsilon)^N C^N$.

We have, therefore, proved

LEMMA 6.1. Let $S = R$

$$S_\infty = R_1 + \sum_{N=1}^{\infty} S_1S_2\ldots S_NR_{N+1}$$

and

$$P = P_1 + \sum_{N=1}^{\infty} S_1S_2\ldots S_NP_{N+1}$$

with ϕ_i chosen as above. Then, both sums converge as operators on L^p for $1 \leqslant p \leqslant \infty$, and

$$P\mathcal{L} = \phi_1 + S_\infty$$

S_∞ is Infinitely Smoothing

We shall first show that S_∞ is bounded from L^p to S_k^p for any k. This is true for any single term $S_1S_2\ldots S_jR_{j+1}$ since R_{j+1} is bounded from L^p to S_k^p and each S_i is bounded on S_k^p. By (4.8) we may thus exclude the terms $S_1S_2\ldots S_jR_{j+1}$ for $j < k$. For any k-tuple $(i_1,\ldots,i_k)$, the operator

$$X_{i_1}\ldots X_{i_k}S_1\ldots S_k$$

is bounded from L^p to L^p by (4.9), since each X_j is of local degree $\leqslant 1$ and therefore any X_jS_ℓ is bounded on L^p by (4.7) and (4.9). Thus, we need only show that

$$\sum_{N=k+2}^{\infty} S_{k+1}\ldots S_{N-1}R_N$$

is bounded on L^p, which follows as in the proof of Lemma 6.1, since each term has norm $\leqslant d(\varepsilon)^{N-k}C^{N-k}C(N+1)^{4r}$, with $Cd(\varepsilon) < 1$.

The proof that S_∞ is bounded from L^p to L_k^p for all k follows immediately from the above and that easy fact that $S_{rk}^p \subset L_k^p$. Finally, to prove that S_∞ is smoothing with respect to the Lipschitz spaces, it suffices to apply Sobolev's Lemma to show that for any α, k can be chosen so large that the inclusion map $L_k^2 \to \Lambda_\alpha$ is bounded. Since the inclusion $L_{loc}^\infty \to L_{loc}^2$ is bounded, the proof is complete. We have, thus, proved

LEMMA 7.1. S_∞ is smoothing of infinite order.

The Limit $P = P_1 + S_1P_2 + S_1S_2P_3 + \ldots$

As above, we begin with the S_k^p spaces.

LEMMA 8.1. For any k,

$$P^{(k)} = \sum_{N=k}^{\infty} S_1S_2\ldots S_{N+3}P_{N+4}$$

is bounded from L^p to S_{k+2}^p.

Proof. Since

$$X_{i_1}X_{i_2}\ldots X_{i_{k+2}}S_1S_2\ldots S_{k+2}$$

is bounded from L^p to L^p, the lemma follows from the convergence, in L^p norm, of the operator

$$\sum_{N=k}^{\infty} S_{k+3}S_{k+4}\ldots S_{N+3}P_{N+4}$$

From the lemma, and the fact that each term is bounded from S_k^p to S_{k+2}^p, it follows that P is bounded from S_k^p to S_{k+2}^p. For the usual Sobolev spaces L_α^p, choose $k > r\alpha + 2$. Then, $S_k^p \subset L_{\alpha+2/r}^p$. Lemma 8.1 then shows the sum $S_1S_2\ldots S_NP_{N+1} + S_1S_2\ldots S_{N+1}P_{N+2} + \ldots$ to be bounded from $L^p \subset L_\alpha^p$ to $S_k^p \subset L_{\alpha+2/r}^p$. Since each term is bounded from L_α^p to $L_{\alpha+2/r}^p$, so is P. The argument for the Lipschitz spaces is similar: each term is bounded from Λ_α to $\Lambda_{\alpha+2/r}$ and the infinite sum is bounded from $L^\infty \subset L^2$ to $L_k^2 \supset \Lambda_{\alpha+2/r}$.

The Transposes P^t and S_∞^t

Since the right parametrix is obtained as the transpose of the left parametrix for $\mathcal{L}^t$, we must prove that the operators P and S_∞ constructed above have appropriately smoothing transposes. These involve

$$S_i^tf(\eta) = -\int K_i(\Theta(\eta,\xi))\phi_i(\xi)f(\xi)d\xi$$

$$P_i^tf(\eta) = \int \phi_{i+1}(\eta)K_2(\Theta(\eta,\xi))\phi_i(\xi)f(\xi)d\xi$$

$$R_i^tf(\eta) = -\int\{\phi_{i+1}'(\eta)k(\Theta(\eta,\xi)) + K_1(\Theta(\eta,\xi)\phi_{i+1}''(\eta)\}\phi_i(\xi)f(\xi)d\xi$$

and

$$P^t = P_1^t + P_2^tS_1^t + P_3^tS_2^tS_1^t + \dots + P_N^tS_{N-1}^t \dots S_1^t + \dots$$

$$S_\infty^t = R_1^t + R_2^tS_1^t + R_3^tS_2^tS_1^t + \dots + R_N^tS_{N-1}^t \dots S_1^t + \dots$$

The convergence in L^p norms of these infinite sums of operators is proved as before. However, the smoothing properties are proved differently; the reason is that the terms

$$X_{i_1} \dots X_{i_{k+2}} R_N^tS_{N-1}^t \dots S_1$$

and

$$X_{i_1} \dots X_{i_k} P_N^tS_{N-1}^t \dots S_1^t$$

may entail $k + 2$ derivatives on ϕ_N.

To handle this, we choose the ϕ_j with slightly more care. We shall still have $\phi_j(\xi) = 1$ for

$$\rho(\xi_0, \xi) \leq \varepsilon \sum_{\ell=1}^{j} \frac{1}{\ell^2}$$

and $\phi_j(\xi) = 0$ for

$$\rho(\xi_0, \xi) > \varepsilon \sum_{\ell=1}^{j+1} \frac{1}{\ell^2}$$

However, we shall require that ϕ_j also satisfy

$$\left| \left(\frac{\partial}{\partial \xi} \right)^\alpha \phi_j \right| \leq CC^{|\alpha|}(|\alpha|!)^2(j + 1)^{2|\alpha|r} \tag{9.1}$$

This may be accomplished by defining ϕ_j by (6.5) as before, but requiring the function ψ_0 used in the construction to be (for example) of Gevrey class two, i.e., satisfying

$$\left| \left(\frac{\partial}{\partial x} \right)^\alpha \psi_0 \right| \leq CC^{|\alpha|}(|\alpha|!)^2$$

Assuming the cutoff functions have been defined as above, we now prove the smoothing properties.

LEMMA 9.1. S^t_∞ is smoothing of infinite order.

Proof. As before, it suffices to prove that S^t_∞ is bounded from L^p to L^p_k for all k, $1 < p < \infty$. Now,

$$X_{i_1} X_{i_2} \cdots X_{i_k} R^t_N S^t_{N-1} \cdots S^t_1$$

is a finite sum of operators with j derivatives falling on the cutoff functions ϕ'_{N+1} and ϕ''_{N+1} of R_N in the η variable, and $k - j$ derivatives (in η) falling on the kernels $k(\Theta(\eta,\xi))$ and $K_2(\Theta(\eta,\xi))$ of R^t_N. By (4.5)

$$\left| X^\eta_{i_1} X^\eta_{i_2} \cdots X^\eta_{i_{(k-j)}} k(\Theta(\eta,\xi)) \right| \leqslant C(k,j) \left| \rho(\xi,\eta) \right|^{-Q-k+j-1} \tag{9.2}$$

where $C(k,j)$ is a constant depending on k and j.

If $\rho(\xi,\eta) \geqslant \varepsilon/2(N+1)^2$ on the support of $\phi'_{N+1}(\eta)\phi_N(\xi)$,

$$\rho(\xi,\eta)^{-Q-k+j+1} \leqslant \left(\frac{2C(N+1)^2}{\varepsilon}\right)^{k+Q-j-1} \leqslant \left(\frac{2C(N+1)^2}{\varepsilon}\right)^{k+Q} \tag{9.3}$$

Using (9.2), (9.3) and (9.1),

$$\left\| X^\eta_{i_1} X^\eta_{i_2} \cdots X^\eta_{i_k} R^t_N \right\|_{L^p} \leqslant C'(k) \frac{(N+1)^{2(k+Q)}}{\varepsilon^{k+Q}} (k!)^2 (N+1)^{2kr}$$

where $\| \ \|_{L^p}$ denotes operator norm, and $C'(k)$ is independent of N. Therefore,

$$\left\| X^\eta_{i_1} X^\eta_{i_2} \cdots X^\eta_{i_k} R^t_N S^t_{N-1} \cdots S^t_1 \right\|_{L^p} \leqslant C''(k) \frac{(N+1)^{(2k+2Q+2kr)}}{\varepsilon^{k+Q}} CC^N d(\varepsilon)^N$$

so, as in Section 7, it suffices to choose $d(\varepsilon) < C^{-1}$.

Finally, we must show

LEMMA 9.2. P^t is smoothing of order two.

Proof. As before, it suffices to show that, for any $(i_1, \ldots, i_{k+2})$,

$$\| X_{i_1} \cdots X_{i_{k+2}} P^t_{N+1} S^t_N S^t_{N-1} \cdots S^t_{N-k} \|_{L^p}$$

$$= \| X_{i_1} \cdots X_{i_{k+2}} \phi_{N+2} P^t_0 \phi_{N+1} S^t_0 \phi_N S^t_0 \cdots S^t_0 \phi_{N-k} \|_{L^p}$$

is bounded by a polynomial in N, where $P_0 g(\xi) = \phi_0(\xi)\int k(\Theta(\eta,\xi))\phi_0(\eta)g(\eta)d\eta$ and $S_0 g(\xi) = \phi_0(\xi)\int k_1(\Theta(\eta,\xi))\phi_0(\eta)g(\eta)d\eta$, $\phi_0 \in C_0^\infty(\{\xi:\rho(\xi_0,\xi) < \varepsilon\})$, and is identically one near the support of ϕ_i for all i. Using (4.9) to pass X's across P_0^t and the S_0^t's, we obtain $c(k)$ terms, the worst of which, namely the one with all derivatives on ϕ_{N+2}, has L^p norm bounded by $c_k(k!)^2(N + 2)^{2r}$.

The Kernels of S_∞ and S'_∞

We now prove that the kernels of S_∞ and S'_∞ are functions in $C_0^\infty(M)$. In Lemmas 7.1 and 9.1, we have shown that S_∞ and its transpose S'^t_∞ are both infinitely smoothing operators. The same is true of S'_∞ and S'^t_∞. Hence, the result follows from a more general result. Lacking an explicit reference, we give a simple proof.

PROPOSITION 10.1. Let $s(x,y)$ be compactly supported and in L^1, separately in x and y and set $Tf(x) = \int_{R^m} s(x,y)f(y)dy$. Suppose that for $f \in L^2$, and all α,

$$|D^\alpha Tf| \leq C_\alpha \|f\|_{L^2}$$

$$|D^\alpha T^*f| \leq C_\alpha \|f\|_{L^2}$$

Then, $s(x,y) \in C_0^\infty(R^m \times R^m)$.

Proof. Let $(1 - \Delta_z)^t$ denote the pseudodifferential operator with symbol $(1 + |\zeta|^2)^{t/2}$, t real, with $|\ |$ the Euclidean norm.

It suffices to show that, for some ℓ, $(1 - \Delta_x)^{-\ell}(1 - \Delta_y)^{-\ell}s(x,y) \in C^\infty$. Choose ℓ even, so that $(1 - \Delta_x)^{-\ell}\delta_{x=x_0} \in L^2_{loc}(R^m)$. Then,

$$|D_x^\alpha \int s(x,y)(1 - \Delta_y)^{-\ell}\delta_{y=y_0}dy| = |(D_x(1 - \Delta_y)^{-\ell}s(x,y))(x,y_0)|$$

$$\leq C_\alpha \quad \text{uniformly in } y_0$$

and, similarly,

$$|(D_y^\beta(1 - \Delta_x)^{-\ell}s(x,y))(x_0,y)| \leq C_\beta \quad \text{uniformly in } x_0$$

But, then, it follows that, for any α', β'

$$|D_x^{\alpha'} D_y^{\beta'}(1 - \Delta_x)^{-\ell}(1 - \Delta_y)^{-\ell}s(x,y)| \leq C_{\alpha'\beta'}$$

which completes the proof.

REFERENCES

Beals, R. Characterization of pseudodifferential operators and applications. Preprint.

Boutet de Monvel, L. Hypoelliptic operators with double characteristics and related pseudodifferential operators. Comm. Pure Appl. Math., 27 (1974), pp. 585-639.

Folland, G. Subelliptic estimates and function spaces on nilpotent Lie groups. Arkiv for Matematik, 13, No. 2 (1975), pp. 161-207.

Folland, G. and Kohn, J. J. The Neumann Problem for the Cauchy-Riemann Complex. Ann. of Math. Studies, No. 75, Princeton Univ. Press, Princeton (1972).

Folland, G. and Stein, E. M. Estimates for the $\bar{\partial}_b$ complex and analysis on the Heisenberg group. Comm. Pure Appl. Math., 27 (1974), pp. 429-522.

Grusin, V. V. On a class of hypoelliptic operators. Math. U.S.S.R. Sbornik, 12, No. 3, (1970), pp. 458-476.

Hörmander, L. Hypoelliptic second order differential equations. Acta Math., 119 (1967), pp. 147-171.

Kohn, J. J. Boundaries of complex manifolds. Proc. Conf. on Complex Manifolds, Minneapolis (1964), pp. 81-94.

Rothschild, L. P. and Stein, E. M. Hypoelliptic differential operators and nilpotent groups. Acta Math., 137 (1976), pp. 247-320.

Sjöstrand, J. Parametrices for Pseudodifferential Operators with Multiple Characteristics. Arkiv for Matematik, 12, No. 1 (1974), pp. 85-130.

Stein, E. M. Singular Integrals and Differentiability Properties of Functions. Princeton Univ. Press, Princeton, New Jersey (1970).

University of Wisconsin
Madison, Wisconsin
Institute for Advanced Study
Princeton, New Jersey

RAYLEIGH WAVES IN LINEAR ELASTICITY AS A PROPAGATION OF SINGULARITIES PHENOMENON

Michael E. Taylor*

Introduction

We consider here the propagation of elastic waves in a bounded medium. We assume our medium is isotropic, and that the displacement u satisfies the equation of linear elasticity

$$\frac{\partial^2}{\partial t^2} u = (\lambda + \mu)\text{grad div } u + \mu\Delta u \tag{1.1}$$

where λ and μ are certain scalar quantities called the "Lamé constants." We assume λ and μ are positive. Then, as is well known, there are two sound speeds, $\sqrt{\mu}$ and $\sqrt{\lambda + 2\mu}$, associated respectively with shear waves (s-waves) and pressure waves (p-waves). Now, it has long been observed that a discontinuous impulse on the surface ∂K of the body K gives rise to a third singular wave, traveling along the boundary at a third, slower speed. This wave is called a Rayleigh wave, and it is of considerable importance in seismology. For example, if the impulse is caused by a sudden break near the surface of the earth, giving rise to an earthquake, the p-waves and s-waves dissipate rapidly, having amplitudes that vary inversely as the square of the distance from the epicenter, while the Rayleigh waves only go down like one over the distance.

In the case ∂K is flat, the propagation of Rayleigh waves has been analyzed in some detail. See Landau and Lifschitz (1970), Love (1944) and Rayleigh (1885). The purpose of this paper is to give a rigorous treatment of the singularity which travels along ∂K, at Rayleigh sound speed, in the case when ∂K is curved. For simplicity in exposition, we carry out the calculations in two space dimensions, but a similar approach will work in three space dimensions. In such a case, we can evidently have a phenom-

*Research partially supported by N. S. F. Grant GP34260.

enon which would not occur in the case of a flat boundary. Namely, a point source on ∂K can give rise to caustics, on which the Rayleigh wave would have fairly large amplitude.

We use the method of geometrical optics and the calculus of Fourier integral operators to analyze the singularities of a solution to (1.1), assumed to satisfy free boundary conditions on ∂K, namely the normal components of the stress tensor, shall vanish:

$$\sum_i n_i \sigma_{ij} = 0 \quad \text{on} \quad \partial K \tag{1.2}$$

where the stress tensor is

$$\sigma_{ij} = \lambda(\text{div } u)\delta_{ij} + \mu \; \frac{\partial u_i}{\partial x_j} + \frac{\partial u_j}{\partial x_i}$$

and n is the normal to ∂K. The method of geometrical optics reduced the problem to studying a certain pseudodifferential operator P on $R \times \partial K$, similar to one studied in Majda and Osher (1975) and Taylor (1975) for analyses of reflection of singularities. Now, the results of these two papers dealt with the case when P was elliptic, or at least hypoelliptic, and, as we will see, the operator P in our case has characteristics which are simple, and near which $\det P$ is real valued. This, together with the fact that t is monotonic on each null bicharacteristic of P, will allow us to construct the appropriate parametrix and analyze the singularities. Aside from the physical interest of this problem, I think this additional case in the analysis of propagation of singularities in domain with boundary is very interesting mathematically.

In Section 2, we give a brief account of the basic existence and uniqueness theory for solutions to the Cauchy problem for (1.1) and (1.2). This will serve to acquaint the reader who is familiar with the general theory of partial differential equations, with the peculiarities of the equations of linear elasticity, and also to give a concise description of the tools needed to justify our geometrical optics construction of Section 3. The solvability and analysis of the pseudodifferential equations obtained in Section 3 can be obtained as a special case of the work of Duistermaat and Hörmander (1972), but, in fact, a simpler construction of global parametrices will suffice (due to the monotonicity of t along null bicharacteristics). We will construct such parametrices in Section 4, both in order to make this exposition more selfcontained and to point out the possibility of constructing global parametrices without necessarily using all the global machinery of Duistermaat and Hörmander (1972) and Hörmander (1971). In Section 5, we put this phenomenon of Rayleigh waves into a general context, complementing our work in 1975.

Finally, let us remark that the phenomenon of Rayleigh waves is connected to the failure of the Kreiss-Sakamoto condition for (1.1), (1.2) in the "elliptic region." In Taylor (1976) we treated the diffraction problem for first order systems satisfying the Kreiss condition (which generalizes to higher order systems with no difficulty) and since that analysis is micro-local and the system (1.1), (1.2) mear the "characteristic variety" in $R \times \partial K$ (over which the grazing rays pass) does satisfy the Kreiss-Sakamoto condition, it follows that we obtain a complete analysis of the singularities of solutions to (1.1), (1.2), with no restriction on the wave front set of the initial data such as we introduce in Section 4 to avoid grazing rays, provided ∂K is convex with respect to the null bicharacteristics of (1.1). Thus, much of the scattering theory developed by Lax and Phillips (1967) for the acoustical equation, and also much of the analysis of Majda and Taylor (to appear), goes through for the scattering of elastic waves off a convex obstacle.

I am grateful to N. Zitron for bringing the problem of scattering of elastic waves to my attention, and R. Burridge and Ed Reiss for some useful conversations and references to the literature.

Basic Existence and Uniqueness, and Smoothness of Solution

In order to analyze the initial value problem, prescribing u and $\frac{\partial}{\partial t} u$ at $t = 0$, for solutions to (1.1) and (1.2), we consider the operator

$$Lu = (\lambda + \mu)\text{grad div } u + \mu\Delta u$$

Note that any such L is symmetric. Furthermore, the boundary condition (1.2) makes L a symmetric operator, i.e., $(Lu,v) = (u,Lv)$ for all smooth u and v on K, with bounded support, which satisfy the boundary condition (1.2).

LEMMA 2.1. L is elliptic and the boundary condition (1.2) coercive.

Proof. The ellipticity of L is obvious. To check the coerciveness of (1.2), we can assume K is a half space, and check the Lopatinski condition. Thus, if K is defined by $x_1 \geqslant 0$, and if $u(x_1)$ is a bounded solution to

$$L\left(\frac{1}{i}\frac{\partial}{\partial x_1},\alpha_2,\alpha_3\right)u(x_1) = 0$$

(obtained by replacing $1/i(\partial/\partial x_2)$ by α_2 and $1/i(\partial/\partial x_3)$ by α_3 in the formula for L) such that the boundary condition

$$\sum n_i\sigma_{ij}\left(\frac{1}{i}\frac{\partial}{\partial x_1},\alpha_2,\alpha_3\right)u(x_1) = 0 \quad \text{at} \quad x_1 = 0$$

is satisfied, we need to show that $u(x_1) \equiv 0$, provided $(\alpha_2,\alpha_3) \neq 0$ in R^2. We may suppose

$$u = E\,e^{i\alpha_1 x_1} + \tilde{E}\,e^{i\tilde{\alpha}_1 x_1}$$

with $E,\tilde{E} \in C^3$. First of all,

$$L\left(\frac{1}{i}\frac{\partial}{\partial x_1},\alpha_2,\alpha_3\right)u(x_1) = 0$$

implies

$$[(\lambda+\mu)ME + \mu(\alpha_1^2+\alpha_2^2+\alpha_3^2)E]e^{i\alpha_1 x_1} + [(\lambda+\mu)\tilde{M}\tilde{E} + \mu(\tilde{\alpha}_1^2+\alpha_2^2+\alpha_3^2)\tilde{E}]e^{i\tilde{\alpha}_1 x_1} = 0 \tag{2.1}$$

where

$$M = \begin{pmatrix} \alpha_1^2 & \alpha_1\alpha_2 & \alpha_1\alpha_3 \\ \alpha_1\alpha_2 & \alpha_2^2 & \alpha_2\alpha_3 \\ \alpha_1\alpha_3 & \alpha_2\alpha_3 & \alpha_3^2 \end{pmatrix}$$

and $\tilde{M}$ is given by the same expression, with α_1 replaced by $\tilde{\alpha}_1$. Thus, $-u/(\lambda+\mu)(\alpha_1^2 + \alpha_2^2 + \alpha_3^2)$ is required to be an eigenvalue of M, and $-u/(\lambda+\mu)(\tilde{\alpha}_1^2 + \alpha_2^2 + \alpha_3^2)$ must be an eigenvalue of M. It is easy to see that the eigenvalues of M are 0,0, and $\alpha_1^2 + \alpha_2^2 + \alpha_3^2$, so (2.1) yields $\alpha_1^2 + \alpha_2^2 + \alpha_3^2 = 0$ and similarly $\tilde{\alpha}_1^2 + \alpha_2^2 + \alpha_3^2 = 0$, so

$$\alpha_1 = i\sqrt{\alpha_2^2 + \alpha_3^2}$$

and

$$\tilde{\alpha}_1 = -i\sqrt{\alpha_2^2 + \alpha_3^2}$$

The boundedness hypothesis implies $\tilde{E} = 0$, so we are left with

$$u = E\, e^{-\eta x_1}$$

where

$$\eta = \sqrt{\alpha_2^2 + \alpha_3^2}$$

Now, the boundary condition $\sigma_{11} = \sigma_{12} = \sigma_{13}$ at $x_1 = 0$ implies

$$\begin{aligned} -(\lambda + 2\mu)\eta E_1 + i\lambda\alpha_2 E_2 + i\lambda\alpha_3 E_3 &= 0 \\ -\eta E_2 + i\alpha_2 E_1 &= 0 \\ -\eta E_3 + i\alpha_3 E_1 &= 0 \end{aligned} \tag{2.2}$$

but, if $E \neq 0$, we see from (2.2) that η must satisfy the equation

$$\det \begin{pmatrix} -(\lambda + 2\mu)\eta & i\lambda\alpha_2 & i\lambda\alpha_3 \\ i\alpha_2 & -\eta & 0 \\ i\alpha_3 & 0 & -\eta \end{pmatrix} = 0$$

which reduces to $(\lambda + 2\mu)\eta^2 + \lambda(\alpha_2^2 + \alpha_3^2) = 0$. But, since $\lambda, \mu > 0$, this is not possible, so the Lopatinsky condition is verified.

If we suppose K is a bounded domain, with smooth boundary, it follows from standard elliptic theory that L_s, defined on those $u \in C^\infty(K)$ satisfying the appropriate boundary conditions, has a unique positive selfadjoint extension, which we will also denote L, and then the unique solvability of the Cauchy problem, given $u(0) \in \mathcal{D}(L)$, $u_t(0) \in L^2(K)$ for (1.1), (1.2) is an exercise in spectral theory.

So much for existence. We now want to show that a function u which solves such a mixed problem, with smooth error, must differ from the exact solution by a smooth function. This is, in fact, a standard consequence of the coerciveness of L, and we sketch the argument briefly.

It suffices to show that, if $u = 0$ for $t < 0$, and

$$\left(\frac{\partial^2}{\partial t^2} - L\right) u = f \quad \text{on} \quad R \times K \tag{2.3}$$

$$\sum n_i \sigma_{ij} = g \quad \text{on} \quad R \times \partial K \tag{2.4}$$

where $f \in C^\infty(R \times \overline{K})$, $g \in C^\infty(R \times \partial K)$, then $u \in C^\infty(R \times \overline{K})$. First, since $R \times \partial K$ is noncharacteristic for $(\partial^2/\partial t^2) - L$, we may use the formal Cauchy-Kovalevsky process and Borel's theorem to solve (2.3) to infinite order at $R \times \partial K$, with (2.4) satisfied (and you could specify both u and $\frac{\partial u}{\partial v}$ on $R \times \partial K$). Thus, it suffices to show that solutions to (2.3) and (2.4) with $u = 0$ for $t > 0$ are smooth, assuming that $g = 0$ and that f vanishes to infinite order on $R \times \partial K$. In such a case, write down u using Duhamel's principle:

$$u(t) = \int_0^t \frac{\sin((t - s)\sqrt{-L})}{\sqrt{-L}} f(s)ds \tag{2.5}$$

Here, L is the selfadjoint operator on $L^2(K)$ with domain $\mathcal{D}(L)$ specified by the boundary condition (1.2). The operator $(1/\sqrt{-L} \sin((t-s)\sqrt{-L})$ is defined by the spectral theorem and is a bounded family of operators on $L^2(K)$, and also on each Hilbert space $\mathcal{D}(L^k)$, $h = 1,2,3,\ldots$. Now, since $f(s) \in C^\infty(R \times \overline{K})$ and vanishes on $R \times \partial K$ to infinite order, it follows that

$$f(s) \in \bigcap_{k=1}^{\infty} \mathcal{D}(L^k)$$

Hence, by (2.5), we see that $u(t)$ is a smooth function of t taking values in $\mathcal{D}(L^k)$, for each k. Now, since the boundary condition (1.2) is coercive for L, it follows that $\mathcal{D}(L^k) \subset H^{2k}(K)$. Thus, u is C^∞ on $R \times \overline{K}$, as desired.

Construction of Parametrices

We construct an approximation to (1.1) satisfying the inhomogeneous boundary condition

$$\sum n_i \sigma_{ij} = f_j \quad \text{on} \quad R \times \partial K \tag{3.1}$$

where $f_j \in \mathcal{E}'(R \times \partial K)$ vanishes for $t < 0$. We suppose $K \subset R^2$ for convenience in calculation. We look for the unique outgoing solution, i.e., we require that $u = 0$ for $t < 0$. We will suppose that $WF(f_j)$ is contained in a small conic neighborhood of a point $(x_0,t_0,\xi_0,\tau_0) \in T^*(R \times \partial K)$. We look for an approximate solution in the form

$$u = \int a(t,x_1,x_2,\zeta)e^{i\phi(t,x_1,x_2,\zeta)}\hat{F}(\zeta)d\zeta \tag{3.2}$$

$$+ \int b(t,x_1,x_2,\zeta)e^{i\psi(t,x_1,x_2,\zeta)}\hat{G}(\zeta)d\zeta$$

where $\zeta \in R^2$, F and G are scalar valued distribution to be determined from the boundary condition (3.1), a and b are vector valued amplitudes and ϕ and ψ are certain plane functions, satisfying the eikonal equation of geometrical optics.

$$\phi_t = (\lambda + 2\mu)^{1/2}|\nabla_x\phi| \tag{3.3}$$

$$\psi_t = \mu^{1/2}|\nabla_x\psi| \tag{3.4}$$

Away from the characteristic variety in $T^*(R \times \partial K)$, the boundary $R \times \partial K$ is noncharacteristic for each of these eikonal equations. We would like to specify that both ϕ and ψ equal some given function $\gamma(t,x,\xi)$ on $R \times \partial K$. There are three cases:

(i) Over $((t,x),\nabla_{tan}\gamma(t,x,\xi)) \in T^*(R \times \partial K)$ pass four rays. This is the hyperbolic region. The eikonal equations (3.3) and (3.4) can both be solved exactly here.

(ii) Over $((t,x),\nabla_{tan}\gamma(t,x,\xi)) \in T^*(R \times \partial K)$ pass two rays. This is the "mixed region." Here, (3.4) can be solved exactly, but (3.3) cannot. However, one can solve (3.3) to infinite order on $R \times \partial K$, which will suffice for the construction of the parametrix. We demand that $\operatorname{Im} \phi \geqslant 0$.

(iii) Over $((t,x),\nabla_{tan}\gamma(t,x,\xi)) \in T^*(R \times \partial K)$ pass no rays. This is the "elliptic region." Here, (3.3) and (3.4) can both be solved to infinite order on $R \times \partial K$, with $\operatorname{Im} \phi \geqslant 0$, $\operatorname{Im} \psi \geqslant 0$ everywhere.

In each case, there are also the associated transport equations for the amplitudes a and b, which are treated similarly.

It follows that (3.2) satisfies (1.1) up to a smooth error. We can prescribe $a(t,x,\zeta)$ and $b(t,x,\zeta)$ for $x \in \partial K$ as long as they are respectively in the $\lambda + 2\mu$ and μ eigenspaces of the symbol $L(t,x,\tau,\xi)$ of $(\lambda + 2\mu)\text{grad div} + \mu\Delta$, evaluated respectively at $(\tau,\xi) = (\phi_t,\phi_x)$ and $(\tau,\xi) = (\psi_t,\psi_x)$. Naturally, we require that these vectors be nonzero on ∂K, so they span R^2 there. Satisfying the boundary condition (3.1) leads to a pseudodifferential equation on $R \times \partial K$ for $\binom{F}{G}$ in terms of

$$\begin{pmatrix} f_1 \\ f_2 \end{pmatrix}$$

which we derive as follows. The left hand side of (3.1) is

$$\lambda(\mathrm{div}\ u)n_j + \mu \sum_i n_i \left(\frac{\partial u_j}{\partial x_i} + \frac{\partial u_i}{\partial x_j} \right)$$

Thus, the vector

$$\sum_i n_i \sigma_{ij} \Big|_{\partial K \times R}$$

is given by

$$T\begin{pmatrix} F \\ G \end{pmatrix} = \int [A(t,x,\zeta)\hat{F}(\zeta) + B(t,x,\zeta)\hat{G}(\zeta)] e^{i\gamma(t,x,\zeta)} d\zeta \tag{3.5}$$

with A and B vector valued symbols of order 1. Their principal parts, homogeneous of degree 1 in ζ, are given by

$$A_1(t,x,\zeta) = i\lambda(\nabla_x\phi \cdot a_0)n + i\mu((\nabla_x\phi \cdot n)a_0 + (a_0 \cdot n)\nabla_x\phi) \tag{3.6}$$

$$B_1(t,x,\zeta) = i\lambda(\nabla_x\psi \cdot b_0)n + i\mu((\nabla_x\psi \cdot n)b_0 + (b_0 \cdot n)\nabla_x\psi) \tag{3.7}$$

We now consider solving the system

$$T\begin{pmatrix} F \\ G \end{pmatrix} = \begin{pmatrix} f_1 \\ f_2 \end{pmatrix} \bmod C^\infty \tag{3.8}$$

given $f_j \in \mathcal{E}'(R \times \partial K)$, supported in $t > 0$, where we demand that F and G vanish for $t < 0$. If coordinates $z = (t,x)$ are chosen on $\partial K \times R$ such that $\gamma(t,x,\zeta) = z \cdot \zeta = x \cdot \xi + t\tau$, then (3.8) is a pseudodifferential equation for

$$\begin{pmatrix} F \\ G \end{pmatrix} \text{ in terms of } \begin{pmatrix} f_1 \\ f_2 \end{pmatrix}$$

We will show that T has the following behavior.

LEMMA 3.1. In the hyperbolic and mixed regions T is elliptic. In the elliptic region, the real valued symbol $\det \sigma_T$ has a simple zero on a hypersurface in $T^*(R \times \partial K)$. On this surface,

$$\frac{\partial}{\partial \tau} \det \sigma_T \neq 0$$

As we will see in Section 4, this leads to the following

LEMMA 3.2. The system (3.8) has a unique solution, mod C^∞, which vanishes for $t < 0$, given f_j supported in $\{t > 0\}$. Then WF(F) and WF(G) are contained in the set $\sum$: the union of $S = WF(f_1) \cup WF(f_2)$ and the set of null bicharacteristics for $\det \sigma_T$ passing over S, travelling in the positive t direction.

Plugging this result into (3.2), we immediately obtain our main result on propagation of singularities for solutions to (1.1) and (3.1).

THEOREM 3.3. Let u be the unique solution to (1.1) and (3.1), vanishing for $t < 0$, given $f_j \in \mathcal{E}'(R \times \partial K)$ vanishing for $t < 0$. Assume $WF(f_j)$ avoids the characteristic variety. Then, in $R \times \text{int } K$, WF(u) is contained in the set of null bicharacteristics of L passing over $WF(f_1) \cup WF(f_2) \subset T^*(R \times \partial K)$, going in the positive t direction, as long as these bicharacteristics do not pass over $R \times \partial K$ again. The solution u is smooth up to ∂K except at the singular supports of f_1, f_2 and at the image in $R \times \partial K$ of the set $\sum$ described in Lemma 3.2. If we consider $u|_{R\times\partial K}$ and $(\partial/\partial\nu)u|_{R\times\partial K}$, the wave front sets of these distributions are contained in $\sum$.

It is the set $\sum$ which forms the wave front set of the Rayleigh wave produced by u. If the null bicharacteristics mentioned in Theorem 3.3 do pass over $R \times \partial K$ again, propagation and reflection of singularities results continue to hold, as described in Taylor (1975), as long as they do not pass over the characteristic variety. Note that such rays cannot pass over the elliptic region, so no further Rayleigh waves are produced. We turn now to the proof of Lemma 3.1.

In order to compute $\det \sigma_T = \det(A_1 B_1)$, where $(A_1 B_1)$ is the 2×2 matrix whose columns are A_1 and B_1, at a point (t_0, x_0, τ, ξ), we may as well suppose Euclidean coordinates are chosen so that $x_0 = 0$ and the plane $\{x_1 = 0\}$ is tangent to ∂K at x_0, with $x_1 \geqslant 0$ on K. At x_0, the eikonal equations solved by ϕ and ψ can be rewritten, with $z = (x_2, t)$,

$$\phi_{x_1} = \lambda_1(x_1, z, \nabla_z \phi)$$

$$\psi_{x_1} = \mu_1(x_1, z, \nabla_z \psi)$$

where, if we write $L(t,x,\tau,\xi)$ for the principal symbol of L, λ_1 and μ_1 are the roots of $\det L(t,x,\tau,\lambda_1,\xi_2) = 0$. Then, the vectors a_0 and b_0 are picked so that

$$L(t,x,\tau,\lambda_1,\xi_2)a_0 = 0$$

$$L(t,x,\tau,\mu_1,\xi_2)b_0 = 0$$

A straightforward computation yields the following formulae:

$$\lambda_2^2 = \frac{1}{\mu}\tau^2 - \xi_2^2 \quad \text{and} \quad \mu_1^2 = \frac{1}{\lambda + 2\mu}\tau^2 - \xi_2^2 \tag{3.9}$$

$$a_0 = \begin{pmatrix} \xi_2 \\ -\lambda_1 \end{pmatrix} \quad \text{and} \quad b_0 = \begin{pmatrix} \mu_2 \\ \xi_1 \end{pmatrix} \tag{3.10}$$

and combining these formulae with (3.6) and (3.7) yields

$$(A_1 B_1) = C \begin{pmatrix} \lambda_1\xi_2\left(1 - \frac{\lambda}{\lambda + 2\mu}\right) & \mu_1^2 + \frac{\lambda}{\lambda + 2\mu}\xi_2^2 \\ \xi_2^2 - \lambda_1^2 & 2\mu_1\xi_2 \end{pmatrix}$$

and C is a nonvanishing scalar. Thus, $\det \sigma_T$ is a nonvanishing multiple of

$$2\mu\lambda_1\xi_2^2 + \mu(2\xi_2^2 - \frac{1}{\mu}\tau^2)^2 = p(\xi_2,\tau) \tag{3.11}$$

Note that (ξ_2,τ) is the fibre variable of $T^*(R \times \partial K)$, which is divided into three regions (excluding the characteristic variety).

I. $|\tau| > (\lambda + 2\mu)^{1/2}|\xi_2|$ (hyperbolic region)

Here λ_1^2 and μ_1^2 are positive, so the roots λ_1 and μ_1 are real. Representing outgoing waves, they must have the same sign.

II. $\mu^{1/2}|\xi_2| < |\tau| < (\lambda + 2\mu)^{1/2}$ $_2$ (mixed region)

Here $\lambda_1^2 > 0$, but $\mu_1^2 < 0$, so λ_1 is real but μ_1 is pure imaginary.

III. $|\tau| < \mu^{1/2}|\xi_2|$ (elliptic region)

Here both λ_1^2 and μ_1^2 are negative, so all roots are pure imaginary.

The assertion of Lemma 3.1 is that $p(\xi_2,\tau)$ is nonvanishing in regions

I and II, and has a simple zero in region III, where $(\partial/\partial\tau)p \neq 0$. The behavior of $p(\xi_2,\tau)$ in regions I and II is easy to investigate. In region I, we have $p(\xi_2,\tau) > 0$; in region II, the term $4\mu\lambda_1\mu_1\xi_2^2$ is imaginary and the term $\mu(2\xi_2^2 - \frac{1}{\mu}\tau^2)^2$ is real and again $p(\xi_2,\tau) \neq 0$. Note that, in region III, $p(\xi_2,\tau)$ is also real valued. To simplify the analysis of $p(\xi_2,\tau)$ there, let $\xi_2 = 1$ and $T = \tau^2$. Since $p(\xi_2,\tau) = 0$ in region III is equivalent to $16\lambda_1^2\mu_1^2\xi_2^4 = (2\xi_2^2 - \frac{1}{\mu}\tau^2)^4$, this condition becomes

$$16(\frac{1}{\mu}T - 1)(\frac{1}{\lambda + 2\mu}T - 1) = (\frac{1}{\mu}T - 2)^4$$

Mutliplying this out and replacing T by $s = \frac{1}{\mu}T$ yields

$$s^4 - 8s^3 + (24 - 16\frac{\mu}{\lambda + 2\mu})s^2 - 16(1 - \frac{\mu}{\lambda + 2\mu})s = 0$$

which, upon division by s, reduces to the cubic equation

$$q(s) = s^3 - 8s^2 + (24 - 16\frac{\mu}{\lambda + 2\mu})s - 16(1 - \frac{\mu}{\lambda + 2\mu}) = 0$$

This is the equation which occurs in the analysis of Rayleigh waves in the half space case (see Love, 1944) and the location of its zero is easy. Our assertion boils down to showing that, for $0 < s < 1$, $q(s)$ has exactly one zero, where $q'(s) \neq 0$. (Outside this interval, $q(s)$ may have other zeros, but these do not correspond to zeros of $p(\xi_2,\tau)$ due to our having squared the equation $4\mu\lambda_1\mu_1\xi_2^2 = -\mu(2\xi_2^2 - \frac{1}{\mu}\tau^2)$.) In fact, we readily see that

$$q(0) = -16(1 - \frac{\mu}{\lambda + 2\mu}) < 0$$

while $q(1) = 1 > 0$, so $q(s)$ has at least one zero in this interval. Meanwhile, $q'(s)$ is easily seen to be strictly positive on $[0,1]$, so this proves our assertion.

Solution of the Boundary Equation

In this section, we give a construction for a parametrix for the pseudodifferential equation

$$Tu = f \quad (\text{mod } C^\infty) \tag{4.1}$$

where we assume T is a $k\times k$ matrix of pseudodifferential operators, which we may suppose to be of order zero, on a manifold $R \times X$ with coordinates (t,x), such that

$$p(t,x,\tau,\xi) = \det \sigma_T \quad \text{is real, and} \tag{4.2}$$

$$\frac{\partial}{\partial\tau} p \neq 0 \quad \text{where} \quad p = 0 \tag{4.3}$$

We assume that $f = 0$ for $t < 0$ and demand that

$$u \in C^\infty \quad \text{for} \quad t < 0 \tag{4.4}$$

We construct a <u>global</u> approximate solution, under the assumption that X is compact. This is a special case of a construction of Duistermaat and Hörmander (1972), which used somewhat heavier machinery.

Let ${}^{co}T$ be the cofactor matrix of T, so ${}^{co}T\,T = P + Q$ where $\sigma_P = p(t.x,\tau,\xi)$ and $Q \in PS(-1)$. Thus, (4.1) implies ${}^{co}T\,Tu = {}^{co}T\,f = g$, or

$$(P + Q)u = g$$

Letting $\Lambda \in PS(1)$ be some scalar first order elliptic operator on $X \times \mathbb{R}$ with real symbol, this is equivalent to

$$(q + B)u = \tilde{g} \qquad (\text{mod } C^\infty) \tag{4.5}$$

where $\tilde{g} = \Lambda g$, $q = \Lambda P \in PS(1)$, $B = \Lambda Q \in PS(0)$. The hypothesis $f = 0$ for $t < 0$ implies $\tilde{g} \in C^\infty$ for $t < 0$.

To solve (4.5), let us proceed momentarily on a formal level. We have

$$\begin{aligned} u &= (iq + iB)^{-1}(i\tilde{g}) && \text{(formally)} \\ &= i\int_0^\infty e^{is(q+B)}\tilde{g}\,ds && \text{(formally)} \\ &= i\int_{-\infty}^0 e^{is(q+B)}\tilde{g}\,ds && \text{(formally)} \end{aligned} \tag{4.6}$$

Now, (4.6) is not well defined. However, $e^{is(q+B)}\tilde{g} = w(s)$ solves the hyperbolic equation

$$\begin{aligned} \frac{\partial}{\partial s} w &= (q + B)w \\ w(0) &= \tilde{g} \end{aligned} \tag{4.7}$$

and we can construct a solution $(\text{mod } C^\infty)$ to this via the method of geometrical optics, for $|s|$ small, and then exploit the group properties to represent such an approximate solution for arbitrary s. Let us denote such an approximate solution to (4.7) by $e_{GO}^{is(q+B)}\tilde{g}$. Then, as is well known, the wave front set of $e_{GO}^{is(q+B)}\tilde{g}$ is obtained from $WF(\tilde{g})$ by following the Hamiltonian flow generated by H_q for s units of "time". Since we are assuming (4.3), it follows that $\dot{t}$ is bounded away from zero on each null bicharacteristic strip. Let $\dot{t} > 0$ on $S_+ \subset \gamma(q)$, the set of characteristics of q, and let $\dot{t} < 0$ on $S_- \subset \gamma(q)$. Write the identity operator on $\mathcal{D}'(R \times X)$ in the form $I = P_+ + P_- + P_0$ where σ_{P_+} is supported in a small conic neighborhood Γ^+ of S_+ with $\dot{t} > 0$ on Γ_+, $\sigma_{P_+} = 1$ on a smaller conic neighborhood of S_+; and likewise, σ_{P_-} is supported on a small conic neighborhood Γ_- of S_- with $\dot{t} < 0$ on Γ_-, and $\sigma_{P_-} = 1$ on a smaller conic neighborhood of S_-. Note that $q + B$ is elliptic on the support of P_0, so if we write $\tilde{g} = P_+\tilde{g} + P_-\tilde{g} + P_0\tilde{g}$, it is easy to solve

$$(q + B)u_0 = P_0\tilde{g} \quad (\text{mod } C^\infty)$$

With these facts in mind, we construct the actual parametrix for (4.5), replacing the formal calculation (4.6). Pick a $T < \infty$, and we desire to solve (4.5) for $t < T$. Choose a $T_0 < \infty$ such that, for all $\zeta \in WF(\tilde{g})$, if $|s| > T_0$, the image of ζ under $C(s)$, where $C(s)$ is the flow on $T^*(R \times X)$ generated by the Hamiltonian vector field H_q, has t coordinate outside the interval $t \in [0,T]$. We set $\psi \in C_0^\infty(R)$ equal to 1 for $|s| \leqslant T_0$, and take

$$u = u_0 + i\int_0^\infty \psi(s)e_{GO}^{is(q+B)}P_+\tilde{g}\,ds - i\int_{-\infty}^0 \psi(s)e_{GO}^{is(q+B)}P_-\tilde{g}\,ds \tag{4.7}$$

It is a simple matter to show that u verifies (4.5) (mod C^∞), for $t < T$, and u is smooth for $t < 0$. It remains to show that such a u solves (4.1) (mod C^∞). Indeed, rewrite (4.5) as $\Lambda^{co}T(Tu - f) = 0$ or, since Λ is elliptic,

$${}^{co}T(Tu = f) = 0 \quad (\text{mod } C^\infty)$$

Apply T to both sides of this, noting that $T^{co}T = P + \tilde{Q}$, P as before, $\tilde{Q} \in PS(-1)$. Thus,

$$(P + \tilde{Q})(Tu - f) = 0 \quad (\text{mod } C^\infty)$$

while $Tu - f$ is smooth for $t < 0$. Now, propagation of singularities results for solutions to $(P + \tilde{Q})w = 0 \pmod{C^\infty}$ yield that $Tu - f$ is smooth for all $t < T$, as desired. Thus, we have our parametrix. Furthermore, the standard propagation of singularities results show that $WF(u)$ is contained in $WF(g)$ union the null bicharacteristic strips of p which pass over $WF(g)$, so we have Lemma 3.2 of the previous section.

We emphasize that the construction described here uses only the local theory of Fourier integral operators.

Generalities

In this final section, we put the phenomenon, analyzed in the previous section for equations of linear elasticity, into a general framework, and also mention some additional phenomena that could occur for general systems. This section is complementary to our work on reflection of singularities (1975).

Thus, we consider a ($k \times k$ matrix of) first order pseudodifferential operators $G(y) = G(y,x,D_x)$ acting on $R_+ \times X$ and consider solutions to the boundary value problem for $(u = u(y) = u(y,x))$

$$\frac{\partial}{\partial y} u = G(y)u \tag{5.1}$$

$$Bu(0) = f \tag{5.2}$$

where B is a pseudodifferential operator on X; say $B \in PS(0)$. The hypothesis that all bicharacteristics intersecting $\partial(R_+ \times X)$ do so transversely and that G has simple characteristics implies that the principal symbol of $G(y)$ is similar to a matrix of the form

$$\begin{pmatrix} i\lambda_1 & & & & \\ & \ddots & & & \\ & & i\lambda_j & & \\ & & & A & \\ & & & & B \end{pmatrix}$$

where $\lambda_\nu(y,x,\xi)$ are real valued (scalar), the spectrum of $A(y,x,\xi)$ has negative real part and the spectrum of $B(y,x,\xi)$ has real positive part. The complete decoupling procedure described in Taylor (1975) implies that any solution $u(y)$ of (5.1), $u(y) \in C([0,y_0),\mathcal{D}'(x))$, can be written in the form

$$u(y) = U(y)\begin{pmatrix} w_1 \\ \cdot \\ \cdot \\ \cdot \\ w_j \\ w_+ \\ w_- \end{pmatrix}$$

for some elliptic $U(y) \in PS(0)$, where w_ν solve the equations

$$\frac{\partial}{\partial y} w_\nu = i\mu_\nu(y,x,D)w_\nu$$

and w_+ and w_- solve, respectively

$$\frac{\partial}{\partial y} w_+ = a(y,x,D)w_+$$

$$\frac{\partial}{\partial y} w_- = b(y,x,D)w_-$$

and, furthermore, the principal symbol of μ_ν is λ_ν, the principal symbol of $a(y,x,D_x)$ is A, and the principal symbol of $b(y,x,D)$ is B. Actually, there might be global topological obstructions to the construction of $U(y)$, but these can be avoided if one microlocalizes appropriately. The details are described in Taylor (1975).

We see that the boundary condition (5.2) is equivalent to

$$BU(0)w(0) = f \tag{5.3}$$

The reflection of singularities phenomenon we consider is described simply as follows. Suppose we know that u is smooth in a conic neighborhood of the rays $\gamma_1,\ldots,\gamma_\ell$ $(0 \leqslant \ell \leqslant j)$ passing over $(x_0,\xi_0) \in T^*(X)$, where γ_ν is a null bicharacteristic strip associated to $(\partial/\partial y) - i\lambda_\nu$. Note that this is equivalent to the smoothness (up to the boundary $y = 0$) of $w_1,\ldots,w_\ell$. More generally, suppose we know the nature of the singularities of h near $\gamma_1,\ldots,\gamma_\ell$, i.e., suppose we know $w_1,\ldots,w_\ell$, mod C^∞. We want to construct a parametrix for $u(y)$ which, in particular, will tell us the nature of the singularities of $w_{\ell+1},\ldots,w_j$, and also the nature of the boundary regularity of w_+. (Note that, since w_+ and w_- solve elliptic evolution equations which are forward and backward, respectively, they are automatically C^∞ inside $(0,y_0) \times X$, and w_- is smooth up to the boundary $y = 0$.) This goal is achieved in Taylor (1976), granted the following hypothesis. (Here, let $w_\nu = P_\nu w$, $w_+ = P_+ w$, $w_- = P_- w$.)

$$\text{Given a knowledge of } w_1(0),\ldots,w_\ell(0) \text{ and of } w_-(0), \text{ the system (5.3) is an elliptic system for } w_{\ell+1}(0),\ldots,w_j(0),w_+(0) \tag{5.4}$$

More generally, as mentioned in Taylor (1975), the system was assumed to be hypoelliptic. However, in the cases we have run across (e.g., Taylor, 1976) hypoelliptic equations seem to occur naturally for a number of grazing ray problems, but in the nongrazing case, one has to work to contrive such a problem (omitting such problems as the $\bar{\partial}$ Neumann problem, where reflection of singularities is not the issue.) We are now in a position to generalize this result as follows.

THEOREM 5.1. One can construct a solution to

$$\frac{\partial}{\partial y} u = Gu \qquad (\text{mod } C^\infty)$$

$$Bu(0) = f \qquad (\text{mod } C^\infty)$$

given $f \in \mathcal{E}'(X)$, with the property that u is smooth along the rays $\gamma_1,\ldots,\gamma_\ell$, provided that, for specified $w_1(0),\ldots,w_\ell(0)$, $w_-(0) \in C^\infty(X)$, we can solve the system

$$BU(0)w(0) = f \qquad (\text{mod } C^\infty)$$

for $w_{\ell+1}(0),\ldots,w_j(0),w_+(0)$.

If we can deduce that $WFw_\nu(0) \subset \Gamma_\nu$, $WFw_+(0) \subset \Gamma_+$, where Γ_ν ($\nu = \ell+1,\ldots,j$) and Γ_+ are closed conic subsets of $T^*(X)$ obtained from $WF(f)$ by some process, it follows that $WF(u)$ is smooth except along those rays passing over Γ_ν and in $\gamma(\frac{\partial}{\partial y} - i\lambda_\nu)$. Furthermore, $w_+(y)$ is smooth up to the boundary $y = 0$ except at points $x \in X$ such that $(x,\xi) \in \Gamma_+$ for some ξ.

The proof the Theorem 5.1 is the same as the proof in the special case where (5.4) is satisfied. The context in which such a situation arises is when $\frac{\partial}{\partial y} - G$ comes from reducing a hyperbolic equation to a first order system of pseudodifferential operators, the time variable being one of the x variables, say $t = x_1$. In such a case, typically the $\gamma_1,\ldots,\gamma_\ell$ are the null bicharacteristics on which t is decreasing (as they leave the boundary) and $\gamma_{\ell+1},\ldots,\gamma_j$ are the null bicharacteristics on which t is increasing. Granted appropriate energy estimates, the approximate solution constructed via Theorem 5.1 will differ from the exact solution by a smooth error (recall the argument at the end of Section 2), so the description of singularities given in Theorem 5.1 will hold for the exact solution.

In such cases as the equations of linear elasticity in three space variables, we need to suppose the w_ν may be vector valued, not merely scalar valued (though the principal symbol of $\mu_\nu(y,x,D_x)$ must be scalar.) Of course, this does not affect the above discussion at all.

In the case of the equations of linear elasticity considered in Sections 2 and 3, the boundary value problem (5.3) fails to be elliptic only in a region where all the eigenvalues of $G_1(y,x,\xi)$ have nonzero real part (the "elliptic region" mentioned in Section 3), so $j = 0$. Thus, in that example, the reflection of singularities phenomenon is exactly as described in our previous paper (1975), except for the Rayleigh waves, which travel along the boundary. No extra singularities propagate into the interior. Now, it is easy to concoct a boundary value problem for which this additional phenomenon will occur. In fact, for vector valued u and v, consider the system

$$\frac{\partial^2}{\partial t^2} u - (\lambda + \mu)\text{grad div } u - \mu\Delta u - L_1 u = 0 \tag{5.5}$$

$$\frac{1}{c^2}\frac{\partial^2}{\partial t^2} v - \Delta v = 0 \tag{5.6}$$

with boundary conditions of the form

$$\sum n_j \sigma_{ij} = f \quad \text{on} \quad \partial\Omega \quad \text{for} \quad u \tag{5.7}$$

$$v = u \quad \text{on} \quad \partial\Omega \tag{5.8}$$

We suppose $f = 0$ for $t < 0$ and solve, assuming $u,v = 0$ for $t < 0$. Note that (5.5), (5.7) is precisely the boundary value problem considered in Sections 2 and 3. So solve it. The equation for v is coupled to that for u via the boundary condition (5.8). Having solved for u, we obtain v by solving the Dirichlet problem for the wave equation, (5.6),(5.8). If the sound speed c in (5.6) is picked to be <u>less</u> than the propagation speed of the Rayleigh waves, it follows from the propagation of singularities results for the Dirichlet problem that v picks up singularities along rays going into Ω in the positive t direction passing over $WF(u|_{\partial\Omega})$, which includes the wave fronts of the Rayleigh waves. I do not know whether there is a physical process for which (5.5)-(5.8) is a model. It would be interesting to find boundary value problems for physical processes for which this additional propagation of singularities phenomenon does occur.

If one mixes grazing rays phenomena in with Rayleigh wave phenomena, additional situations are encountered. Some of these can be handled by the techniques of Melrose (1975), and Taylor (1976), and some need further study. We will not try to say anything about this here.

REFERENCES

Achenbach, J. <u>Wave Propagation in Elastic Solids</u>. North Holland Publishing Company (1973).

Duistermaat, J. <u>Fourier Integral Operators</u>. Courant Inst. of Math. Sci., N. Y. U. Lecture Notes (1973).

Duistermaat, J. and Hörmander, L. Fourier Integral Operators, II. <u>Acta Math</u>., Vol. 128 (1972), pp. 183-259.

Hörmander, L. Fourier Integral Operators, I. <u>Acta Math</u>., Vol. 127 (1971), pp. 79-183.

Hörmander, L. On the existence and regularity of solutions of linear partial differential equations. <u>L'Enseignment Math</u>., Vol 17 (1971), pp. 99-163.

Landau, L. and Lifschitz, E. <u>Theory of Elasticity</u>. Second Edition, Addison Wesley (1970).

Lax, P. and Phillips, R. <u>Scattering Theory</u>. Academic Press, New York (1967).

Love, A. <u>A Treatise on the Mathematical Theory of Elasticity</u>. Fourth Edition, Dover, New York (1944).

Majda A., and Osher, S. Reflection of singularities at the boundary. <u>Comm. Pure Appl. Math</u>., Vol. 28 (1975), pp. 479-499.

Majda, A., and Taylor, M. Inverse scattering problems for transparent obstacles, electromagnetic waves, and hyperbolic systems. <u>Comm. P. D. E.</u>, To appear.

Majda, A., and Taylor, M. The asymptotic behavior of the diffraction peak in classical scattering theory. <u>Comm. Pure and Appl. Math</u>. To appear.

Melrose, R. Microlocal parametrices for diffractive boundary value problems. <u>Duke Math. J</u>., Vol. 42 (1975), pp. 605-635.

Lord Rayleigh. On waves propagated along the plane surface of an elastic solid. Proc. London Math. Soc., 17 (1885), pp. 4-11.

Taylor, M. Reflection of singularities of solution to systems of differential equations. Comm. Pure Appl. Math., Vol 28 (1975), pp. 457-478.

Taylor, M. Grazing rays and reflection of singularities of solutions to wave equations. Comm. Pure Appl. Math., Vol 29 (1976) pp. 1-38

Taylor, M. Grazing rays and reflection of singularities of solutions to wave equations, part II (systems). Comm. Pure Appl. Math., Vol 29 (1976)

Rice University
Houston, Texas

FOURIER INVERSION PROBLEMS ON LIE GROUPS AND A CLASS OF PSEUDODIFFERENTIAL OPERATORS

Joseph A. Wolf*

The Classical Plancherel Formula

Let G be a unimodular separable locally compact groups and, to avoid technicalities, suppose that G is of type I. The unitary dual $\hat{G}$, the set of all equivalence classes $[\pi]$ of irreducible unitary representations π of G, has a standard Borel structure. Given $\pi \in [\pi] \in \hat{G}$, let $\mathcal{H}_\pi$ denote the representation space, and also let π denote the corresponding $*$-representation of $L^1(G)$,

$$\langle\pi(f)u,v\rangle_{\mathcal{H}_\pi} = \int_G f(x)\langle\pi(x)u,v\rangle_{\mathcal{H}_\pi} dx \quad \text{for} \quad f \in L^1(G), \quad u,v \in \mathcal{H}_\pi$$

Segal's Plancherel Theorem (1950), which extends earlier results on compact groups and on abelian groups, says: there is a unique Borel measure μ on $\hat{G}$ such that

(i) if $f \in L^1(G) \cap L^2(G)$, then $\pi(f)$ is a Hilbert-Schmidt operator on $\mathcal{H}_\pi$ for μ-almost-all $[\pi] \in \hat{G}$ and

(ii)

$$\|f\|^2_{L^2(G)} = \int_{\hat{G}} \|\pi(f)\|^2_{HS} d\mu[\pi] \tag{1.1}$$

Here μ is the "Plancherel measure" and (1.1) is the "Plancherel Formula."

If f is of the form $h^* * h$, $h^*(x) = \overline{h(x^{-1})}$ so that $\pi(h^*) = \pi(h)^*$, then

$$\|h\|^2_{L^2(G)} = f(1_G)$$

and $\|\pi(h)\|^2_{HS} = \text{trace } \pi(f)$. Then, (1.1) becomes a Fourier inversion formula

$$f(1_G) = \int_{\hat{G}} \text{trace } \pi(f) d\mu[\pi] \tag{1.2}$$

*Research supported in part by N. S. F. Grant MCS76-01692.

When G is not Unimodular

Let us now drop the unimodularity condition of G. Let dx denote right Haar measure,

$$\int_G f(xg)dx = \int_G f(x)dx \quad \text{for} \quad f \in C_0(G)$$

and let $\delta = \delta_G$ denote the modular function

$$\delta(g)\int_G f(gx)dx = \int_G f(x)dx \quad \text{for} \quad f \in C_0(G)$$

Set $f^g(x) = f(g^{-1}xg)$. Then, for $[\pi] \in G$,

$$\text{trace } \pi(f^g) = \text{trace } \int_G f(g^{-1}xg)\pi(x)dx = \delta(g)\text{trace } \pi(f)$$

while, of course, $f^g(1_G) = g(1_G)$. So, the Plancherel Formula (1.2) does not make sense.

The solution to this problem is to insert an operator that is semiinvariant of type δ_G, either the infinitesimal operators D_π, $[\pi] \in \hat{G}$, specified up to scalar multiple by

$$\pi(g)D_\pi\pi(g)^{-1} = \delta(g)D \quad \text{on} \quad \mathcal{H}_\pi$$

or a global operator D on $L^2(G)$ such that

$$D^g(f) = \delta(g)D(f) \quad \text{where} \quad D^g(f) = D(f^{g^{-1}})$$

The relation is that, in some suitable sense, $D_\pi = \pi(D)$.

We illustrate this with the case of the Heisenberg group with scale. The Heisenberg group of dimension $2n + 1$ is $N = \mathbb{R} + \mathbb{R}^n + \mathbb{R}^n$ with

$$(z,y,x)(z',y',x') = (z + z' + x \cdot y' - y \cdot x', y + y', x + x')$$

and the Heisenberg group with scale is $G = N \cdot A$, $A \cong \mathbb{R}^+$, with

$$(z,y,x,a)(z',y',x',a') = (z + a^2a' + ax \cdot y' - ay \cdot x', y + ay', x + ax', aa')$$

Write Z for the center $\{(z,0,0): z \in \mathbb{R}\}$ of N. Recall that $\hat{N}$ consists of (i) the unitary characters

$$\chi_f(z,y,x) \to e^{if(y,x)}$$

where f is a linear functional on $\mathbb{R}^n + \mathbb{R}^n$ and (ii) the infinite dimensional representation classes for the

$$\gamma_\lambda = \mathrm{Ind}_Q^N((z,y,0) \mapsto e^{i\lambda z}), \quad \lambda \neq 0, \quad Q = \{(z,y,x) \in N; x = 0\}$$

So Mackey's "little group method" tells us that $\hat{G}$ consists of the equivalence classes of (i) the trivial representation, (ii) the $\mathrm{Ind}_N^G(\chi_f)$, $\|f\| = 1$, and (iii) the two representations $\pi_\lambda = \mathrm{Ind}_N^G(\gamma_\lambda)$, $\lambda = \pm 1$. The Plancherel formula will use only these $\pi_{\pm 1} = \mathrm{Ind}_Q^G((z,y,0,1) \mapsto e^{\pm iz})$.

G has right Haar measure $d(z,y,x,a) = a^{-1}dzdydxda$ where dy, dx are Lebesgue measure on $\mathbb{R}^n$. The Hilbert space for π_λ is the space of all measureable $f: G \to \mathbb{C}$ such that $f((z,y,0,1)g) = e^{i\lambda z}f(g)$ and $\int |f(0,0,x,a)|^2 a^{-1}dxda < \infty$, with π_λ given by $[\pi_\lambda(g')f](g) = f(gg')$. We view π_λ as a representation on $L^2(\mathbb{R}^n \times \mathbb{R}^+, a^{-1}dxda)$ by setting $\phi_f(x,a) = f(0,0,x,a)$, so

$$[\pi_\lambda(z',y',x',a')\phi_f](x,a) = f((0,0,x,a)(z',y',x',a'))$$

$$= f(a^2z' + ax\cdot y', ay', x + ax', aa')$$

$$= f((a^2z' + a^2x'\cdot y' + 2ax\cdot y', ay', 0, 1)(0m0mx + ax', aa'))$$

$$= e^{i\lambda(a^2z'+a^2x'\cdot y'+2ax\cdot y')}\phi_f(x + ax', aa')$$

If $\psi \in C_0^\infty(G)$, now

$$[\pi_\lambda(\psi)\phi_f](x.a) = \int\psi(z',y',x',a')[\pi_\lambda(z',y',x',a')\phi_f](x,a)a'^{-1}dz'dy'dx'da'$$

$$= \int K_\psi(x,a;x'',a'')\phi_f(x'',a'')dx''da''$$

where $x'' = x + ax'$, $a'' = aa'$, and

$$K_\psi(x,a;x'',a'') = \int\psi(z',y',x',a')e^{i\lambda(a^2z'+z^2x'\cdot y'+2ax\cdot y')}\,\frac{a^{-n}}{a''}\,dz'dy'$$

$$= (2\pi)^{n+1/2}\psi(\cdot,\cdot,\frac{x''-x}{a}, \frac{a''}{a})^\wedge(\lambda a^2, \lambda a(x''+x))\,\frac{a^{-n}}{a''}$$

so

$$\text{trace } \pi_\lambda(\psi) = \int K_\psi(x,a;x,a)dxda$$

$$= (2\pi)^{n+1/2}\int\psi(\cdot,\cdot,0,1)^\wedge(\lambda a^2, 2ax)a^{-(n+1)}dxda$$

$$= (2\pi)^{n+1/2}2^{-(n+1)}\int\psi(\cdot,\cdot,0,1)^\wedge(\lambda b, x)b^{-(n+1)}dxdb$$

Here a and $b = a^2$ run from 0 to ∞, so for the appropriate real constant c, we have the Plancherel formula

$$\operatorname{trace} \pi_+(D\psi) + \operatorname{trace} \pi_-(D\psi) = \psi(1_G) \qquad D = \left(ic \frac{\partial}{\partial z}\right)^{n+1} \tag{2.1}$$

In the case $n = 0$, where G is the "ax + b group," the Plancherel formula (2.1) is due independently to A. Hohari (1961) and C. C. Moore (1971 and 1973). The general (2.1) was known to Moore, and probably also to Tatsuuma in 1972 and to Pukánszky in 1971.

The idea of using semiinvariants D_π or D was first suggested by Dixmier's work (1952) on quasi-Hilbert algebras and the Tomita-Takesaki theory (1967 and 1970) of modular Hilbert algebras. It then developed along several lines. Kohari's method (1961) for the $ax + b$ group was perfected by Tatsuuma (1972), who defined the D_π as corresponding to multiplication by the modular function δ_G, and showed that the Plancherel formula for G need only involve representations induced from the kernel of δ_G. Somewhat more direct approaches were taken by Kleppner and Lipsman (1972 and 1973) for the global operator and by Duflo and Moore (1976) for the infinitesimal operators. In the case where G is of type I, the Duflo-Moore result says: There exist a positive standard Borel measure μ on $\hat{G}$, and measurable fields $\{(\pi, D_\pi) : [\pi] \in \hat{G} \setminus (\mu\text{-null set})\}$ where D_π is a nonzero selfadjoint operator on $\mathcal{H}_\pi$ semiinvariant of type δ_G, such that

if $f \in L^1(G) \cap L^2(G)$, then $D_\pi^{1/2}\pi(f)$ is Hilbert-Schmidt a.e. $(\hat{G},\mu)$; if $f \in C_0^\infty(G)$, then $D_\pi^{1/2}\pi(f)D_\pi^{1/2}$ is of trace class a.e. $(\hat{G},\mu)$ (2.2)

$f \to D_\pi^{1/2}\pi(f)$ extends to an isometry of $L^2(G)$ onto $\int_{\hat{G}} \mathcal{H}_\pi \otimes \mathcal{H}_\pi^* d\mu[\pi]$ intertwining the left (resp. right) regular representation with $\int_{\hat{G}} \pi \otimes 1 \, d\mu[\pi]$ (resp. $\int_{\hat{G}} 1 \otimes \pi^* d\mu[\pi]$) (2.3)

the D_π are unique up to scalars depending on π and the quantity $D_\pi^{1/2} d\mu[\pi]$ is unique up to a scalar multiple that depends only on normalization of Haar measure (2.4)

The infinitesimal operators D_π have been computed (Duflo and Rais, to appear, and Charbonnel, 1975) for simply connected solvable Lie groups of type I. Also, for simply connected solvable Lie groups, Pukánszky,

(1971) showed that the global operator may be realized as the quotient of of two elements in the center of the universal enveloping algebra of the nilradical. But the only cases in which the global operator is known explicitly, besides (2.1) above, are the cases studied by Keene (to appear), by Keene, Lipsman and Wolf (to appear) and by Lipsman and Wolf (to appear). There, there remain some interesting analytic problems, and that is what I want to discuss in the remainder of this paper.

Maximal Parabolic Subgroups of Unitary Groups

Let $\mathbb{F}$ be one of the fields $\mathbb{R}$ (real), $\mathbb{C}$ (complex) or $\mathbb{Q}$ (quaternion). Given integers $p,q \geqslant 0$, we denote

$$\begin{aligned} &\mathbb{F}^{p,q}: (p+q)\text{-tuples over } \mathbb{F} \quad \text{with } \langle x,y\rangle = \sum_{i}^{p} x_a \bar{y}_a - \sum_{p+1}^{p+q} x_b \bar{y}_b \\ &U(p,q;\mathbb{F}): \text{all } \mathbb{F}\text{-linear } \langle,\rangle\text{-isometries of } \mathbb{F}^{p,q} \end{aligned} \tag{3.1}$$

Here, $U(p,q;\mathbb{R})$ is the indefinite orthogonal group $O(p,q)$;$U(p,q:\mathbb{C})$ is the usual indefinite unitary group $U(p,q)$; and $U(p,q;\mathbb{Q})$ is the indefinite symplectic group $Sp(p,q)$.

A subspace $E \subset \mathbb{F}^{p,q}$ is <u>totally isotropic</u> if $\langle E,E\rangle = 0$. The $U(p,q;\mathbb{F})$ have an important class of subgroups, the <u>parabolic subgroups</u>, which are the normalizers of nested sequences $0 \neq E_1 \subsetneqq \cdots \subsetneqq E_k$ of totally isotropic subspaces,

$$P_{E_1,\ldots,E_k} = \{g \in U(p,q;\mathbb{F}); gE_\ell = E_\ell \quad \text{for} \quad 1 \leqslant \ell \leqslant k\}$$

The conjugacy class of

$$P_{E_1,\ldots,E_k}$$

is determined by the dimension sequence $\dim_{\mathbb{F}} E_1 < \ldots < \dim_{\mathbb{F}} E_k$. Thus, $U(p,q;\mathbb{F})$ has $\min(p,q)$ conjugacy classes of maximal parabolic subgroups,

$$P_E = \{g \in U(p,q;\mathbb{F}): gE = E\} \quad E \text{ nonzero totally isotropic in } \mathbb{F}^{p,q} \tag{3.2}$$

We know (Wolf, 1976) the structure of P_E and $\hat{P}_E$. The group P_E is a semidirect product $N\cdot(M \times A)$ - its Langlands decomposition is MAN - as follows. To describe the nilradical N, we let $\mathbb{F}^{r\times s}$ denote the $r\times s$ matrices over $\mathbb{F}$, we let $\mathrm{Im}: \mathbb{F}^{s\times s} \to \mathbb{F}^{s\times s}$ and $\mathrm{Re}: \mathbb{F}^{s\times s} \to \mathbb{F}^{s\times s}$ denote the projections $z \mapsto \frac{1}{2}(z - z^*)$ and $z \mapsto \frac{1}{2}(z + z^*)$ where $z^* = {}^t\bar{z}$, and we let $\mathbb{F}^{s\times(t,u)}$ denote $\mathbb{F}^{s\times(t+u)}$ with the "hermitian" map

$$\mathcal{H}: \mathbb{F}^{s\times(t,u)} \times \mathbb{F}^{s\times(t,u)} \to \mathbb{F}^{s\times s} \quad \text{by} \quad \mathcal{H}((v_0,w_0),(v,w)) = v_0v^* - w_0w^*$$

To describe $M \times A$, we need

$$GL'(s;\mathbb{F}) = \{g \in GL(s;\mathbb{F}): \ g \text{ preserves Lebesgue measure on } \mathbb{F}^s\}$$

Here note $GL(s:\mathbb{F}) = GL'(s;\mathbb{F}) \times \mathbb{R}^+$ where $\mathbb{R}^+$ is the multiplicative group of positive real numbers. Now, if $s = \dim_{\mathbb{F}} E$,

$$\begin{aligned}
&\text{(i)} \quad N = \operatorname{Im}\mathbb{F}^{s\times s} + \mathbb{F}^{s\times(p-s,q-s)} \text{ with } (z,v)(z',v') \\
&\qquad = (z + z' + \operatorname{Im}\mathcal{H}(v,v'), v + v') \\
&\text{(ii)} \quad A = \mathbb{R}^+ \text{ and } M = GL'(s;\mathbb{F}) \times U(p-s,q-s;\mathbb{F}) \\
&\text{(iii)} \quad A \times M \text{ acts on } N \text{ by } (a,\gamma,g):(z,v) \to (a^2\gamma z\gamma^*, a\gamma v g^*)
\end{aligned} \tag{3.3}$$

For example, if $\mathbb{F} = C$ and $s = 1$, then N is the Heisenberg groups of real dimension $2p + 2q - 3$; if $\mathbb{F} = \mathbb{R}$, $s = 1$, $p = 2$, and $q = 4$, then $U(p,q;\mathbb{F})$ is the conformal group $O(2,4)$, $N\cdot M$ is its Poincaré subgroup and P_E is the Poincaré group with scale.

In any case, N really is just a fancy sort of Heisenberg group. N <u>has center</u> $Z = [N,N] = \operatorname{Im}\mathbb{F}^{s\times s}$ <u>except in the cases</u>

$$\begin{aligned}
&p = q = s: \text{ here } N = \operatorname{Im}\mathbb{F}^{s\times s}, \text{ commutative} \\
&s = 1 \text{ and } \mathbb{F} = \mathbb{R}: \text{ here } N = \mathbb{R}^{p-1,q-1}, \text{ commutative}
\end{aligned} \tag{3.4}$$

In any case, N is 2-step nilpotent and $\hat{N}$ comes directly out of the Kirillov theory.

A class, $[\pi] \in \hat{N}$, is called <u>square integrable</u> (mod Z) if its matrix coefficients $f_{\xi,\eta}(n) = \langle\xi,\eta(n)\eta\rangle$ satisfy $|f_{\xi,\eta}| \in L^2(N/Z)$. If $\hat{N}$ has a square integrable representation, one knows (see Moore and Wolf, 1973) that Plancherel-almost-all classes in $\hat{N}$ are square integrable and that those classes correspond, using Kirillov theory, to the coadjoint orbits $Ad^*(N)\cdot\phi$ of the form $Ad^*(N)\cdot\phi = \{\psi \in n^*: \psi|_z = \phi|_z\}$, where n and z are the Lie algebras of N and Z, and $*$ denotes real linear dual space. In our case, one can see (Wolf, 1976) that N has square integrable representations in all cases except

$$\mathbb{F} = \mathbb{R} \quad s \text{ odd} \quad s > 1 \quad (p - s) + (q - s) > 0 \tag{3.5}$$

Leaving (3.5) aside, the coadjoint orbits giving square integrable representations are the coadjoint orbits of the linear functionals

$$\begin{aligned}&s = 1 \text{ and } \mathbb{F} = \mathbb{R}: \phi_w(v) = \mathrm{Re}\langle v,w\rangle \quad w \in \mathbb{R}^{p-1,q-1}\\&p = q = s: \phi_z(z_0) = \mathrm{trace}\ \mathrm{Re}(z_0 z^*) \quad z \in \mathrm{Im}\mathbb{F}^{s\times s}\\&\text{otherwise: } \phi_z(z_0,v_0) = \mathrm{trace}\ \mathrm{Re}(z_0 z^*) \quad z \in \mathrm{Im}\mathbb{F}^{s\times s} \text{ nonsingular}\end{aligned} \tag{3.6}$$

According to the Mackey machine, the corresponding representations of P_E obtained by the little-group method are sufficient for harmonic analysis in $L^2(P_E)$. Here we may take $w \neq 0$ in the $s = 1$, $\mathbb{F} = \mathbb{R}$ case, and may take z of maximal possible rank in the $p = q = s$ case.

The Case of Real Rank One

Keene (to appear) and Keene, Lipsman and Wolf (to appear) studied the case of parabolic subgroups of the real rank one unitary groups $U(p,1;\mathbb{F})$. There, $s = 1$ and $P_E = NAM$ is given by

$$\begin{aligned}&N = \mathrm{Im}\mathbb{F} + \mathbb{F}^{p-1} \text{ with } (z_0,v_0)(z,v) = (z_0 + z + \mathrm{Im}\langle v_0,v\rangle, v_0 + v)\\&A = \mathbb{R}^+ \text{ and } M = \{\gamma \in \mathbb{F} : |\gamma| = 1\} \times U(p-1;\mathbb{F})\\&A \times M \text{ acts on } N \text{ by } (a,\gamma,g):(z,v) \to (a^2\gamma z\bar{\gamma}, a\gamma v g^*)\end{aligned} \tag{4.1}$$

Square integrable representations of N are associated to the linear functionals $\lambda \in z^* - \{0\}$, the corresponding class $[\gamma_\lambda] \in \hat{N}$ being characterized by its central character

$$\gamma_\lambda(zn) = e^{i\lambda(\log z)}\gamma_\lambda(n) \quad \text{for } z \in Z,\ n \in N$$

If $\mathbb{F} = \mathbb{R}$, γ_λ is just the unitary character $z \to e^{i\lambda(z)}$ on $\mathbb{R}^{p-1}$. If $\mathbb{F} = \mathbb{C}$, $[\gamma_\lambda]$ is the infinite dimensional representation class of the Heisenberg group N with central character $e^{i\lambda}$, and one may view

$$\gamma_\lambda = \mathrm{Ind}_Q^N((z,v) \to e^{i\lambda(z)}) \qquad Q = \mathrm{Im}\mathbb{C} + \mathbb{R}^{p-1} \subset N \tag{4.2}$$

as in Section 1. If $\mathbb{F} = \mathbb{Q}$, the situation is similar.

A acts on $\{[\gamma_\lambda]:\lambda \in z^* - \{0\}\}$ by $a\cdot[\gamma_\lambda] = [\gamma_{a^{-1}\lambda}]$ for $\mathbb{F} = \mathbb{R}$ and $a\cdot[\gamma_\lambda] = [\gamma_{a^{-2}\lambda}]$ for $\mathbb{F} \neq \mathbb{R}$, so the

$$\eta_\nu = \mathrm{Ind}_N^{NA}(\gamma_\lambda) \qquad \lambda \in z^* - \{0\} \tag{4.3}$$

are irreducible, and $[\eta_\lambda] = [\eta_{\lambda'}]$ exactly when $\lambda = r\lambda'$ for some $r > 0$. If we denote

$$S: \text{ unit sphere in } z^* \tag{4.4}$$

then, <u>the generic representations of</u> NA <u>are</u> $\{[\eta_\lambda]:\lambda \in S\}$.

Conjugation by $m \in M$ commutes with induction from N to NA, so it sends $[\gamma_\lambda]$ to $[\gamma_{\mu(m)*\lambda}]$ and $[\eta_\lambda]$ to $[\eta_{\mu(m)*\lambda}]$ where μ is the representation of M on z. The latter is given by

$$\begin{array}{l} \mathbb{R}:M = \{\pm 1\} \times O(p-1) \quad \mu \text{ is nontrivial scalar on } \{\pm 1\} \\ \quad \text{and } \mu \text{ is the usual representation of } O(p-1) \\ \quad \text{on } \mathbb{R}^{p-1} \\ \mathbb{C}:M = \{t \in \mathbb{C}: |t| = 1\} \times U(p-1) \quad \text{and} \quad \mu \text{ is trivial} \\ \mathbb{Q}:M = \{t \in \mathbb{Q}: |t| = 1\} \times Sp(p-1) \quad \mu \text{ is the 3-dimen-} \\ \quad \text{sional representation of } \{t \in \mathbb{Q}: |t| = 1\} \cong SU(2) \\ \quad \text{and } \mu \text{ is trivial on } Sp(p-1) \end{array} \tag{4.5}$$

Now, the action of M on z^* and the $[\gamma_\lambda]$, $\lambda \in S$, satisfies

$$\begin{array}{l} \mathbb{R}:M \text{ is transitive with isotropy } M_\lambda \cong \{\pm 1\} \times O(p-2) \\ \mathbb{C}:M \text{ fixes the 2 points of } S, \text{ so isotropy } M_\lambda = M \\ \mathbb{Q}:M \text{ is transitive with } M_\lambda \cong \{t \in \mathbb{C}: |t| = 1\} \times Sp(p-1) \end{array} \tag{4.6}$$

One can check that η_λ extends to a linear representation $\tilde{\eta}_\lambda$ of NAM. Now, apply Mackey's little-group method. If $\mathbb{F} \neq \mathbb{C}$, fix $\lambda_1 \in S$ and set $M_1 = M_{\lambda_1}$; then, the generic representations of $P_E = NAM$ are the

$$\pi_\tau = \mathrm{Ind}_{NAM_1}^{NAM} (\widetilde{\eta_{\lambda_1}} \otimes \tau) \qquad [\tau] \in \hat{M}_1 \tag{4.7}$$

If $\mathbb{F} = \mathbb{C}$, then $S = \{\lambda_1, \lambda_{-1}\}$ and the generic representations of NAM are the

$$\pi_\tau^+ = \widetilde{\eta_{\lambda_1}} \otimes \tau \quad \text{and} \quad \pi_\tau^- = \widetilde{\eta_{\lambda_{-1}}} \otimes \tau \qquad [\tau] \in \hat{M} \tag{4.8}$$

The final ingredient of the Plancherel formula on P_E is the Laplacian $\Delta = -\sum \partial^2/\partial z_i^2$ on the Euclidean vector group Z. View Δ as an operator on P_E using the diffeomorphic splittings $\mathrm{Im}\mathbb{F} \times \mathbb{F}^{p-1} \times A \times M$. Set

$$k = \dim_{\mathbb{R}} Z \qquad \ell = \dim_{\mathbb{R}} N/Z \qquad r = k + \frac{1}{2}\ell \tag{4.9}$$

Then, for certain constants c_i the pseudodifferential operators

$$D_{NA} = c_1 \Delta^{r/2} \text{ on } NA \qquad D_{NAM} = c_2 \Delta^{r/2} \text{ on } NAM \tag{4.10}$$

come into the Plancherel formula as follows (see Keene, Lipsman and Wolf, to appear.)

For NA: if $f \in C_c^\infty(NA)$, then

(i) $D_{NA}f \in L_1(NA)$, so the $\eta_\lambda(D_{NA}f)$ are defined

(ii) each $\eta_\lambda(D_{NA}f)$ is of trace class

(iii) trace $\eta_\lambda(D_{NA}f)$ is a C^∞ function of $\lambda \in S$, and

(iv) we have

$$f(1_{NA}) = \int_S \text{trace } \eta_\lambda(D_{NA}f)d\sigma(\lambda) \tag{4.11}$$

where σ is the standard volume element on S.

For NAM: if $f \in C_c^\infty(NAM)$, then

(i) $D_{NAM}(f) \in L_1(NAM)$, so the $\pi_\tau^{(\pm)}(f)$ are defined

(ii) each $\pi^{(\pm)}(f)$ is of trace class, and

(iii)

$$\begin{aligned} f(1_p) &= \sum_{\tau\in\hat{M}_1} (\dim \tau)\text{trace } \pi_\tau(D_{NAM}f) \qquad \mathbb{F} \neq \mathbb{C} \\ f(1_p) &= \frac{1}{2} \sum_{\tau\in\hat{M}} (\dim \tau)\text{trace } \{\pi_\tau^+(D_{NAM}f) + \pi_\tau^-(D_{NAM}f)\} \qquad \mathbb{F} = \mathbb{C} \end{aligned} \tag{4.12}$$

The group theoretic and measure theoretic aspects of the proof use methods of Duflo (1972) and Kleppner and Lipsman (1972 and 1973), but they are not our concern here. Rather, we are concerned with the first conclusion

$$D:C_0^\infty(G) \to L_1(G) \qquad G = NA \text{ or } NAM \tag{4.13}$$

which is needed simply in order to make sense of the transformations

$$f \mapsto \eta(D_{NA}f) \qquad f \mapsto \pi(D_{NAM}f)$$

that appear in the Plancherel formulae (4.11) and (4.12).

There is nothing to (4.13) when D is differential. But Keene's discovery (to appear) that D may be strictly pseudodifferential (e.g., when $\mathbb{F} = \mathbb{Q}$, so $r/2 = \frac{3}{2} + (p - 1))$ raises the question of L_1 estimates of the form (4.13). In the real rank one case here, where D is a multiple of a positive power of Δ, this is not so serious (see Keene, Lipsman and Wolf, to appear). But the next examples will show that (4.13) can be very tricky and, in fact, still open.

The Estimate $DC_0(G) \subset L^1(G)$

If we compare the Plancherel formula (2.1) with the case $\mathbb{F} = \mathbb{C}$ of (4.11), we see that the global operator D is not quite unique. Indeed, D can be replaced by any densely defined invertible operator on $L^2(G)$, say D', that also is semiinvariant of type δ_G ($G = NA$ or NAM). Then the corresponding local operators $\pi(D') = a_\pi \cdot \pi(D)$, a_π nonzero scalars, for Plancherel almost all $[\pi] \in \hat{G}$. This leaves a lot of freedom with the solvable groups NA. For example, if $z: Z \to \mathbb{R}^k$ is a linear coordinate on the vector group Z, then (4.11) can be recast as

$$f(1_{NA}) = \int_S \text{trace } \eta_\lambda(D'f)d\sigma'(\lambda)$$

for an appropriate measure σ' on S, where $D' = (\partial/\partial z_1)^r$, $r = \dim_{\mathbb{R}} Z + \frac{1}{2}\dim_{\mathbb{R}} N/Z$ as in Section 4.

The matter is somewhat different for the full parabolic $P_E = NAM$. In order to manage the group theoretic aspects of the proof of the Plancherel formula, and also for esthetic reasons, one wants that D be M-invariant.

Let us see what this means in the case $s = 1$. There,

$$N = \text{Im}\mathbb{F} + \mathbb{F}^{p-1,q-1} \qquad M = \{\gamma \in \mathbb{F} : |\gamma| = 1\} \times U(p-1,q-1;\mathbb{F})$$

If $\mathbb{F} \neq \mathbb{R}$, then $Z = \text{Im}\mathbb{F}$, $U(p-1,q-1;\mathbb{F})$ acts trivially on Z, and $\{\gamma \in \mathbb{F} : |\gamma| = 1\}$ is $\mathbb{R}$-irreducible on Z. If we require that $D = D_{NAM}$ be an operator on Z, it follows that D must be the Z-Fourier-transform of a radial function homogeneous of degree $\dim_{\mathbb{R}} Z + \frac{1}{2}\dim_{\mathbb{R}} N/Z$, and if we further require positivity, then D must essentially be the

$$\tfrac{1}{2}(\dim_{\mathbb{R}} Z + \tfrac{1}{2}\dim_{\mathbb{R}} N/Z)$$

power of the Laplacian on Z. In this case, things go more or less as for (4.12) with no serious problem in the L_1 estimate (4.13)

If $\mathbb{F} = \mathbb{R}$, the situation changes. There, $Z = \mathbb{R}^{p-1,q-1}$ and $M = \{\pm 1\} \times O(p-1,q-1)$ acts on it in the vector representation. With the Z-Fourier-transform and the positivity requirement as above, $D = D_{NAM}$ must be a positive multiple of the $\frac{1}{2}\dim Z$ power of a wave operator,

$$D = c|\square|^{k/2} \qquad \square = \sum_1^k \varepsilon(j)\partial^2/\partial x_j^2 \qquad k = (p-1) + (q-1) \tag{5.1}$$

where $(x_1,\ldots,x_k)$ is a linear coordinate on $Z = \mathbb{R}^{p-1,q-1}$ in which $\langle x,x'\rangle = \sum \varepsilon(j)x_j x_j'$, $\varepsilon = \pm 1$. There, one can show† that $|\square|^{\sigma}C_0^\infty(Z)$ is con-

†Information received from R. Prosser and C. Fefferman.

tained in $L^1(Z)$ for $\sigma > 2[\frac{k}{4}] + 2$, but is not contained in $L^1(Z)$ for $\sigma < \frac{1}{2}k - 1$ and $\min(p,q) > 1$. There, the question of (4.13), where $\sigma = k/2$, remains open.

More generally, assuming $s = \dim_{\mathbb{F}} E$ even in case $\mathbb{F} = \mathbb{R}$, that is, leaving aside the case (3.5), where N does not have square integrable representations, and also the case $s = 1$, $\mathbb{F} = \mathbb{R}$ just discussed, one takes (see Lipsman and Wolf, to appear) the operators $D = D_{NA}$ and D_{NAM} in the form

$$Df = cF^{-1}(|p(\zeta)|^t \cdot F(f)) \qquad c > 0 \tag{5.2}$$

as follows. ζ is the coordinate on the real dual vector space Z^*, F is Fourier transform on the real vector space Z, p is the real polynomial function on $Z = \mathrm{Im}\mathbb{F}^{s\times s}$ such that $z \in Z$ sends Lebesgue measure $d\nu$ on $\mathbb{F}^s$ to $|p(z)|d\nu$, and t is the positive real number such that $|p(z)|^t$ is homogeneous of degree $r = \dim_{\mathbb{R}} Z + \frac{1}{2}\dim_{\mathbb{R}} N/Z$. If $s > 1$, then $p(z)$ is considerably more complicated than the quadratic polynomial $\sum\varepsilon(j)z_j^2$ implicit in (5.1), and the problem of proving an L_1 a priori estimate (4.13) seems even stickier.

Avoiding the Estimate-Possibility and Disadvantages

Lipsman and I recently (to appear) managed, without an estimate of the type (4.13), to prove a Plancherel formula for all the cases in which N has square integrable representations. The formula is of the form

$$f(1_p) = \sum_i \int_{\hat{M}_i} \mathrm{trace}\,\pi_\tau^{(i)}(Df)\,d\nu_i(\tau) \tag{6.1}$$

where the sum runs over the (finite) set of MA-orbits on the space of square integrable classes in $\hat{N}$, where M_i is the M-stabilizer of an element $\lambda_i \in z^*$ corresponding to the ith orbit, where $\pi_\tau^{(i)}$ is defined as in (4.7) for $[\tau] \in \hat{M}_i$, and where $D + D_{NAM}$ is given by (5.1) or (5.2). The formula applies to functions $f \in C^\infty(P_E)$ whose Z-Fourier-transform $F(f) \in C_c^\infty((Z^*\backslash(\text{zeros of } p)) \times N/Z \times A \times M)$. Of course, one can enlarge the domain of D in the obvious way, using Schwartz functions f such that $F(f)$ vanishes, in the Z^* variable, to sufficiently high order at the zeros of $p(\zeta)$. Still, this is not completely satisfactory for harmonic analysis on P_E, and it would be very good to know whether the estimate (4.13) holds for the pseudodifferential operators D of (5.1) and (5.2)

In summary, we have the Fourier inversion formula (6.1) for a moderately large class of nonunimodular groups, but its utility is limited by our lack of exact knowledge of the class of functions to which it applies. This lack would be remedied by some special cases of a problem in psuedo-differential equations. The problem: Let $p(z)$ be a polynomial on $\mathbb{R}^k$, homogeneous of degree d, and let $D_{p,t}$ denote the pseudodifferential operator on $\mathbb{R}^k$ given by

$$D_{p,t}(f)^\wedge(\zeta) = |p(\zeta)|^t \hat{f}(\zeta) \qquad t > 0$$

For which values of t does $D_{p,t}$ keep $C_0^\infty(R^k)$ inside $L^1(\mathbb{R}^k)$?

REFERENCES

Bernat, P., et al. Représentations des Groupes de Lie Résolubles. Dunod, Paris (1972).

Charbonnel, J. La formule de Plancherel pour un groupe de Lie résoluble connexe. Thesis, Univ. de Paris (1975).

Dixmier, J. Algèbres quasi-unitaires. Comm. Math. Helv., 26 (1952), pp. 275-322.

Duflo, M., and Moore, C. C. On the regular representation of a nonunimodular locally compact group. J. Funct. Anal., 21 (1976), pp. 209-243.

Duflo, M., and Raïs, M. Sur l'analyse harmonique sur les groupes de Lie résolubles. To appear.

Keene, F. W. Square integrable representations and a Plancherel theorem for parabolic groups. Trans. Amer. Math. Soc. To appear.

Keene, F. W., Lipsman, R. L., and Wolf, J. A. The Plancherel formula for parabolic subgroups. Israel Math. J. To appear.

Kleppner, A., and Lipsman, R. L. The Plancherel formula for group extensions. Ann. Sci. École Norm. Sup., 5 (1972), pp. 459-516.

Kleppner, A., and Lipsman, R. L. The Plancherel formula for groups extensions, II. Ann. Sci École Norm. Sup., 6 (1973), pp. 103-132.

Kohari, A. Harmonic analysis on the group of linear transformations of the straight line. Japan Acad. Proc., 37 (1961), pp. 250-254.

Lipsman, R. L., and Wolf, J. A. The Plancherel formula for parabolic subgroups of the classical groups. To appear.

Moore, C. C. A Plancherel formula for non-unimodular groups. Address at International Conf. on Harm. Anal., Univ. Md. (1971).

Moore, C. C. Representations of solvable and nilpotent groups and harmonic analysis on nil- and solvmanifolds. Proc. Symp. Pure Math., 24 (1973), pp. 3-44.

Moore, C. C., and Wolf, J. A. Square integrable representations of nilpotent groups. Trans. Amer. Math. Soc., 185 (1973), pp. 445-462.

Pukánszky, L. Unitary representations of solvable Lie groups. Ann. Sci. Ecole Norm. Sup., 4 (1971), pp. 464-608.

Segal, I. E. An extension of Plancherel's formula to separable unimodular groups. Ann. of Math., 52 (1950), pp. 272-292.

Takesaki, M. Tomita's Theory of Modular Hilbert Aglebras and its Applications. Springer Lecture Notes in Math., 128 (1970).

Tatsuuma, N. Plancherel formula for nonunimodular locally compact groups. J. Math., Kyoto Univ., 12 (1972), pp. 179-261.

Tomita, M. Standard forms of von Neumann algebras. 5th Func. Anal. Symp. of Math. Soc. of Japan, Sendai (1967).

Wolf, J. A. Unitary representations of maximal parabolic subgroups of the classical groups. Memoirs, Amer. Math. Soc., 180 (1976).

University of California
Berkeley, California

HARMONIC MAPS BETWEEN RIEMANNIAN MANIFOLDS

Shing-Tung Yau

Introduction

This talk outlines joint work with R. Schoen (1976, and to appear), which was in progress at the time of the conference. Since then, we have found out more about harmonic maps. However, we decided to write the report only on those materials that we knew during the conference.

Let M and N be two Riemannian manifolds with metrics ds_M^2 and ds_N^2 respectively. Let f be a C^1 map from M into N. Then, the energy of f is defined to be

$$E(f) = \frac{1}{2} \int \operatorname{tr}_{ds_M^2}(f^*ds_N^2)$$

Harmonic maps are those stationary maps with respect to this functional. (When M is not compact, we use only compactly supported deformation.)

Since S. Bochner (1940) formally introduced the concept of harmonic maps, Eells and Sampson (1964) were the first to lay the foundation for the theory of harmonic maps. They proved that, if N is compact and has non-positive curvature, then any smooth map from a compact manifold M into N is homotopic to a harmonic map. R. Hamilton (1975) was able to solve the boundary value problem for harmonic maps where M is allowed to have boundary. K. Uhlenbeck (1970) was able to introduce a perturbation method to give a new proof of the theorem of Eells and Sampson. R Schoen (thesis) was able to improve and simplify Hamilton's theorem to the case where N is merely required to be complete. Hildebrandt, Kaul and Wideman (1975) were able to relax the curvature condition on N. However, they had to restrict the size of N so that the radial function was convex. Only recently, Eells and Wood (1976) were able to demonstrate that there is no degree one harmonic map from the two dimensional torus to the two sphere. This shows that some kind of restriction must be imposed on N for the existence of harmonic maps in a given homotopic class. (From the recent experience with this class of maps, it is likely that the homotopic properties of N play an important role.)

In my research with R. Schoen, we considered the properties and the applications of harmonic maps which should be considered as important as the existence. In the following, we discuss those different aspects of our research.

Univalence of Harmonic Maps

Let M and N be two compact manifolds with negative curvature. Suppose the fundamental groups of M and N are isomorphic. Then, it is a natural conjecture that M is diffeomorphic to N.

Since both M and N are $K(\pi,1)$ and have the same fundamental group, it is well known that M must be homotopic to N. Using the theorem of Eells and Sampson, one can then find a harmonic map from M to N which is a homotopic equivalence. If we were able to prove that this map is one to one with nonzero Jacobian, then we would be able to prove the above conjecture. Since Hartman (1967) had proved that the above harmonic map is unique in its homotopic class, one should expect that this harmonic map will enjoy more properties than one can see at first sight. It is therefore not unreasonable to believe the univalence of the map. On the other hand, the existence of the harmonic map is either proved by heat equation methods or by the variational method. The proof of the univalence of the harmonic map is not clear a priori.

R. Schoen and the author (to appear) were able to prove that, if N is a compact two dimensional Riemannian manifold with nonpositive curvature and M is another compact two dimensional Riemannian manifold with the same genus as N, then any degree one map from M into N is homotopic to a unique harmonic diffeomorphism. We were also able to prove a similar theorem when N has boundary which is convex. The latter theorem is considerably harder than the previous one. In the course of our proof, we derive a new local formula for harmonic maps. After integrating this formula, one obtains a global formula similar to the Hurwitz formula for harmonic maps. This last global formula was also obtained by Eells and Woods (1976) as a demonstration for the nonexistence of degree one harmonic maps from torus into the two sphere. However, they found their formula by pure topological consideration.

A Nonlinear Bochner Method

R. Schoen and the author (1976) applied the theory of harmonic maps to study the topology of complete manifolds with nonnegative Ricci curvature and stable minimal hypersurfaces of Euclidean space. We found the following new topological obstruction for the above two classes of manifolds.

Let Ω be a smooth compact domain in the above manifold. Let $\Omega/\partial\Omega$ be the C.W. complex obtained by collapsing the boundary of Ω to a point. Let N be any compact manifold with nonpositive curvature. Then, there is no nontrivial homomorphism from $\pi_1(\Omega/\partial\Omega)$ to $\pi_1(N)$.

As an example, we can see from the above assertion that, if M is any manifold with dimension $\geqslant 2$, then the open manifold obtained by deleting a point from $S^1 \times M$ admits no complete metric with nonnegative Ricci curvature.

The proof of the above assertion can be sketched as follows. We construct a harmonic map h with finite energy from the complete manifold with nonnegative Ricci curvature into N so that h induces the same map as the given homomorphism from $\Omega/\partial\Omega$ to N. The energy density of the harmonic map h satisfies some differential equation depending on the curvature. In our case, it turns out to be subharmonic. Since there is no nonnegative L^2 subharmonic function defined on a complete Riemannian manifold (see Yau, 1976), and the volume of any complete manifold with nonnegative Ricci curvature is infinite (see Yau, 1976), the map h must be a constant map and the corresponding algebraic homomorphism on $\pi_1(\Omega/\partial\Omega)$ must be trivial.

Existence of Incompressible Minimal Surfaces with Positive Genus

Let M be a compact Riemannian manifold with $\pi_2(M) = 0$. Let G be a subgroup of $\pi_1(M)$ which is isomorphic to the fundamental group of a compact surface Σ_g with genus $g \geqslant 1$. Then, R. Schoen and the author (to appear) proved that there exists a branched minimal immersion h from Σ_g into M so that $h_*:\pi_1(\Sigma_g) \to \pi_1(M)$ is injective and $h_*(\pi_1(\Sigma_g)) = G$. This was proved independently by Sachs and Uhlenbeck by a somewhat different method.

Our method of proof runs as follows. For each Riemannian metric on Σ_g, we can use a method of C. B. Morrey (1948) to prove the existence a harmonic map from Σ_g into M which satisfies the above condition on the fundamental group and which minimizes energy among the above maps. Then, we vary the conformal structure on Σ_g and obtain a harmonic map which minimizes energy even when we change the conformal structure on Σ_g. In proving the existence here, we need to use the Teichmüller theory and we have to minimize over the Teichmüller space. Then it is well known that the map that we found in the above is a branched minimal immersion. When M is three dimensional, it is possible to show the nonexistence of branch points using a method of Osserman and Gulliver. For a more detailed discussion of this paragraph, see Schoen and Yau (to appear).

Obstruction for Group Actions

Finally, we show here how one can exploit the uniqueness of harmonic maps to study groups acting smoothly on a manifold. For simplicity, we shall only state a special case of a result obtained by R. Schoen and the author.

Let N be a compact n-dimensional manifold with nonpositive curvature. Let $A(N)$ be the group of affine diffeomorphisms of N, i.e., the subgroup of $\mathrm{Diff}(N)$ which preserves connections of N. Let M be any compact smooth manifold such that, for some smooth map f from M into N, the induced map $f^*: H^n(N) \to H^n(M)$ is nontrivial and $f_*(\pi_1(M)) = \pi_1(N)$. Let G be any finite group that acts smoothly on M. Suppose for each $g \in G$, we can find an (not necessarily unique or nontrivial) element $\tilde{g}$ in $P(N)$ such that $f \circ g$ is freely homotopic to $\tilde{g} \circ f$. Then, there is a surjective harmonic map h homotopic to f and a homomorphism $\gamma: G \to P(N)$ so that $h \cdot g = \gamma(g) \cdot h$ for all $g \in G$.

Let $\mathrm{Ker}(\gamma)$ be the kernel of the homomorphsim γ and $\bar{G}$ be the subgroup of G consisting of elements g so that $f \circ g$ is freely homotopic to f. Then we can prove that $\bar{G}/\mathrm{Ker}(\gamma)$ is a finite abelian subgroup which is a product of m cyclic groups with m not greater than the first Betti number of M and the rank of the center of $\pi_1(N)$.

If $\dim M = \dim N$, then we can prove that $\mathrm{Ker}(\gamma)$ is a finite group whose order divides the degree of f. In particular, if the degree of f is one, $\mathrm{Ker}(\gamma) = (1)$. In this case, we can also prove that the group $\bar{G}$ acts freely on M.

There are a number of interesting consequences of the above theorem (see Schoen and Yau, to appear). One of them is the following corollary. Let M be an n-dimensional compact manifold. Suppose there are homology classes $\alpha_1, \ldots, \alpha_n \in H_1(M, Z)$ such that if $\tilde{\alpha}_1, \ldots, \tilde{\alpha}_n \in H^1(M, Z)$ are the cohomology classes dual to $\alpha_1, \ldots, \alpha_n$, then $\tilde{\alpha}_1 \cup \ldots \cup \tilde{\alpha}_n([M]) = 1$. Then, a consequence of the above theorem implies that any finite group acting smoothly on M is isomorphic to a subgroup of the semidirect product of $S\ell(n, Z)$ with R^n.

REFERENCES

Bochner, S. Harmonic surfaces in Riemann metric. Trans. Amer. Math. Soc., 47 (1940), pp. 146-154.

Eells, J., and Sampson, H. Harmonic mappings of Riemannian manifolds. Amer. J. Math., 86 (1964), pp. 109-160.

Eells, J., and Wood, J. Restrictions on harmonic maps of surfaces. Topology, 15 (1976), pp. 263-266.

Hartman, P. On homotopic harmonic maps. Canad. J. Math., 19 (1967), pp. 673-687.

Hamilton, R. Harmonic maps of manifolds with boundary. Lecture Notes in Math., No. 471. Springer-Verlag, Berlin, Heidelberg, New York (1975).

Hildebrandt, Kaul and Wideman. Harmonic mappings into Riemannian manifolds with non-positive sectional curvature. Math. Scand., 37 (1975).

Morrey, C. B. The problem of Plateau on a Riemannian manifold. Ann. of Math., 49 (1948), pp. 807-851.

Schoen, R. Thesis.

Schoen, R., and Yau, Shing-Tung. Harmonic maps and the topology of stable hypersurfaces and manifolds with non-negative Ricci curvature. Comm. Math. Helv., 39 (1976), pp. 333-341.

Schoen, R., and Yau, Shing-Tung. On univalent harmonic maps between surfaces. To appear in Invent. Math.

Schoen, R., and Yau, Shing-Tung. Harmonic maps, group actions and the topology of manifolds with non-positive curvature. To appear.

Schoen, R., and Yau, Shing-Tung. Harmonic maps and incompressible minimal surfaces in a compact Riemannian manifold. To appear.

Uhlenbeck, K. Harmonic maps; a direct method in the calculus of variations. Bull. Amer. Math. Soc., 76 (1970), pp. 1082-1087.

Yau, Shing-Tung. Some function-theoretic properties of complete Riemannian manifolds and their applications to geometry. Indiana Math. J., 25 (1976), pp. 659-670.

University of California
Los Angeles, California

AN EXAMPLE OF L_2 COHOMOLOGY AND HODGE THEORY FOR NONCOMPACT MANIFOLDS

Steven Zucker

Introduction and Statement of Results

The results and conjectures contained herein are an analog in the subject of Riemannian geometry of a theorem in transcendental algebraic geometry (see Zucker, to appear). We will restate that theorem in this section; it is not essential, however, that the reader find the statement intelligible in order to comprehend the work at hand. While working on "Hodge theory with degenerating coefficients", my faith in the result was sustained by the simplest case, which coincides with the springboard for the problem on Riemannian manifolds. It is this example that I wish to present, and then discuss directions of generalization.

Let $\overline{M}$ be a compact orientable surface (which may then be given a complex structure and be regarded as a Riemann surface), and let M be the open submanifold obtained by deleting from $\overline{M}$ finitely many points. We endow M with a Riemannian metric such that, in the punctured neighborhood of a deleted point of $\overline{M}$, the metric is asymptotically the Poincaré metric of the punctured disc Δ^*:

$$(ds)^2 = r^{-2}\log^{-2}r[(dr)^2 + (rd\theta)^2] \qquad (*)$$

in local polar coordinates. The L_2 Hodge theory related the following two observations.

(i) Let $A^*_{(2)}(M)$ denote the complex of global $\mathbb{C}$-valued $\mathcal{C}^\infty$ differential forms ϕ on M for which ϕ and $d\phi$ are square summable in the given metric. Then, the intrinsic L_2 cohomology groups

$$H^i_{(2)}(M) = H^i(A^*_{(2)}(M),d)$$

are isomorphic to $H^i(\overline{M},\mathbb{C})$. This follows from the existence of an L_2 Poincaré Lemma for Δ^*. Heuristically, the usual generator $d\theta$ (or $z^{-1}dz$) for $H^1(\Delta^*,\mathbb{C})$ is not an L_2 1-form, and no "extraneous" cohomology arises. (Of course, we may use $\mathbb{R}$-coefficients as well.)

(ii) One considers the square summable harmonic forms $\mathcal{H}^i_{(2)}(M)$ for M, which we now view as a Riemann surface. Ignoring the L_2 conditions,

$$\mathcal{H}^0(M) = \mathbb{C} \quad \mathcal{H}^1(M) = \mathcal{H}^{1,0} \oplus \overline{\mathcal{H}^{1,0}}$$

where $\mathcal{H}^{1,0}$ denotes the space of holomorphic differentials. Imposing the L_2 requirement now, one checks directly, using the local form (*), that the constants are in L_2; moreover, any L_2 holomorphic 1-form on M actually has removable singularities, i.e., extends holomorphically to $\overline{M}$. Thus, we have recovered the usual harmonic forms for $\overline{M}$ in $\mathcal{H}^0_{(2)}(M)$ and $\mathcal{H}^1_{(2)}(M)$, which represent $H^0(\overline{M},\mathbb{C})$ and $H^1(\overline{M},\mathbb{C})$ respectively. Generating $\mathcal{H}^2_{(2)}(M)$, we have the Kähler class of M, which is now a 2-form with singularities at the points of $\overline{M} - M$. The Hodge structure on $H^*(\overline{M},\mathbb{C})$ is seen as arising from the L_2 cohomology of M. We formalize the above in stating:

THEOREM 1.1. Let $\overline{M}$ be a compact surface, M the complement of finitely many points in $\overline{M}$, metrized as above [(*)]. Then,

$$H^i(\overline{M},\mathbb{C}) \simeq H^i(A^*_{(2)}(M),d) \simeq \mathcal{H}^i_{(2)}(M)$$

We will see that these isomorphisms are not "accidental".

For the complex analytic situation, let $\Omega^p_{\overline{M}}$ denote the sheaf of germs of holomorphic p-forms on $\overline{M}$. Define $A^{p,q}_{(2)}(M)$ to be the space of global C^∞ forms ϕ on M of type (p,q) with both ϕ and $\bar\partial\phi$ square summable, and let $\mathcal{H}^{p,q}_{(2)}(M)$ be the subset of $A^{p,q}_{(2)}(M)$ consisting of harmonic differentials.

THEOREM 1.2. With notation as above,

$$H^q(\overline{M},\Omega^p_{\overline{M}}) \simeq H^q(A^{p,*}_{(2)}(M),\bar\partial) \simeq \mathcal{H}^{p,q}_{(2)}(M)$$

In addition,

$$\mathcal{H}^i_{(2)}(M) = \bigoplus_{p+q=i} \mathcal{H}^{p,q}_{(2)}(M)$$

Before continuing, we state the algebraic geometer's generalization of Theorems 1.1 and 1.2 (for definitions, see Schmid, 1973).

THEOREM 1.3. (see Zucker, to appear.) Let M be a nonsingular algebraic curve over $\mathbb{C}$, let V be a locally constant sheaf of complex vector spaces on M which underlies a polarizable variation of Hodge structure of weight m, and let $j: M \to \overline{M}$ denote the inclusion of M in its smooth completion. Then there is a natural polarizable Hodge structure of weight $m + i$ on $H^i(\overline{M}, j_*V)$.

The situation $M \subset \overline{M}$ is precisely that of Theorems 1.1 and 1.2. The Hodge decomposition of $H^i(\overline{M}, j_*V)$ is deduced from the L_2 V-valued differential forms on M, using a metric with form (*). One passes back to Theorem 1.1 by taking $V = \mathbb{C}$ (with trivial variation of Hodge structure). In general, V acquires nonremovable singularities at the points of $\overline{M} - M$ (see Schmid, 1973), so working on M becomes essential.

We will prove Theorem 1.1 in the next two sections. It should be remarked that our present method of proof is somewhat different from the one we give in "Hodge theory with degenerating coefficients." It is the one, however, which figures to be easier to generalize in the long run. Theorem 1.2 is proved in much the same way, so we will omit its proof. Since the class of manifolds covered by Theorems 1.1 and 1.2 is rather narrow, one should regard these results as indicators of something more general. I feel strongly that the Hodge theoretic assertions should be a consequence of the intrinsic Riemannian geometry of M, obtainable without reference to $\overline{M}$. We will discuss conjectural generalizations in Section 4.

Reduction to Local Questions

We let $A^i_{(2)}$ denote the sheaves (defined on $\overline{M}$) determined by the presheaf

$$A^i_{(2)}(U) = \{\phi : \phi \text{ is a } C^\infty \text{ i-form on } U \cap M \text{ with } \phi|_U \in A^i_{(2)}(U)$$
$$\text{and } d\phi|_U \in A^{i+1}_{(2)}(U)\}$$

It is clear that

$$A^i_{(2)}(M) = \Gamma(\overline{M}, A^i_{(2)})$$

and that $A^i_{(2)}$ is a fine sheaf on $\overline{M}$). The L_2 Poincaré Lemma asserts

PROPOSITION 2.1. $A^*_{(2)}$ is a resolution of the constant sheaf $\mathbb{C}$ on $\overline{M}$.

We will use the Hilbert space methods of a priori estimates. Define

$L^i_{(2)}$ to be the sheaf on $\overline{M}$ determined by the presheaf

$$L^i_{(2)}(U) = \{\phi : \phi \text{ is an } L_2 \text{ i-form on } U, \text{ and } d\phi \text{ exists weakly as an } L_2 \text{ (i + 1)-form on } U\}$$

Note that we may identify $\Gamma(\overline{M}, L^i_{(2)})$ as the domain of the weak extension of d for the given L_2 spaces. The sheaves $L^i_{(2)}$ admit partitions of unity, so by regularity theory, Proposition 2.1 will follow from

PROPOSITION 2.2. $L^*_{(2)}$ is a resolution of $\mathbb{C}$ on $\overline{M}$.

We can define sheaves $A^{p,q}_{(2)}$ and $L^{p,q}_{(2)}$ in a similar fashion, with reference to the $\bar{\partial}$-operator, and obtain complex analytic analogs of Propositions 2.1 and 2.2.

We will prove Proposition 2.2 in the next section. Assuming the result, we can derive Theorem 1.1 in a straightforward manner. Of course, the first half comes by standard sheaf theory. To prove the second part, one must know that the Laplacian has closed range. It is equivalent to know that d has closed range; but this is automatic, since the first isomorphism, for all i the image of d has finite codimension in $\ker d$. It follows that $H^i(\overline{M},\mathbb{C})$ is isomorphic to the space of L_2 harmonic forms which are in the domain of the operator $dd^* + d^*d$. To see that this is all of $H^i_{(2)}(M)$, we invoke Chernoff's criterion (1973): since M is metrically complete, the Laplacian has a unique closure as an unbounded operator. Viewing it as the weak extension, we obtain the desired result.

The a priori Estimate

We need to derive an estimate locally on $\overline{M}$

$$\|d\phi\|^2 + \|d^*\phi\|^2 \geq C\|\phi\|^2 \quad \text{for some } C > 0 \tag{3.1}$$

for all i-forms, $i > 0$, in Domain(d) $\cap$ Domain(d^*), for this will imply the exactness for d in degrees greater than zero.

Given any point of $\overline{M}$, we may choose coordinates in which M appears locally as either Δ or Δ^*. We have (3.1) for Δ (Neumann problem). For the more interesting case Δ^*, we will verify (3.1) on a space of forms dense in the graph norm $(\|\phi\|^2 + \|d\phi\|^2 + \|d^*\phi\|^2)^{1/2}$. We will do only the case $i = 1$, the other being similar (and easier). Since Proposition 2.2 is insensitive to equivalent changes of norm, we may assume that the metric is exactly the Poincaré metric in $\Delta_A^* = \{z \in \mathbb{C} : 0 < |z| \leq A < 1\}$.

LEMMA 3.1. $S = \{\phi = f dr + g d\theta : f, g \in C_0^1(\Delta_A^*)$, and $f = 0$ when $|z| = A\}$ is dense in the graph norm for $d \oplus d^*$.

Proof. One combines the smoothing and approximation methods of Andreotti and Vesentini (1965) in the graph norm (allowing truncation near the origin) and Hörmander (1965) (allowing approximation by C^1 forms which satisfy the boundary conditions).

Expanding f and g as Fourier series

$$f = f_n(r)e^{in\theta} \qquad g = g_n(r)e^{in\theta}$$

(with summation over n understood), we have, if $\phi \in S$,

$$d\phi = (g_n' - inf_n)e^{in\theta} dr \wedge d\theta$$

$$d^*\phi = -([rf_n]' + ing_n r^{-1})e^{in\theta} r \log^2 r$$

$$\|d\phi\|^2 = \int_0^A (|g_n'|^2 + n^2|f_n|^2) r \log^2 r dr - 2\mathrm{Re} \int_0^A inf_n \overline{g}'_n r \log^2 r dr$$

$$\|d^*\phi\|^2 = \int_0^A (|(rf_n)'|^2 + n^2 r^{-2}|g_n|^2) r \log^2 r dr - 2\mathrm{Re} \int_0^A (f_n r)' in\overline{g}_n \log^2 r dr$$

Now, with $0 < \varepsilon < 1$,

$$\|g d\theta\|^2 = \int_0^A |g_n|^2 r^{-1} dr = \int_0^A \left| \int_0^r g_n'(\rho) d\rho \right|^2 r^{-1} dr$$

$$\leq \int_0^A r^{-1} \left(\int_0^r |g_n'(\rho)|^2 \rho |\log|^{1+\varepsilon} \rho d\rho \right) \left(\int_0^r \rho^{-1} |\log|^{-1-\varepsilon} \rho d\rho \right) dr$$

$$= C_1 \int_0^A \left(\int_0^r |g_n'(\rho)|^2 \rho |\log|^{1+\varepsilon} \rho d\rho \right) r^{-1} |\log|^{-\varepsilon} r dr$$

$$= C_1 \int_0^A |g_n'(\rho)|^2 \rho |\log|^{1+\varepsilon} \rho \left(\int_\rho^A r^{-1} |\log|^{-\varepsilon} r dr \right) d\rho$$

$$\leq C_2 \int_0^A |g_n'(\rho)|^2 \rho \log^2 \rho d\rho$$

There is an analogous estimate for $\|f dr\|^2$. Putting them together, we will obtain (3.1) provided we have

$$2\left|\int_0^A [inf_n\bar{g}_n' r \log^2 r + (f_n r)' in\bar{g}_n \log^2 r]dr\right| \leq \int_0^A n^2(|rf_n|^2 + |g_n|^2)r^{-1}\log^2 r dr \qquad (3.2)$$

Integrating the left hand side by parts yields

$$4\left|\int_0^A inf_n\bar{g}_n \log r dr\right|$$

since $\phi \in S$. Using the Schwartz inequality, this is

$$\leq 4n\left(\int_0^A |f_n|^2 r \log^2 r dr\right)^{1/2}\left(\int_0^A |g_n|^2 r^{-1} dr\right)^{1/2}$$

$$\leq 4n\left(\sup_{[0,A]} \log^{-2} r\right)\left(\int_0^A |f_n|^2 r \log^2 r dr\right)^{1/2}\left(\int_0^A |g_n|^2 r^{-1}\log^2 r dr\right)^{1/2}$$

which is less than the right hand side of (3.2), provided A is chosen sufficiently small.

Since the constants are in L_2, Proposition 2.2 now follows.

A Conjectural Generalization

We have just finished proving Theorem 1.1, a result on the L_2 cohomology of a rather restrictive class of manifolds. As an educated guess, I find it indicative of a general phenomenon. By construction, the manifolds M that we have considered satisfy the following three geometric conditions: (a) complete, (b) finite volume, and (c) uniform boundedness (in Riemannian norm) of the curvature tensor and each of its covariant derivatives. Some calculations for the classical parametrix for the Laplacian (see Zucker, to appear) suggest that these differential geometric conditions are sufficient to guarantee that the Laplacian has closed range of finite codimension, yielding finite dimensional L_2 cohomology groups represented by harmonic forms.

The methods of this paper should be adequate to produce a higher dimensional version of Theorem 1.1, in which $\bar{M}$ is compact and M is obtained by deleting D, the union of a finite number of codimension two manifolds meeting transverally. In this case, one can construct metrics on M which grow like the Poincaré metric near D. If $\bar{M}$ is a complex Kähler manifold, and the components of D are complex submanifolds, one can arrange that the metric constructed on M be Kähler, so we would correspondingly generalize Theorem 1.2. Such manifolds comprise the natural setting for generalizing the theorem from "Hodge theory with degenerating coefficients".

REFERENCES

Andreotti, A., and Vesentini, E. Carleman estimates for the Laplace-Beltrami equation of complex manifolds. Pub. Math. IHES, 25 (1965), pp. 81-130.

Chernoff, P. Essential self-adjointness of powers of generators of hyperbolic equations. J. Func. Anal., 12 (1973), pp. 401-414.

Hörmander. L. L^2 estimates and existence theorems for the $\bar{\partial}$ operator. Acta Math., 113 (1965), pp. 89-152.

Schmid, W. Variation of Hodge structure: the singularities of the period mapping. Inven. Math., 22 (1973), ;;. 211-319.

Zucker, S. Hodge theory with degenerating coefficients. To appear.

Zucker, S. Estimates for the classical parametrix for the Laplacian. To appear.

Rutgers University
New Brunswick, New Jersey

INDEX